Fundamentals of Applied Biochemistry and Bioengineering

Fundamentals of Applied Biochemistry and Bioengineering

Contributors

Shaniko Shini, Asad Sultan et al.

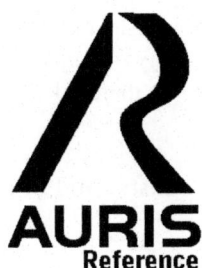

AURIS
Reference

www.aurisreference.com

Fundamentals of Applied Biochemistry and Bioengineering

Contributors: Shaniko Shini, Asad Sultan et al.

Published by Auris Reference Limited

www.aurisreference.com

United Kingdom

Fundamentals of Applied Biochemistry and Bioengineering

ISBN: 978-1-78154-842-4

British Library Cataloguing in Publication Data
A CIP record for this book is available from the British Library

Printed in the United Kingdom

Exclusively distributed by CBS Publishers & Distributors Pvt. Ltd.

Sales & Distribution Rights only for India, Pakistan, Bangladesh, Sri Lanka, Nepal and Bhutan.This book is not to be sold outside these territories.

Contents

List of Abbreviations

ACC	Adaptive Control with Constraints
ANFIS	Adaptive neurofuzzy inference system
ACO	Ant Colony Optimization
ANN	Artificial neural network
CVD	Chemical vapor deposition
CML	Combined machine loading
CNC	Computer numerical control
DE	Design Exploration
DOF	Degrees of Freedom
EDA	Electrical discharge alloying
EDM	discharge machining
EDM	Electrical discharge machining
FEA	Finite Element Analysis
FMS	Flexible Manufacturing System
GEP	Genetic expression programming
GHG	Greenhouse gas
ICDR	International Center for Dielectric Research
LIBWE	Laser-induced back-side wet etching
LCS	Learning Classifier System
LMP	Linear Mathematical Programming
LPCVD	Low pressure chemical vapour deposition
MRR	Material removed ratio
MRE	Mean relative errors
MIP	Mixed Integer Programming
NSERC	Natural Science and Engineering Research Council of Canada
NSFC	Natural Science Foundation of China
PSO	Particle Swarm Optimization
PVD	Physical vapor deposition
PVDF	Polyvinylidene fluoride
QCM	Quartz crystal micro-balance
RBFNs	Radial basis function neural networks
VCM	Varying-coefficient model
XRD	X-ray diffraction

List of Contributors

Shaniko Shini
School of Biomedical Sciences, University of Queensland, St Lucia, QLD 4072, Australia

Asad Sultan
Department of Poultry Science, Khyber Pakhtunkhwa Agricultural University, Peshawar 25120, Pakistan

Wayne L. Bryden
School of Agriculture and Food Sciences, University of Queensland, Gatton, QLD 4343, Australia

Mustafa Türk
Department of Biology, Faculty of Science and Art, Kırıkkale University, Yahsihan, 71450 Kirikkale, Turkey

Zakir M. O. Rzayev
Department of Chemical Engineering, Faculty Engineering, Hacettepe University, Beytepe, 06800 Ankara, Turkey

Gülcihan Kurucu
Department of Chemical Engineering, Faculty Engineering, Hacettepe University, Beytepe, 06800 Ankara, Turkey

Gregory A. Lewbart
North Carolina State University, College of Veterinary Medicine, Raleigh, North Carolina, United States of America

Maximilian Hirschfeld
University San Francisco de Quito, Galapagos Science Center, Puerto Baquerizo Moreno, Galapagos, Ecuador

Judith Denkinger
University San Francisco de Quito, Galapagos Science Center, Puerto Baquerizo Moreno, Galapagos, Ecuador

Karla Vasco
University San Francisco de Quito, Galapagos Science Center, Puerto Baquerizo Moreno, Galapagos, Ecuador

Nataly Guevara
University San Francisco de Quito, Galapagos Science Center, Puerto Baquerizo Moreno, Galapagos, Ecuador

Juan Garcı
Galapagos National Park Service, Puerto Ayora, Galapagos, Ecuador

Juanpablo Munoz
University San Francisco de Quito, Galapagos Science Center, Puerto Baquerizo Moreno, Galapagos, Ecuador

Kenneth J. Lohmann
Department of Biology, University of North Carolina, Chapel Hill, North Carolina, United States of America

Mustafa Türk
Department of Biology, Faculty of Arts and Sciences, Kırıkkale University, Yahşihan, Turkey

Gülten Kahraman
Sarayköy Nuclear Research and Training Center, Turkish Atomic Energy Authority, Ankara, Turkey

Sevda A. Khalilova
Scientific Research Institute of Medicinal Prophylaxis, Ministry of Public Health, Baku, Azerbaijan

Zakir M. O. Rzayev
Institute of Science & Engineering, Division of Nanoscience and Nanomedicine, Hacettepe University Ankara, Turkey
Serpil Oguztüzün
Department of Biology, Faculty of Arts and Sciences, Kırıkkale University, Yahşihan, Turkey

Laura Mejía-Teniente
Facultad de Ingeniería, CA Ingeniería de Biosistemas, Universidad Autónoma de Querétaro Centro Universitario Cerro de las Campanas s/n, Querétaro, Qro, México

Angel María Chapa-Oliver
Facultad de Ingeniería, CA Ingeniería de Biosistemas, Universidad Autónoma de Querétaro Centro Universitario Cerro de las Campanas s/n, Querétaro, Qro, México

Moises Alejandro Vazquez-Cruz
Facultad de Ingeniería, CA Ingeniería de Biosistemas, Universidad Autónoma de Querétaro Centro Universitario Cerro de las Campanas s/n, Querétaro, Qro, México

Irineo Torres-Pacheco
Facultad de Ingeniería, CA Ingeniería de Biosistemas, Universidad Autónoma de Querétaro Centro Universitario Cerro de las Campanas s/n, Querétaro, Qro, México

Ramón Gerardo Guevara-González
Facultad de Ingeniería, CA Ingeniería de Biosistemas, Universidad Autónoma de Querétaro Centro Universitario Cerro de las Campanas s/n, Querétaro, Qro, México

Umberto Lucia
Dipartimento Energia, Politecnico di Torino, Corso Duca degli Abruzzi 24, 10129 Torino, Italy

Pedro Fernandes
Institute for Biotechnology and Bioengineering (IBB), Centre for Biological and Chemical Engineering, Instituto Superior Técnico, Avenue Rovisco Pais, 1049-001 Lisboa, Portugal

Heping Cao
U.S. Department of Agriculture, Agricultural Research Service Southern Regional Research Center U.S.A.

A. Pontini
Department of Neurosensorial Specialties, Institute of Plastic Reconstructive Surgery and Burn Unit, Padova University Hospital, Padova, Italy

M.M. Sfriso
Department of Molecular Medicine, Human Anatomy Section, Padova University Hospi- tal, Padova, Italy

M.I. Buompensiere
Department of Molecular Medicine, Human Anatomy Section, Padova University Hospi- tal, Padova, Italy

V.Vindigni
Department of Neurosensorial Specialties, Institute of Plastic Reconstructive Surgery and Burn Unit, Padova University Hospital, Padova, Italy

F. Bassetto
Department of Neurosensorial Specialties, Institute of Plastic Reconstructive Surgery and Burn Unit, Padova University Hospital, Padova, Italy

Collins Odhiambo
Kenya Medical Research Institute, Kisumu, Kenya

Boaz Oyaro
Kenya Medical Research Institute, Kisumu, Kenya

Richard Odipo
Kenya Medical Research Institute, Kisumu, Kenya

Fredrick Otieno
Kenya Medical Research Institute, Kisumu, Kenya

George Alemnji
U.S. Centers for Disease Control and Prevention (CDC), Bridgetown, Barbados

John Williamson
U.S. Centers for Disease Control and Prevention (CDC), Kisumu, Kenya

Clement Zeh
U.S. Centers for Disease Control and Prevention (CDC), Kisumu, Kenya

Bassem Jaouadi
Laboratory of Microorganisms and Biomolecules, Centre de Biotechnologie de Sfax-University of Sfax, Road of Sidi Mansour Km 6, Tunisia

Badis Abdelmalek
Laboratory of Biochemistry and Industrial Microbiology, Department of Industrial Chemistry, University Saad Dahlab of Blida, Algeria

Nedia Zaraî Jaouadi
Laboratory of Microorganisms and Biomolecules, Centre de Biotechnologie de Sfax-University of Sfax, Road of Sidi Mansour Km 6, Tunisia

Samir Bejar
Laboratory of Microorganisms and Biomolecules, Centre de Biotechnologie de Sfax-University of Sfax, Road of Sidi Mansour Km 6, Tunisia

Lionella Palego
Department of Clinical and Experimental Medicine, University of Pisa, 56126 Pisa, Italy

Laura Betti
Department of Pharmacy, University of Pisa, 56126 Pisa, Italy
Interdepartmental Center of "Nutraceutical Research and Food for Health", University of Pisa, 56124 Pisa, Italy

Alessandra Rossi
Department of Clinical and Experimental Medicine, University of Pisa, 56126 Pisa, Italy

Gino Giannaccini
Department of Pharmacy, University of Pisa, 56126 Pisa, Italy
Interdepartmental Center of "Nutraceutical Research and Food for Health", University of Pisa, 56124 Pisa, Italy

Preface

Biochemistry is the branch of science that explores the chemical processes within and related to living organisms. The principle engineering applications involving the fields of biology and medicine are known as bioengineering. The field addresses various challenges that exist within the biological scientific community and also encompasses all aspects of biomedical engineering and biotechnology. The text *Fundamentals of Applied Biochemistry and Bioengineering* describes recent advances in biochemistry and bioengineering. First chapter focuses on selenium biochemistry and bioavailability. Second chapter evaluates selected blood gas, blood biochemical, and hematology parameters from 28 wild-caught green turtles in two coastal foraging areas adjacent to San Cristóbal Island, Galapagos, Ecuador. In third chapter, the comparative analysis of HeLa cells (cancel cells) and L929 Fibroblast cells (normal cells) has been investigated. The goal of fourth chapter is synthesis and characterization of novel organo-boron amide-ester derivatives by amidolysis of PAA with 2-aminoethyldiphenyl borinate (2-AEPB). Biotechnological approaches for control in crops have been presented in fifth chapter. In sixth chapter, we develop the bioengineering thermodynamic of biological cells, with particular regards to possible control of the cells growth by a control of the ions transport across the cell membrane. The aim of seventh chapter is to provide an updated and succinct overview on the applications of enzymes in the food sector, and of progresses made, namely, within the scope of tapping for more efficient biocatalysts, through screening, structural modification, and immobilization of enzymes. Eighth chapter focuses on bioengineering recombinant diacylglycerol acyltransferases. Bioengineering of vascular conduits has been discussed in ninth chapter. Evaluation of locally established reference intervals for hematology and biochemistry parameters in western Kenya has been outlined in tenth chapter. The aim of eleventh chapter is to provide an overview on the current quest for novel natural bacterial alkaline proteases with special emphasis on the purification and characterization of two enzymes, namely SAPB and KERAB. Structural, nutritional, metabolic, and medical aspects of tryptophan biochemistry in humans have been presented in last chapter.

Chapter 1

SELENIUM BIOCHEMISTRY AND BIOAVAILABILITY: IMPLICATIONS FOR ANIMAL AGRICULTURE

Shaniko Shini[1], Asad Sultan[2] and Wayne L. Bryden[3]

[1]School of Biomedical Sciences, University of Queensland, St Lucia, QLD 4072, Australia

[2]Department of Poultry Science, Khyber Pakhtunkhwa Agricultural University, Peshawar 25120, Pakistan

[3]School of Agriculture and Food Sciences, University of Queensland, Gatton, QLD 4343, Australia

ABSTRACT

Selenium (Se) is an essential trace mineral required for growth, development, immune function, and metabolism. Selenium exerts its biological effects as an integral component of selenoproteins (SePs). Deficiency or low Se status leads to marked changes in many biochemical pathways and a range of pathologies and disorders which are associated with SeP function. Animals, and presumably humans, are able to efficiently utilize nutritionally adequate levels of Se in both organic and inorganic forms. It is now clear that the bioavailability of Se varies depending on the source and chemical form of the Se supplement. There are a range of products available for dietary Se supplementation, however, organic sources have been shown to be assimilated more efficiently than inorganic compounds and are considered to be less toxic and more appropriate as a feed supplement. Yeast enriched with Selenohomoalanthionine (SeHLan) has recently become commercially available, and initial research suggests that it may be an efficacious source for the production of Se enriched animal products.

INTRODUCTION

Selenium (Se) is an essential trace element for animals and humans. It was discovered in 1818 and named Selene after the Greek goddess of the moon. However, it was not until the 1950s that it was recognized as an essential dietary micronutrient for mammals [1]. Selenium presents a nutritional conundrum because it is an essential trace element with a narrow range between dietary adequacy and toxicity. It is this characteristic of Se that allows it to be considered as either a friend, or, a foe.

Selenium exerts its biological effects as an integral component of selenoproteins (SePs) that contain selenocysteine at their active site [2]. Some thirty SePs, mostly enzymes, have been identified, including a series of glutathione peroxidases, thioredoxin reductases, and iodothyronine deiodinases [1,2,3,4]. The majority play important roles in redox regulation, detoxification, immunity and viral suppression [5,6,7]. Deficiency or low Se status leads to marked changes in many biochemical pathways and a range of pathologies associated with defects of SeP function may occur [8,9,10].

Selenium content of soils can vary widely. In areas where soils are low in bioavailable Se, deficiencies can occur in humans and animals consuming plant based foods grown in those soils. Selenium deficiency have been reported in many countries including China, Japan, Korea, and Siberia, Northern Europe, USA, Canada, New Zealand, and Australia [11,12,13]. Within each country there are large regional differences in soil Se status and in some localities there are plants that accumulate Se resulting in Se toxicity (or selenosis) in grazing animals. However, health outcomes are not only dependent on the total Se content of soils, but also on the amount of Se taken up into plants and animals—the bioavailable selenium [11].

Dietary Se supplementation was first permitted some forty years ago [13]. Since then, there have been significant advances in our knowledge of Se metabolism and the important role that Se plays in animal productivity and health [3,6,9]. During this period, Se has become an important addition to dietary supplements for animals [14]. The following pages provide an overview of the nutritional biochemistry of Se, and an update on recent developments in Se bioavailability and the possible impact on Se enrichment of animal products.

BIOCHEMISTRY OF SELENIUM AND SELENOPROTEINS

Selenium exists in four oxidation states: elemental Se (Se^0), selenide (Se^{-2}), selenite (Se^{+4}), and selenate (Se^{+6}) in a variety of inorganic and organic matrices [14]. The natural soluble inorganic forms, selenite and selenate, account for the majority of total global selenium [12]. The ultimate source of all Se is the rocks and soils of our terrestrial environment. Most soils contain from 0.1 to 2 µg/g of Se, but this is not evenly distributed. Some areas of the world have soils rich in Se (100 µg/g), while many areas have deficient soils (none or >0.1 µg/g) [11]. Organically bound selenide compounds are predominantly seleno-amino acids; the principle chemical form of Se in animal tissues is selenocysteine, while selenomethionine predominates in plants [15].

The chemistry of selenium resembles that of sulphur in several respects but these elements are not completely interchangeable in animal systems. Both, sulphur and Se occur in proteins as constituents of amino acids. Sulphur is one of the most prevalent elements in the body and is present in the sulphur-containing amino acids: methionine, cysteine, homocysteine and taurine. Selenium is a trace element and a component of the amino acids selenocysteine, and selenomethionine [2]. Selenocysteine is identical to cysteine except that sulphur is replaced by a Se atom, which is typically ionized at physiological pH when compared with sulphur-containing species.

Selenocysteine (from animal tissues) and selenomethionine (from plants, marine algae, yeast and bacteria) are both sources of Se suitable for synthesis of SePs [16]. Selenocysteine is incorporated into selenium-containing proteins, designated selenoproteins, following the action of a specific codon for selenocysteine residue [3]. Replacement of selenocysteine by cysteine in a SeP usually results in a dramatic decrease of enzymatic activity, confirming that the ionized Se atom is critical for optimum protein function [9]. Significantly, within all cell types there is a specific biosynthesis pathway that facilitates selenocysteine synthesis and its subsequent incorporation into SePs. Cellular Se concentrations are therefore tightly regulated. The regulation of SeP synthesis is central to understanding Se homeostasis and disorders following the failure of homeostasis [5,17,18]. Cellular Se concentration is a key regulator of Se incorporation into SePs and acts mainly at the post-transcriptional level in response to alterations in Se availability [19]. Selenocysteine biosynthesis represents the main regulatory point for SeP synthesis and not absorption as occurs with many nutrients. In contrast, ingested selenomethionine can be nonspecifically incorporated into proteins in place of methionine or converted to selenocysteine by the trans-sulphuration pathway in liver or kidney and may be used as a biological Se pool [6,17].

The biochemistry of Se is different from most other trace elements as it is incorporated in proteins (SePs) at their highest level of complexity and function. Selenoproteins incorporate selenium only in the form of selenocysteine and this occurs during translation in the ribosome using a transfer RNA specific for selenocysteine. Seleno-amino acids (selenocysteine or selenocystine and selenomethionine) are required for the synthesis of selenium-containing peptides and proteins [15]. There are no known human or animal functionally active SePs that contain selenomethionine.

Only proteins that are genetically programmed and perform essential biological functions are classified as SePs [3]. Some of these SePs are enzymes such as the six antioxidant glutathione peroxidases and the three thioredoxin reductases; the three deiodinases are involved in thyroid function by catalyzing the activation and deactivation of the thyroid hormones. [20]. Selenoprotein P, for example, functions as a transporter of selenium between the liver and other organs [4]. The functional characterization of many SePs remains to be delineated.

SELENIUM ABSORPTION, DISTRIBUTION AND METABOLIC FATE

An overview of the metabolism of Se is shown in Figure 1. Absorption of selenium occurs in the small intestine, where both inorganic and organic forms of Se are readily absorbed. Selenite is passively absorbed across the gut wall, while selenate appears to be transported by a sodium-mediated carrier mechanism shared with sulphur [14,21]. Organic forms of Se are actively transported. Selenium is distributed throughout the body from the liver to the brain, pancreas and kidneys. The highest Se concentrations are found in the liver and kidneys but the greatest total concentration occurs in muscle because of their proportion of body weight [22]. However, dose–response studies conducted with chicken and lamb have found that in muscle tissue, the proportion of selenomethionine and selenocysteine was equivalent [23], suggesting that both selenomethionine and selenocysteine-containing proteins are able to be absorbed and transferred to peripheral tissues. However, the mechanism is unclear. Selenium is transported in blood by two SePs; selenoprotein P (SePP1) and extracellular glutathione peroxidase (GSH-Px) [6]. Plasma SePP1 contains more than 50% of circulating Se [6,19]. Only insignificant transitory amounts of free selenomethionine are found in blood. It has also been shown, that dietary supplementation with Se enriched yeast increased blood selenomethionine in a dose dependent manner whereas selenocysteine remained constant [24]. These changes in circulating selenised amino acids may reflect saturation of selenoenzymes and a shift from incorporation of Se into functional SePs to non-specific incorporation of selenomethionine into liver tissue protein. This may indicate that alternative transport mechanisms for selenomethionine exist and need to be clarified through further research. Following protein turnover, the released Se, can be recycled via enterohepatic circulation or excreted. Selenium is eliminated primarily in urine and faeces [17]. The distribution between the two routes varies with the level of exposure and time after exposure.

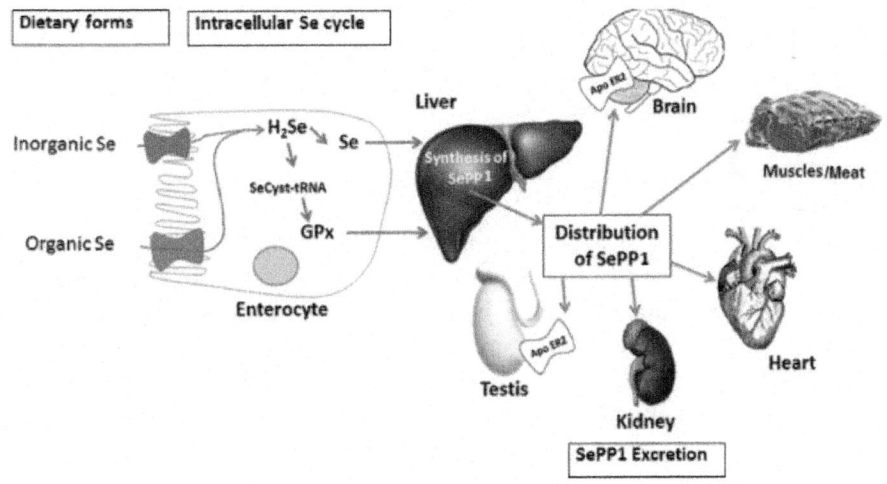

Figure 1: General pathways of Se absorption, hepatic synthesis of SePP1 and distribution to various organs. Adapted from Kumar and Priyadarsini [25].

In ruminants, selenite is the primary compound available for absorption because the reducing conditions within the rumen convert the majority of selenate to selenite [26]. In the rumen, about a third of selenite is converted to insoluble forms that are passed into manure. Of the soluble selenite that reaches the intestine, some 40% will be absorbed, compared to about 80% of selenomethionine [27]. Subsequently, in ruminants, the digestibility of Se from selenite is around 50% compared to about 66% for selenium-yeast [27]. There is no information on the impact of the gut microbiota on the Se requirements of monogastric animals.

Inorganic Se is recognized by the digestive tissues and is absorbed and converted into SePs [27]. In contrast, organic Se (selenomethionine) is not recognized as Se-containing by mammalian cells [28]. Hence, selenomethionine is absorbed and metabolized relative to methionine needs. If selenomethionine is broken down within the cell, Se is released and recognized by the cell as a mineral [29]. It is then processed according to the need for Se. However, if the cell does not break down selenomethionine, it may be inadvertently incorporated into a wide variety of proteins that are not genetically programmed to contain Se [30], rather than be excreted rapidly. Such incorrect incorporation of Se into protein may be toxic [3,31]. As a metabolic safeguard, neither dietary selenocysteine nor selenomethionine is directly incorporated into selenoproteins. All dietary forms of selenium must be metabolized and converted to selenocysteine and selenoproteins under the genetically controlled mechanism within the cell [32]. Much of the absorbed organic Se is transferred into the amino acid pool, where together with the existing intracellular pool it is metabolized by different pathways (Figure 1). From there, it is enzymatically converted in the liver to selenide, which serves as the Se source for selenocysteine synthesis [33].

SELENIUM DEFICIENCY AND REQUIREMENTS

Selenium acts biochemically in mammals and birds in a complimentary manner to vitamin E [34,35,36]. Both nutrients prevent peroxidation of unsaturated fatty acids in cell membranes. Most of the deficiency signs of these nutrients can be explained by their antioxidant properties. The requirement for each is therefore influenced by the dietary concentration of the other [2,37]. For example, the Se requirement of the chick is inversely proportional to dietary vitamin E intake. Thus Se has sparing effect on the requirement for vitamin E and vice versa; Se-dependent peroxidases essentially serve as antioxidants when vitamin E levels are not sufficient to prevent lipid peroxidation. On the other hand, vitamin E neutralizes oxidants before the initiation of chain reactions thus reducing Se loss from the body [16]. The metabolic interrelationship of GSH-Px with vitamin E is particularly evident in deficiency diseases that can be prevented either by vitamin E or Se [1,37]. The other major role for Se is the production of thyroid hormones as a component of the enzyme type 1 iodothyronine deiodinase which converts triiodothyronine to thyroxine [1].

Manifestation of Se deficiency can take many forms and varies between species. Muscular degeneration or white muscle disease occurs to varying degrees in all species [14,38]. In birds, pancreatic fibrosis is an uncomplicated Se deficiency, whereas exudative diathesis (generalised oedema visible under the skin) is responsive to both Se and vitamin E [34]. Pigs with hepatosis diatetica (severe necrotic liver lesions) are responsive to Se supplements [39], while both Se and vitamin E are effective in treating mulberry heart disease (a dietetic microangiopathy) [40]. Reproductive disorders, including retained placenta in dairy cows [41], while lowered disease resistance are observed in all Se deficient species [26,27]. Some species, such as rabbits and horses, seem to be more dependent on vitamin E than Se for their antioxidant protection [42]. This may reflect species differences in dependence on non-selenium containing GSH-Px.

Selenium presents a nutritional conundrum because it is both essential and highly toxic. Adequate Se intake levels have a relatively narrow range between deficiency and toxicity. Dietary Se levels of 0.2–0.3 mg Se/kg diet have shown to prevent deficiency and resultant diseases such as white muscle disease in cattle and sheep, exertional myopathy in horses, hepatosis dietetica in pigs, and exudative diathesis in chickens [22,37]. The tolerable concentrations for

Se in most livestock feed is considered to be 2–5 mg Se/kg diet, although some believe 4–5 mg Se/kg diet can inhibit growth. Animals that consume diets containing 5–40 mg Se/kg diet over a period of several weeks or months suffer from chronic selenosis [37].

There are several approaches to measuring Se status. These include the measurement of changes in plasma Se concentration, measurement of GSH-Px enzyme activity, and absorption/retention studies. The use of stable isotopes of Se has been used in human studies and to determine endogenous forms of selenium in foods [43]. All of these biomarkers are useful indicators of Se status but because of the role of Se in many biochemical pathways, a single indicator may not be an appropriate index of Se status.

DIETARY SELENIUM SUPPLEMENTATION

Soils are the major source of Se for plants, while plants are major sources of Se for animals. To improve Se status of plants grown on deficient soils, Se enriched fertilizers may be used or plant breeding strategies may be applied [13]. Despite these strategies, Se is also routinely added to animal diets to ensure that requirements are met. There has been increased interest recently in dietary supplementation with Se to enrich animal products. The production of selenium-enriched meat, milk and eggs is viewed as an effective and safe way of improving the selenium status of humans [44].

There are a range of products available for dietary Se supplementation (Table 1). Selenium is commonly added to diets as sodium selenite. However, there has been growing interest in dietary addition of organic Se. Organic sources are assimilated more efficiently than inorganic Se and considered to be less toxic and therefore more appropriate as a feed supplement [22,45]. Yeast has become the most popular vehicle for the addition of organic Se because of its rapid growth, ease of culture and high capacity to accumulate Se [46,47,48]. The major product in selenized yeast is selenomethionine.

Selenomethionine was found to be four times more effective than selenite in preventing the characteristic pancreatic degeneration caused by selenium deficiency in chicks [49]. Selenium yeast (selenomethionine) was found to be much more effective than inorganic Se in increasing the Se concentration of cow's milk [31]. This is in accord with many animal studies and human clinical trials that have demonstrated the superior efficacy of L-selenomethionine, in increasing Se muscle content compared to inorganic Se [45,50].

Table 1: Selenium compounds and their application in animal and human nutrition

Name and Content	Nature or Origin	Uses	Reference *
Sodium selenate Sodium selenite Selenase 50 mcg/mL	Synthetic Inorganic	For short-term selenium supplementation; orally in the diet, or by injections for both animals and humans	[12,36,44, 45,49,50]
Sintomin BIOSEL 2000	Inactive dry yeast containing high levels of organic selenium	All animals	[51]

Selenohomolanthionine (SeHLan; 2 hydroxy-4-methylselenobutanoic acid) was recently identified in Japanese pungent radish and has generated much interest as it was less toxic in human cell culture than selenomethionine [53,56]. As shown inFigure 2, differences in metabolism between SeHLan and selenomethionine may, in-part, explain the apparent difference in toxicity. Selenomethionine mimics methionine by sharing the same metabolic pathways and can replace methionine in peptide synthesis, and as noted above, thus causing signs of selenosis [31]. The proposed metabolic pathway for SeHLan (Figure 2) appears to be much less complex; SeHLan is only utilized in the trans-selenation pathway for selenoprotein synthesis and therefore is not expected to interfere with the methionine metabolic pathways. The tissue distribution of these two selenoamino acids may also contribute to differences in toxicity [6]. Both are distributed throughout the body with higher liver and pancreas accumulation of selenomethionine in contrast to SeHLan which preferentially accumulates in the liver and kidneys [19]. At higher doses, selenomethionine has been shown to induce pancreas damage [55,57] whereas SeHLan is excreted by the kidneys without inducing pancreatic damage [53]. Tsuji et al., have reported specific toxicity of SeHLan to the kidneys which could be avoided when it is administered at a lower dose than in their study [53].

Selenomethionine enriched yeast has been used as a feed supplement in animal agriculture for many years. Recently, a yeast product enriched with SeHLan has become available and efficacy studies with growing pigs [58] and broiler chickens [54] have been conducted with these selenoamino acid sources. In the studies both selenomethionine (Sel Plex) and SeHLan (AB Tor-Sel) were compared to sodium selenite. In clean research, experimental conditions, as demonstrated on many occasions, dietary supplementation with both inorganic and organic selenium resulted in similar animal and bird performance. However, tissue accumulation was significantly greater when the organic forms of Se were fed, which is in accord with the literature [31,48,49,50]. Interestingly, the yeast enriched with SeHLan generated significantly higher Se concentrations in muscle tissue of broiler chickens than the selenomethionine enriched product [54]. The implication of this finding in both pig and broilers may imply a greater efficacy of SeHLan in stressful commercial environments. These interesting results should be the subject of further research.

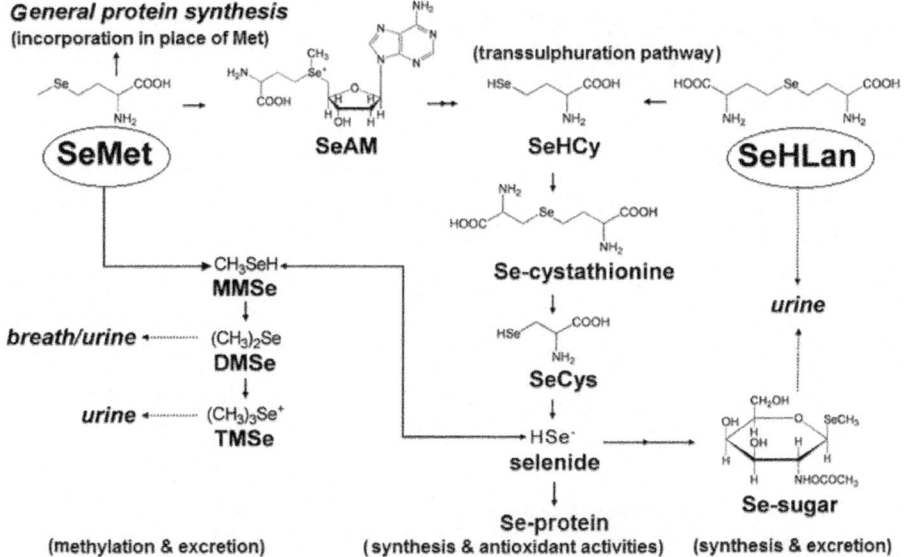

Figure 2: Proposed metabolic pathways for SeHLan and SeMet in animal cells. Adapted from Tsuji et al. 2010 [53].

CONCLUSIONS

Selenium's nutritional essentiality was discovered in the 1950s. It is now clear that the importance of having adequate amounts of Se in the diet is primarily due to the fact that this micronutrient is required for the biosynthesis of selenocysteine as a part of functional selenoproteins. Improved analytical techniques will be required to answer those questions that still remain regarding Se metabolism. Although animals, and presumably humans, are able to efficiently utilize nutritionally adequate levels of Se in both organic and inorganic forms for SeP synthesis, it is clear that the bioavailability of Se varies, depending on the source and chemical form of the Se supplement. Tissue enrichment with Se is greater when organic forms of the micronutrient are fed. Greater tissue reserves of Se may enhance an animal's resilience to stress and disease challenge.

Organic Se, in the form of yeast enriched with selenomethionine, is widely used in animal nutrition. Recently, yeast enriched with SeHLan became commercially available and initial research suggests that it may also be an efficacious source for tissue accumulation of Se. This has obvious implications for the production of Se enriched animal products.

CONFLICTS OF INTEREST

The author declares no conflict of interest.

REFERENCES

1. Brown, K.M.; Arthur, J.R. Selenium, selenoproteins and human health: A review. Public Health Nutr. **2001**, 4, 593–599

2. Bock, A.; Forchhammer, K.; Heider, J.; Leinfelder, W.; Sawers, G.; Veprek, B.; Zinoni, F. Selenocysteine: The 21st amino acid. Mol. Microbiol. **1991**, 5, 515–520

3. Hatfield, D.L.; Gladyshev, V.N. How Selenium Has Altered Our Understanding of the Genetic Code. Mol. Cell. Biol.**2002**, 22, 3565–3576

4. Reeves, M.A.; Hoffmann, P.R. The human selenoproteome: Recent insights into functions and regulation. Cell. Mol. Life Sci. **2009**, 66, 2457–2478

5. Bellinger, F.P.; Raman, A.V.; Reeves, M.A.; Berry, M.J. Regulation and function of selenoproteins in human disease.Biochem. J. **2009**, 422, 11–22

6. Burk, R.F.; Hill, K.E.; Motley, A.K. Selenoprotein metabolism and function: Evidence for more than one function for selenoprotein P. J. Nutr. **2003**, 133, 1517s–1520s.

7. Duntas, L.H. The role of selenium in thyroid autoimmunity and cancer. Thyroid **2006**, 16, 455–460

8. Cardoso, B.R.; Roberts, B.R.; Bush, A.I.; Hare, D.J. Selenium, selenoproteins and neurodegenerative diseases.Metallomics **2015**, 7, 1213–1228

9. Moghadaszadeh,B.;Beggs,A.H.Selenoproteinsandtheirimpactonhuman health through diverse physiological pathways. Physiology **2006**, 21, 307–315

10. Köhrle, J.; Jakob, F.; Contempré, B.; Dumont, J.E. Selenium, the thyroid, and the endocrine system. Endocr. Rev. **2005**,26, 944–984

11. Selinus, O.; Fordyce, F. Selenium deficiency and toxicity in the environment. In Essentials of Medical Geology; Springer Netherlands: Heidelberg, Germany, 2013; pp. 375–416.

12. Oldfield, J.E. Selenium World Atlas: Updated Edition; Selenium-Tellurium Development Association: Grimbergen, Belgium, 2002.

13. WHO. Guidelines on Food Fortification with Micronutrients; World Health Organization: Geneva, Switzerland, 2006.

14. Reilly, C. Selenium in Food and Health; Springer Science Media: New York, NY, USA, 2006.

15. Mangiapane, E.; Pessione, A.; Pessione, E. Selenium and selenoproteins: An overview on different biological systems.Curr. Protein Pept. Sci. **2014**, 15, 598–607

16. Rahmanto, A.S.; Davies, M.J. Selenium-containing amino acids as direct and indirect antioxidants. IUBMB Life **2012**,64, 863–871

17. Patching, S.G.; Gardiner, P.H. Recent developments in selenium metabolism and chemical speciation: A review. J. Trace Elem. Med. Biol. **1999**, 13, 193–214.

18. Steinbrenner, H.; Sies, H. Selenium homeostasis and antioxidant selenoproteins in brain: Implications for disorders in the central nervous system. Arch. Biochem. Biophys. **2013**, 536, 152–157

19. Hill, K.E.; Wu, S.; Motley, A.K.; Stevenson, T.D.; Winfrey, V.P.; Capecchi, M.R.; Atkins, J.F.; Burk, R.F. Production of selenoprotein P (Sepp1) by hepatocytes is central to selenium homeostasis. J. Biol. Chem. **2012**, 287, 40414–40424

20. Labunskyy, V.M.; Hatfield, D.L.; Gladyshev, V.N. Selenoproteins: Molecular pathways and physiological roles.Physiol. Rev. **2014**, 94, 739–777

21. Wastney, M.E.; Combs Jr, G.F.; Canfield, W.K.; Taylor, P.R.; Patterson, K.Y.; Hill, A.D.; Moler, J.E.; Patterson, B.H. A human model of selenium that integrates metabolism from selenite and selenomethionine. J. Nutr. **2011**, 141, 708–717

22. Schrauzer, G.N. Nutritional selenium supplements: Product types, quality, and safety. J. Am. Coll. Nutr. **2001**, 20, 1–4

23. Bierla, K.; Dernovics, M.; Vacchina, V.; Szpunar, J.; Bertin, G.; Lobinski, R. Determination of selenocysteine and selenomethionine in edible animal tissues by 2D size-exclusion reversed-phase HPLC-ICP MS following carbamidomethylation and proteolytic extraction. Anal. Bioanal. Chem. **2008**, 390, 1789–1798

24. Juniper, D.T.; Phipps, R.H.; Bertin, G. Effect of dietary supplementation with selenium-enriched yeast or sodium selenite on selenium tissue distribution and meat quality in commercial-line turkeys. Animal **2011**, 11, 1751–1760

25. Santhosh Kumar, B.; Priyadarsini, K.I. Selenium nutrition: How important is it? Biomed. Prev. Nutr. **2014**, 4, 333–341.

26. Spears, J.W. Trace Mineral Bioavailability in Ruminants. J. Nutr. **2003**, 133, 1506S–1509S.

27. Weiss, W.P. Selenium sources for dairy cattle. In Proceedings of Tri-State Dairy Nutrition Conference, Fort Wayne, IN, USA, May 2005; pp. 61–72.

28. Behne, D.; Kyriakopoulos, A. Mammalian selenium-containing proteins. Ann. Rev. Nutr. **2001**, 21, 453–473

29. López-Alonso, M.R. Trace Minerals and Livestock: Not Too Much Not Too Little. ISRN Vet. Sci. **2012**

30. Combs, G.F. Biomarkers of Selenium Status. Nutrients **2015**, 7, 2209–2236

31. Rayman, M.P. The use of high-selenium yeast to raise selenium status: how does it measure up? Br. J. Nutr. **2004**, 92, 557–573

32. Turanov, A.A.; Xu, X.-M.; Carlson, B.A.; Yoo, M.-H.; Gladyshev, V.N.; Hatfield, D.L. Biosynthesis of Selenocysteine, the 21st Amino Acid in the Genetic Code, and a Novel Pathway for Cysteine Biosynthesis. Adv. Nutr.: An Int. Rev. J.**2011**, 2, 122–128

33. Ma, S.; Caprioli, R.M.; Hill, K.E.; Burk, R.F. Loss of selenium from selenoproteins: Conversion of selenocysteine to dehydroalanine in vitro. J. Am. Soc. Mass. Spectrom. **2003**, 14, 593–600.

34. Combs, G.F.; Scott Jr., M.L. Dietary requirements for vitamin E and selenium measured at the cellular level in the chick. J. Nutr. **1974**, 104, 1292–1296.

35. Scott, M.L.; Noguchi, T.; Combs, G.F., Jr. New evidence concerning mechanisms of action of vitamin E and selenium.Vitam. Horm. **1974**, 32, 429–444.

36. Diplock, A.T. The nutritional and metabollic roles of selenium and vitamin E. Proc. Nutr. Soc. **1974**, 33, 315–322

37. National Research Council (US) Subcommittee on Selenium. Selenium in Nutrition: Revised Edition; National Academies Press: Washington, DC, USA, 1983.

38. Kinoshita, H.; Sugai, K.; Goto, Y.; Nonaka, I. Early onset distal muscular dystrophy. Brain Dev. **1995**, 17, 206–209.

39. Sharp, B.A.; Van Dreumel, A.A.; Young, L.G. Vitamin E, selenium and methionine supplementation of dystrophogenic diets for pigs. Can. J. Comp. Med. **1972**, 36, 398–402.

40. Sharp, B.A.; Young, L.G.; Van Dreumel, A.A. Dietary induction of mulberry heart disease and hepatosis dietetica in pigs. I. Nutritional aspects. Can. J. Comp. Med. **1972**, 36, 371–376.

41. Jovanović, I.B.; Veličković, M.; Vuković, D.; Milanović, S.; Valčić, O.; Gvozdić, D. Effects of Different Amounts of Supplemental Selenium and Vitamin E on the Incidence of Retained Placenta, Selenium, Malondialdehyde, and Thyronines Status in Cows Treated with Prostaglandin F2α; for the Induction of Parturition. J. Vet. Med. **2013**

42. Cheeke, P.R.; Dierenfeld, E.S. Comparative animal nutrition and metabolism; CABI: Wallingford, UK, 2010.

43. Finley, J.W. Bioavailability of selenium from foods. Nutr. Rev. **2006**, 64, 146–151

44. Fisinin, V.I.; Papazyan, T.T.; Surai, P.F. Producing selenium-enriched eggs and meat to improve the selenium status of the general population. Crit. Rev. Biotechnol. **2009**, 29, 18–28

45. Schrauzer, G.N.; Surai, P.F. Selenium in human and animal nutrition: resolved and unresolved issues. A partly historical treatise in commemoration of the fiftieth anniversary of the discovery of the biological essentiality of selenium, dedicated to the memory of Klaus Schwarz (1914–1978) on the occasion of the thirtieth anniversary of his death. Crit. Rev. Biotechnol. **2009**, 29, 2–9.

46. Kieliszek, M.; Błażejak, S.; Gientka, I.; Bzducha-Wróbel, A. Accumulation and metabolism of selenium by yeast cells.Appl. Microbiol. Biotechnol. **2015**, 99, 5373–5382

47. Suhajda, A.; Hegoczki, J.; Janzso, B.; Pais, I.; Vereczkey, G. Preparation of selenium yeasts I. Preparation of selenium-enriched Saccharomyces cerevisiae. J. Trace Elem. Med. Biol. **2000**, 14, 43–47.

48. Ouerdane, L.; and Mester, Z. Production and characterization of fully selenomethionine-labeled Saccharomyces cerevisiae. J. Agric. Food Chem. **2008**, 56, 11792–11799

49. Cantor, A.H.; Langevin, M.L.; Noguchi, T.; Scott, M.L. Efficacy of selenium in selenium compounds and feedstuffs for prevention of pancreatic fibrosis in chicks. J. Nutr. **1975**, 105, 106–111.

50. Payne, R.L.; Southern, L.L. Comparison of inorganic and organic selenium sources for broilers. Poult. Sci. **2005**, 84, 898–902

51. da Silva, ICM; Ribeiro, A.M.L.; Canal, C.W.; Trevizan, L.; Macagnan, M.; Gonçalves, T.A.; Hlavac, N.R.C.; de Almeida, L.L.; Pereira, R.A. The impact of organic and inorganic selenium on the immune system of growing broilers submitted to immune stimulation and heat stress. Rev. Bras. Cienc. Avic. **2010**, 12, 247–254.

52. Oliveira, T.F.B.; Rivera, D.F.R.; Mesquita, F.R.; Braga, H.; Ramos, E.M.; Bertechini, A.G. Effect of different sources and levels of selenium on performance, meat quality, and tissue characteristics of broilers. J. Appl. Poult. Res. **2014**, 23, 15–22.

53. Tsuji, Y.; Mikami, T.; Anan, Y.; Ogra, Y. Comparison of selenohomolanthionine and selenomethionine in terms of selenium distribution and toxicity in rats by bolus administration. Metallomics **2010**, 2, 412–418

54. Celi, P.; Selle, P.H.; Cowieson, A.J. Effects of organic selenium supplementation on growth performance, nutrient utilisation, oxidative stress and selenium tissue concentrations in broiler chickens. Anim. Prod. Sci. **2014**, 54, 966–971.

55. Suzuki, K.T.; Doi, C.; Suzuki, N. Metabolism of 76Se-methylselenocysteine compared with that of 77Se-selenomethionine and 82Se-selenite. Toxicol. Appl. Pharmacol. **2006**, 217, 185–95

56. Anan, Y.; Ogra, Y. Toxicological and pharmacological analysis of selenohomolanthionine in mice. Toxicol. Res. **2013**, 2, 115–122.

57. Kaufman, N.; Klavins, J.V.; Kinney, T.D. Pancreatic damage induced by excess methionine. Arch. Pathol. **1960**, 70, 331–337.

58. Henman, D.J.; (Riverlea Pty Ltd., Corowa, Australia). Personal communication, 2014.

Chapter 2

BLOOD GASES, BIOCHEMISTRY, AND HEMATOLOGY OF GALAPAGOS GREEN TURTLES (CHELONIA MYDAS)

Gregory A. Lewbart[1], Maximilian Hirschfeld[2], Judith Denkinger[2], Karla Vasco[2], Nataly Guevara[2], Juan Garcia[3], Juanpablo Munoz[2], Kenneth J. Lohmann[4]

[1] North Carolina State University, College of Veterinary Medicine, Raleigh, North Carolina, United States of America

[2] University San Francisco de Quito, Galapagos Science Center, Puerto Baquerizo Moreno, Galapagos, Ecuador

[3] Galapagos National Park Service, Puerto Ayora, Galapagos, Ecuador

[4] Department of Biology, University of North Carolina, Chapel Hill, North Carolina, United States of America

ABSTRACT

The green turtle, *Chelonia mydas,* is an endangered marine chelonian with a circum-global distribution. Reference blood parameter intervals have been published for some chelonian species, but baseline hematology, biochemical, and blood gas values are lacking from the Galapagos sea turtles. Analyses were done on blood samples drawn from 28 green turtles captured in two foraging locations on San Cristóbal Island (14 from each site). Of these turtles, 20 were immature and of unknown sex; the other eight were males (five mature, three immature). A portable blood analyzer (iSTAT) was used to obtain near immediate field results for pH, lactate, pO_2, pCO_2, HCO_3^-, Hct, Hb, Na, K, iCa, and Glu. Parameter values affected by temperature were corrected in two ways: (1) with standard formulas; and (2) with auto-corrections made by the iSTAT. The two methods yielded clinically equivalent results. Standard laboratory hematology techniques were employed for the red and white blood cell counts and the hematocrit determination, which was also compared to the hematocrit values generated by the iSTAT. Of all blood analytes, only lactate concentrations were positively correlated with body size. All other values showed no significant difference between the two sample locations nor were they correlated with body size or internal temperature. For hematocrit count, the iSTAT blood analyzer yielded results indistinguishable from those obtained with high-speed centrifugation. The values reported in this study provide baseline data that may be useful in comparisons among populations and in detecting changes in health status among Galapagos sea turtles. The findings might also be helpful in future efforts to demonstrate associations between specific biochemical parameters and disease.

INTRODUCTION

The green turtle (*Chelonia mydas*), also known as the black turtle in the Pacific Ocean, is a marine chelonian inhabiting oceans throughout the world [57]. The green turtle is currently listed as Endangered on the IUCN Red List, and no commercial use is permitted under CITES Appendix I. Major threats to green turtle populations include habitat destruction, pollution, disease, consumption of meat and eggs by local populations, fishing gear entanglement, and consumption of plastics and other anthropogenic materials [16], [37], [42], [43], [59]. Health assessments of green turtles may therefore have implications for wildlife biology and species conservation. Considerable research on natural history has been performed in this species and studies on the health parameters of green turtles, while still relatively limited, have increased dramatically in the last 5 years[1], [4], [5], [8], [17], [18], [22], [27], [30], [31], [39], [40], [45],[50]. A recent review summarizes the health of wild sea turtles and methods of assessment, including blood parameters [20].

Biochemical and hematology parameters, as measured in peripheral blood, are a useful diagnostic tool in animal health management [2], [3]. To determine the significance of alterations in biochemical and hematological values, it is essential to establish species-specific (or at least taxon-specific) normal values for parameters of interest. Reference intervals for certain chelonian species, including sea turtles [2], [5], [6], [8]–[10], [12], [15], [18], [19], [24],[33], [34], [39], [46], [60], have been widely investigated. Additionally, wild sea turtle blood gas values have been reported and analyzed [28], [55]. However, few studies have combined blood gas, biochemistry, and hematology parameters from the same geographic subpopulation of sea turtles. The present study evaluates selected blood gas, blood biochemical, and hematology parameters from 28 wild-caught green turtles (*Chelonia mydas*) in two coastal foraging areas adjacent to San Cristóbal Island, Galapagos, Ecuador. The current study appears to be the first health assessment study on Galapagos sea turtles and on Western Hemisphere green turtles south of the Equator.

MATERIALS AND METHODS

Ethics Statement

This study was performed as part of a population health assessment approved and supported by the Galapagos National Park Service (Permit # PC-35-12 to J. Denkinger) and approved by the Universidad San Francisco de Quito ethics and animal handling protocol. All handling and sampling procedures were consistent with standard vertebrate protocols and veterinary practices.

Turtle Capture and Sampling

Turtles were captured in shallow water within 100 m from the shore by swimmers using standard snorkel, mask, and fins. The animals were carried to shore and placed on the beach. Turtles were aligned with their heads facing the water so that the slight incline of the beach assisted efforts to obtain blood samples from the dorsal jugular vein. A few turtles were sampled while still partially in water when a receding tide exposed rocks, which prevented turtles from being carried safely onto the sandy beach.

Blood samples were obtained within approximately 5 minutes of capture. Fourteen turtles were captured, examined, and sampled from La Loberia Beach (0° 55′ 40″ S, 89° 36′ 43″ W) on 30 June, 2013 and 14 turtles were similarly captured and sampled on 1 July, 2013 at a second site at Punta Carola (0° 53′ 26″ S, 89° 34′ 46″ W). Both study sites are shallow bays of similar size, sheltered from wave exposure, and were identified as important foraging and resting areas where individual turtles show very high site fidelity [16]. To avoid capturing the same individual more than once, a line of white zinc oxide ointment was applied to each turtle's carapace after a blood sample had been obtained and before the animal was released. The line was clearly visible underwater and remained on for at least a day, but disappeared soon after (Hirschfeld, personal observation).

In addition to obtaining a blood sample from each turtle, photographs were taken of each side of the head, as well as the carapace. These images can be used to identify individuals in the future [35], [44], [47].

Blood Sample Collection and Handling

All turtles were manually restrained and blood samples of approximately 2.5 mLs were obtained from either the left or right dorsal jugular sinus using a heparinized 22 gauge needle attached to a 3.0 mL syringe. The blood was then immediately divided into sub-samples. Some subsamples were used for making blood films on clean glass microscope slides, some were stored on ice in sterile plastic vials for future analyses, and others were loaded into the CG-8+ and CG-4+ iSTAT cartridges within 10 minutes of sample collection.

Blood Gas and Biochemistry Parameters

The blood gas, electrolyte, and biochemistry results were obtained using an iSTAT Portable Clinical Analyzer (Heska Corporation, Fort Collins, Colorado, USA) with CG8+ and CG4+ cartridges. The iSTAT is a portable, handheld, battery-operated electronic device with the ability to measure a wide variety of blood gas, chemistry, and basic hematology parameters with only a few drops (0.095 mL) of whole, non-coagulated blood. The following parameters were measured and recorded: pH, lactate, pO_2, pCO_2, HCO_3^-, Hct, Hb, Na, K, iCa, and glucose.

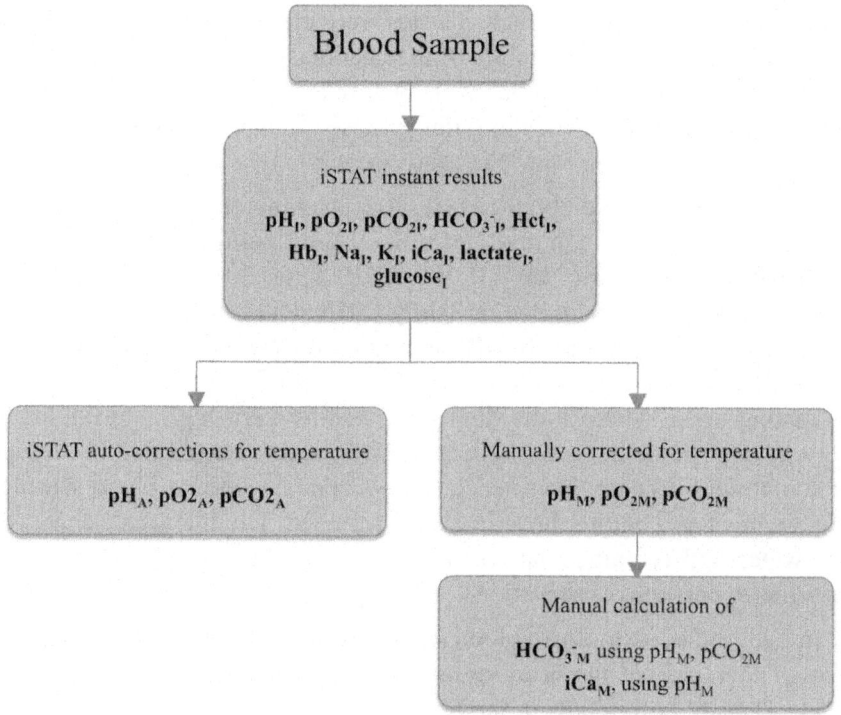

Figure 1: Flow chart of the iSTAT sample handling and calculations.

doi:10.1371/journal.pone.0096487.g001.

The iSTAT device analyzed the blood at 37°C then corrected pH, pO_2, and pCO_2 for body temperature once this information was entered. Because the validity of the iSTAT temperature corrections has been questioned by some authors [13], [28], we also manually calculated an independent set of corrections for pH, pO_2, pCO_2, iCa, and HCO_3^- based on the turtle cloacal temperature (Ti) at the time of sampling [5], [38], using the equations listed below. Both sets of values (i.e., those derived from auto-corrections from the iSTAT and those derived from independent calculations) are reported in Table 1. The methodology for obtaining the various measured and calculated values is summarized in Figure 1. For clarity, instant values obtained from the iSTAT are denoted by an 'I' subscript. Those that were auto-corrected for temperature by entering the turtle's body temperature into the iSTAT are denoted by 'A' subscript. Values that were manually corrected for temperature using the equations shown below are denoted by 'M' subscript.

Table 1: Mean, standard deviation, and range for blood gas and blood biochemical values for wild Galapagos green turtles

Analyte	n	Mean	SD	Min	Max
$HCO_3^-{}_I$ (mmol/L)	24	43.8	5.8	35.6	58.2
$HCO_3^-{}_M$ (mmol/L)	28	41.1	5.6	33.0	54.4
pH_A	28	7.477	0.085	7.273	7.626
pH_M	28	7.441	0.075	7.256	7.568
pCO_{2A} (mmHg)	24	46.5	7.1	32.4	65.4
pCO_{2M} (mmHg)	28	49.0	9.2	32.4	68.3
pO_{2A} (mmHg)	28	22	5	14	32
pO_{2M} (mmHg)	28	53	10	36	72
Na_I (mmol/L)	28	148	3	143	153
K_I (mmol/L)	28	3.4	0.5	2.7	4.3
iCa_I (mmol/L)	27	0.87	0.14	0.64	1.18
iCa_M (mmol/L)	27	0.79	0.12	0.57	1.06
Glu_I(Umg/dl)	28	60	9	46	82
Hct_I(L/L)	28	0.24	0.05	0.17	0.38
Hb_I (g/L)	28	80	16	58	129
Lac_I (mmol/L)	28	3.73	2.44	0.8	8.73

'I' subscript denotes values obtained through the instant iSTAT analysis, 'M' subscript indicates values manually corrected for temperature using standard equations (see Results and Figure 1), and 'A' subscript indicates values that were auto-corrected for temperature by the iSTAT after a turtle's cloacal temperature was entered into the iSTAT.
doi:10.1371/journal.pone.0096487.t001

$$pH_M = 0.014(\Delta T_{37\text{-}25}) + 0.005(\Delta T_{25\text{-}Ti})$$

$$+ pH_I [\text{at temperatures} < 25°C].$$

Anderson et al. (2011) [5] simplifies the above for Ti values below 25°C:

$$pH_M = 0.168 + 0.005(\Delta T_{25\text{-}Ti}) + pH_I [\text{at temperatures} < 25°C].$$ (We note that a typographical error in [5] shows 0.0168 instead of 0.168.).

Because pH has an effect on iCa calculation, the formula below was used in the current study[5], [21], [32]: $iCa_M = iCa_I (1 + 0.53[pH_I - pH_M])$

The following formulas [5] were used to correct pCO$_2$ and pO$_2$ for temperature: $pO_{2M} = pO_{2I}(10^{-0.0058\Delta T})$ and $pCO_{2M} = pCO_{2I}(10^{-0.019\Delta T})$

For the calculation of bicarbonate the following adaptation of the Henderson-Hasselbach Equation was used [36], [51]: $HCO_{3\text{-}M} = \alpha CO_2 \times pCO_{2M} \times 10^{(pHM - pKa)}$

The corresponding αCO_2 and pKa values were calculated using the formulas published by Stabenau and Heming (1993) [51].

Hematology

Heparinized whole blood was stored on ice immediately after collection and refrigerated overnight. Total erythrocyte and leukocyte counts (Table 2) were obtained within 24 hours using Natt Herick's stain and a Neubaeuer hemocytometer [11]. Hematocrit was determined using high-speed centrifugation of blood-filled microhematocrit tubes. Differential white blood cell counts were conducted by examining 100 white blood cells on a peripheral smear stained with Wright-Giemsa stain.

Table 2: Mean, standard deviation, and range for manually analyzed hematology values of wild Galapagos green turtles

Analyte	n	Mean	SD	Min	Max
Hct (L/L)	28	0.236	0.048	0.17	0.38
WBC (x10^9/L)	28	6.58	4.02	1.76	22.4
Heterophils (%)	28	16.4	6.6	8.0	35.0
Lymphocytes (%)	28	50.5	7.72	33.0	67.0
Monocytes (%)	28	12.0	5.73	6.0	32.0
Azurophils (%)	28	0.04	0.19	0	1.0
Eosinophils (%)	28	20.8	5.99	5.0	31.0
Basophils (%)	28	0.36	0.56	0	2.0
Heterophils (x10^9/L)	28	1.06	0.66	0.23	2.59
Lymphocytes (x10^9/L)	28	3.42	2.24	0.97	119
Monocytes (x10^9/L)	28	0.79	0.60	0.18	2.92
Azurophils (x10^9/L)	28	0.001	0.01	0	0.03
Eosinophils (x10^9/L)	28	1.41	1.07	0.37	5.16
Basophils (x10^9/L)	28	0.02	0.04	0.00	0.13

doi:10.1371/journal.pone.0096487.t002

Turtle Measurements and Body Temperature

A flexible measuring tape was used to determine curved carapace lengths (CCL). The sex of an immature sea turtle cannot be determined with confidence on the basis of an external examination [8]. Green turtles in the Galapagos have the slowest reported growth rate for this species [26]. Although the smallest nesting female recorded for the archipelago is 60.7 cm CCL, two peaks for nesting females exist at 80 and 95 cm CCL [59]. Furthermore, laparoscopic gonad assessment of green turtles has shown size at maturity is highly variable [20]. Therefore, we used a relatively large size of 80 cm CCL as a threshold to distinguish immature from mature individuals. Adults were sexed on the basis of the external sexual dimorphism of adult green turtles, while some smaller individuals could also be sexed as males by their evident secondary sexual characteristics [54], [57]. An EBRO® Compact J/K/T/E Thermocouple Thermometer was used to obtain all temperature readings (model EW-91219-40; Cole-Parmer, Vernon Hills, Illinois, USA 60061). Core body temperatures were recorded from the cloaca using the probe T PVC epoxy tip 24GA×3 ft in length.

Statistical Analysis

First, the iSTAT blood chemistry results of all overlapping values of the CG8+ and CG4+ cartridges (all analytes except lactate) were compared using paired t-tests. The iSTAT results for hematocrit were compared (paired t-test) to the results of manually determined hematocrit. Subsequently, we grouped values for each blood analyte by foraging site and compared them using Student's t-test. Group sizes for different age classes (mature/immature) as well as sex (male/female) were too small to perform statistical analysis. Linear regressions were used to examine the relationship between the turtles' body size or body temperature and the measured blood chemistry analytes. A standard alpha level of p=0.05 was used for all statistical tests using R statistical software, version 3.0.2 (R Development Core Team).

RESULTS

Turtle Demographics and Health Status

A total of 28 green turtles, 23 immature (CCL<80 cm) and 5 mature (CCL>80 cm) were sampled. Half of the animals (n=14) were captured at each study location. The mean CCL was 68.0 cm; values ranged from 41.8 cm CCL for the smallest immature to 84.6 cm CCL for the largest adult male. All of the mature sea turtles (n=5) were males and an additional three males of 79, 76.5 and 71.1 cm CCL were identified. Internal body temperature varied little among individuals, with a mean of 20.6°C and ranging between 19.4°C and 22.5°C. All 28 sea turtles appeared clinically healthy, had no barnacle or excessive algae growth on their carapaces and were bright, alert, and responsive. Many were observed feeding just prior to capture. One turtle had a large piece of monofilament fishing line on the right front flipper causing a strangulating lesion at the elbow that was deemed significant but had not compromised blood flow to the distal limb. The ligature was cut free of the turtle and discarded.

Blood Biochemical Analysis

Values obtained from the CG4+ and CG8+ cartridges were consistent and not statistically different; thus, all overlapping iSTAT values reported are from the first cartridge run for all samples (CG8+). Table 1 displays the biochemistry, blood gas, and hematology results. In a few samples the iSTAT blood analyzer was unable to calculate values for some of the parameters resulting in a slightly smaller n. The hematocrit figures generated by the iSTAT were not significantly different from the manually determined values. There were no statistical differences between the two sampling sites for any values of blood gas, electrolyte, biochemical, and hematology parameters. In addition, none of these values, with one exception, were correlated either with body size or body temperature.

The only blood analyte found to be positively correlated ($r^2=0.172$, $p<0.05$) with body length (CCL) was lactate (Figure 2). Male green turtles appeared to have a higher lactate concentration than the rest of the unsexed and immature animals, but low sample size and uncertainty of determining sex of smaller individuals prevented a comparative analysis.

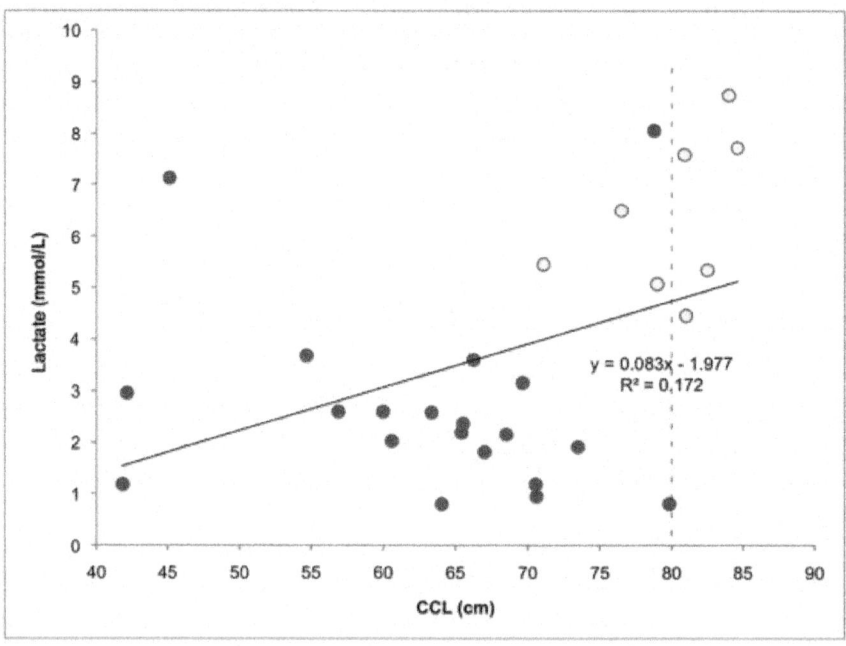

Figure 2: Scatter plot with linear regression line showing the correlation between sea turtle body size and blood lactate levels.

Closed circles indicate data for immature turtles of undetermined sex. Open circles indicate turtles that could be identified as male based on secondary sexual characteristics. The dotted line indicates the size threshold (80 cm CCL) used to classify individuals as immature or mature.

DISCUSSION

Previous studies have reported baseline blood gas, biochemical, and hematology parameters for some chelonian species, including sea turtles. Clinicians desire species-specific baseline values for the parameters commonly measured by commercial blood gas and chemistry analyzers. When working with reptiles, species specificity of health data is especially important, due to the diverse environmental conditions that exist in different habitats. This study reports the first set of blood gas, biochemistry, and hematology values in Galapagos green turtles and, indeed, in any subequatorial Western Hemisphere green turtle population.

Field sampling of green turtles was performed after the nesting season, when a resident subpopulation appears to remain in the Galapagos [16], [26], [48]. In addition, the two foraging areas sampled are frequented by resident turtles, some of which have been observed repeatedly for up to 5 years [16]. Thus, the turtles sampled in our study were most likely resident turtles, although the possibility that some of the turtles were transient visitors rather than residents cannot be ruled out.

The small sample size (n=28) in the present study precluded the calculation of formal reference intervals, a process that would require a minimum of 120 individuals [23]. However, given the limited sample size available, and the lack of previously published data regarding biochemical, blood gas, and hematology parameters in resident Galapagos green turtles of the eastern tropical Pacific subspecies, these results provide a useful starting point for clinicians and researchers. All 28 turtles were judged to be clinically healthy and their blood parameters support this assessment.

In general, most of the blood parameters we recorded were similar to those reported previously for healthy green turtles [1], [17], [22], [30], [40]. For example, Harms et al. (2009) [30] reported iSTAT values obtained from five green turtles kept in captivity for several years. Relative to these, Galapagos green turtles had slightly lower pO_2, higher pCO_2, and slightly lower lactate. Similarly, Anderson et al. (2009) [5] reported median values for green turtles in North Carolina, U.S.A.; relative to the Galapagos turtles, most values were nearly equivalent, but significant differences existed in hematocrit and blood glucose. The hematology values we observed appear consistent with ranges reported for other clinically healthy green turtles and other chelonian species [1], [5], [8], [17], [25], [45], [56], [58].

In our study, parameter values affected by temperature were corrected based on published, standard formulas; these corrected values sometimes differed from auto-corrected iSTAT values (Table 1) but the differences did not appear to be clinically important. We conclude that iSTAT auto-corrected values are usually sufficient for clinical applications in the field, but also suggest that when accuracy is paramount, investigators studying animals with body temperatures below 37°C should be cautious about relying on auto-corrected iSTAT values for parameters that are influenced by temperature.

Galapagos green turtles in general had higher lactate values than other sea turtles [29], [32],[51], [52]. Only gillnet trapped green turtles [50], loggerhead turtles captured in shrimp trawlers, and cold-stunned Kemp's ridley turtles (sampled after 2 to 3 days of hospitalization) [28], [36], had higher lactate. Loggerheads caught using pound nets had low lactate levels when sampled initially, but lactate concentrations were significantly higher after 30 min [28]. Similarly, Berkson (1966) [7] found no increase in blood lactate of green turtles until 30 to 60 min after forced submergence. Since most voluntary dives are thought to be aerobic and no difference has been found in the dive time of green turtles differing in size, none of these factors readily explain the overall high blood lactate and its increase with body size [41], [49]; Figure 2.

Several factors can influence chelonian biochemistry values, including environmental conditions, age, and sex. Relative to males, females frequently possess higher albumin, calcium, cholesterol, phosphorus, and triglyceride values, which are generally attributed to vitellogenesis [9], [14], [46], [58]. We could not investigate differences between sexes with regard to calcium or other parameters in the Galapagos green turtles due to the difficulty of determining the sex of immature animals in the field [53].

Age also affects some biochemical parameters in sea turtles. Casale et al. (2009) [12] reported that juvenile loggerhead turtles had lower values of albumin, calcium, globulins, hematocrit, total protein, and triglycerides than did adult animals, a difference probably at least partly attributable to egg production in adult females. In green turtles, albumin, total protein, and triglyceride levels were found to increase with body size while calcium, glucose, potassium, and sodium did not [39].

A number of papers have examined differences between clinically healthy and ill or compromised green sea turtles. Aguirre et al. (2009) [1] determined, by examining hematology, biochemistry, and adrenocortical values, that green turtles afflicted with green turtle fibropapillomatosis (GTFP) were immunocompromised and stressed. Similarly, Anderson et al. (2011) [5] reported that many important biochemical parameters (including calcium, glucose, total protein, and some electrolytes) were significantly lower, and uric acid and blood urea nitrogen higher, in cold-stunned green turtles versus healthy, wild individuals.

In a recent study with a very large sample size, Flint et al. (2010) [18] report blood reference intervals (RIs) for 290 Australian green turtles, 211 of which were deemed clinically healthy. All of the 25 turtles judged unhealthy possessed at least one value outside of the RIs. Aside from higher blood glucose levels in the Australian turtles, healthy green turtles in Australia and the Galapagos appear to have similar blood parameters.

In summary, data reported in this study represent an important step toward determining the normal range of values against which future blood gas and biochemistry results in green turtles can be compared. Such assessments are important for health monitoring and disease diagnostics. Because green turtles are an endangered species with importance in the wildlife biology research community and the aquarium/zoo industry, health assessments are important from the standpoint of sustainable conservation and management. These results add to a growing database of knowledge about health management in wild chelonian species. Future research should continue to establish reference values in this species and facilitate comparisons of blood values across age groups and disease states.

ACKNOWLEDGMENTS

We thank Diana Amoguimba, Eduardo Espinoza, Craig Harms, Tillie Laws, Carlos Mena, Philip Page, Kent Passingham, Carlos Valle-Castillo, Galo Quezada, Erich Stabenau, and Stephen Walsh for their support and assistance with this project.

REFERENCES

1. Aguirre AA, Balazs GH, Spraker TR, Gross TS (1995) Adrenal and hematological responses to stress in juvenile green turtles (*Chelonia mydas*) with and without fibropapillomas. Physiol Zool 68(5): 831–854.

2. Aguirre AA, Balazs GH (2000) Plasma biochemistry values of green turtles (*Chelonia mydas*) with and without fibropapillomas in the Hawaiian Islands. Comp Haematol Int 10: 132–137. doi: 10.1007/s005800070004

3. Aguirre AA, Lutz P (2004) Sea turtles as sentinels of marine ecosystem health: is fibropapillomatosis an indicator? Eco Health 1(3): 275–283.

doi: 10.1007/s10393-004-0097-3

4. 4.Anderson NL, Wack RF, Hatcher R, Wack F (1997) Hematology and clinical chemistry reference ranges for clinically normal, captive New Guinea Snapping Turtles (*Elseya novaeguineae*) and the effects of temperature, sex, and sample type. J Zoo Wildl Med 28: 394–403.

5. Anderson ET, Harms CA, Stringer EM, Cluse WM (2011a) Evaluation of hematology and serum biochemistry of cold-stunned green sea turtles (*Chelonia mydas*) in North Carolina, USA. J Zoo Wild Med 42(2): 247–255. doi: 10.1638/2010-0217.1

6. Anderson ET, Minter LJ, EO Clarke III, Mroch RM, Beasley JF, et al.. (2011b) The effects of feeding on hematological and plasma biochemical profiles in green (*Chelonia mydas*) and Kemp's Ridley (*Lepidochelys kempii*) sea turtle. Vet Med Internat doi: 10.4061/2011/890829.

7. Berkson H (1966) Physiological adjustments to prolonged diving in the Pacific green turtle (*Chelonia mydas agassizii*). Comp Biochem Phys 18: 101–119. doi: 10.1016/0010-406x(66)90335-5

8. Bolton AB, Bjorndal KA (1992) Blood profiles for a wild population of green turtles (*Chelonia mydas*) in the Southern Bahamas: Size-specific and sex-specific relationships. J Wild Dis 28(3): 407–413. doi: 10.7589/0090-3558-28.3.407

9. Brenner D, Lewbart G, Stebbins M, Herman D (2002) Health survey of wild and captive Bog Turtles (*Clemmys muhlenbergii*) in North Carolina and Virginia. J Zoo Wild Med 33: 311–316.

10. Metabolic and respiratory status of stranded juvenile loggerhead sea turtles (*Caretta caretta*): 66 cases (2008–2009). J Am Vet Med Assoc 242: 396–401. doi: 10.2460/javma.242.3.396

11. Campbell TW (1995) *Avian Hematology and Cytology*, 2nd Ed. Iowa State University Press, Ames, IA. Pp. 7–11.

12. Casale AB, Camacho M, López-Jurado LF, Juste C, Orós J (2009) Comparative study of hematologic and plasma biochemical variables in Eastern Atlantic juvenile and adult nesting loggerhead sea turtles (*Caretta caretta*). Vet Clin Path 38: 213–218. doi: 10.1111/j.1939-165x.2008.00106.x

13. Chittick EJ, Stamper MA, Beasley JF, Lewbart GA, Horne WA (2002) Medetomidine, ketamine, and sevoflurane for anesthesia of injured loggerhead sea turtles: 13 cases (1996–2000). J Am Vet Med Assoc 221: 1019–1025. doi: 10.2460/javma.2002.221.1019

14. Christopher MM, Berry KH, Wallis IR, Nagy KA, Henen BT, et al.

(1999) Reference intervals and physiologic alterations in hematologic and biochemical values of free–ranging Desert Tortoises in the Mojave Desert. J Wild Dis 35: 212–238. doi: 10.7589/0090-3558-35.2.212

15. Deem SL, Norton TM, Mitchell M, Segars A, Alleman AR, et al. (2009) Comparison of blood values in foraging, nesting, and stranded loggerhead turtles (*Caretta caretta*) along the coast of Georgia, USA. J Wild Dis 45: 41–56. doi: 10.7589/0090-3558-45.1.41

16. **16.**Denkinger J, Parra M, Munoz JP, Carrasco C, Murillo JC, et al. (2013) Are boat strikes a threat to sea turtles in the Galapagos Marine Reserve? Ocean Coastal Mgt 80: 29–35. doi: 10.1016/j.ocecoaman.2013.03.005

17. Flint M, Morton JM, Limpus CJ, Patterson-Kane JC, Murray PJ, et al. (2010a) Development and application of biochemical and haematological reference intervals to identify unhealthy green sea turtles (*Chelonia mydas*). Vet J 185: 299–304. doi: 10.1016/j.tvjl.2009.06.011

18. Flint M, Patterson-Kane JC, Limpus CJ, Murray PJ, Mills PC (2010b) Health surveillance of stranded green turtles in Southern Queensland, Australia (2006–2009): An epidemiological analysis of causes of disease and mortality. Eco Health 7: 135–145. doi: 10.1007/s10393-010-0300-7

19. Flint M, Morton JM, Limpus CJ, Patterson-Kane JC, Murray PJ, et al. (2010c) Reference intervals for plasma biochemical and hematologic measures in loggerhead sea turtles (*Caretta caretta*) from Moreton Bay, Aust J Wild Dis. 46: 731–741. doi: 10.7589/0090-3558-46.3.731

20. Flint M (2013) Free ranging sea turtle health. In: Wyneken J, Lohmann KJ, Musick JA (eds.) *The Biology of Sea Turtles Vol. III*. CRC Press, Inc., Boca Raton, FL. 379–397.

21. **21.**Fogh-Anderson N (1981) Ionized calcium analyzer with a built-in pH correction. Clin Chem 27: 1264–1267.

22. Fong CL, Chen HC, Cheng IJ (2010) Blood profiles from wild populations of green sea turtles in Taiwan. J Vet Med Anim Health 2(2): 8–10.

23. Geffre A, Friedricks K, Harr K, Concordet D, Trumel C, et al. (2009) Reference values: a review. Vet Clin Path 38: 288–298. doi: 10.1111/j.1939-165x.2009.00179.x

24. Gelli D, Ferrari V, Zanella A, Arena P, Pozzi L, et al. (2008) Establishing physiological blood parameters in the loggerhead sea turtle (*Caretta caretta*). Eur J Wild Res 55: 59–63. doi: 10.1007/s10344-008-0214-7

25. Gibbons PM, Klaphake E, Carpenter JW (2013) Reptiles. In: Exotic Animal Formulary,Fourth Edition (Carpenter J, Marion C eds.), Elsevier

Saunders, St. Louis, MO, pp. 84–170.

26. Green D (1993) Growth rates of wild immature green turtles in the Galapagos Islands, Ecuador. J Herp 27(3): 338–341. doi: 10.2307/1565159

27. Hamann M, Schauble CS, SimonT, Evans S (2006) Demographic and health parameters of green sea turtles *Chelonia mydas* foraging in the Gulf of Carpentaria, Australia. Endang Spec Res 2: 81–88. doi: 10.3354/esr002081

28. Harms CA, Mallo KM, Ross PM, Segars A (2003) Venous blood gases and lactates of wild loggerhead sea turtles (*Caretta caretta*) following two capture techniques. J Wild Dis 39(2): 366–374. doi: 10.7589/0090-3558-39.2.366

29. Harms CA, Eckert SA, Kubis SA, Campbell M, Levenson DH, et al. (2007) Field anesthesia of leatherback sea turtles (*Dermochelys coriacea*). Vet Rec 161: 15–21. doi: 10.1136/vr.161.1.15

30. Harms CA, Eckert SA, Jones TT, Dow Piniak WE, Mann DA (2009) A technique for underwater anesthesia compared with manual restraint of sea turtles undergoing auditory evoked potential measurements. J Herp Med Surg 19(1): 8–12.

31. Hasbún CR, Lawrence AJ, Naldo J, Samour JH, Al-Ghais SM (1998) Normal blood chemistry of free-living green sea turtles, *Chelonia mydas*, from the United Arab Emirates. Comp Haematol Int 8(3): 174–177. doi: 10.1007/bf02642510

32. Innis CJ, Tlusty M, Merigo C, Weber ES (2007b) Metabolic and respiratory status of cold-stunned Kemp's ridley sea turtles (*Lepidochelys kempii*). J Comp Physiol B 177: 623–630.

33. Innis CJ, Ravich JB, Tlusty MF, Hoge MS, Wunn DS, et al. (2009) Hematologic and plasma biochemical findings in cold-stunned Kemp's Ridley Turtles: 176 cases (2001–2005). J Am Vet Med Assoc 235: 426–32. doi: 10.2460/javma.235.4.426

34. Innis CJ, Merigo C, Dodge K, Tlusty M, Dodge M, et al. (2010) Health evaluations of leatherback turtles (*Dermochelys coriacea*) in the Northwestern Atlantic during direct capture and fisheries gear disentanglement. Chelon Conserv Biol 9(2): 205–222. doi: 10.2744/ccb-0838.1

35. 35.Jean C, Ciccione S, Talma E, Ballorain K, Bourjea J (2010) Photo-identification method for green and hawksbill turtles- First results from Reunion. Indian Ocean Turt News 11: 8–13.

36. Keller KA, Innis CJ, Tlusty MF, Kennedy AE, Bean SB, et al. (2012)

Metabolic and respiratory derangements associated with death in cold-stunned Kemp's ridley turtles (*Lepidochelys kempii*): 32 cases (2005–2009). J Am Vet Med Assoc 240: 317–323. doi: 10.2460/javma.240.3.317

37. Koch V, Nichols WJ, Peckham H, De La Toba V (2006) Estimates of sea turtle mortality from poaching and bycatch in Bahia Magalena, Baja California Sur, Mexico. Biol Conserv 128: 327–334. doi: 10.1016/j.biocon.2005.09.038

38. Kraus DR, Jackson DC (1980) Temperature effects on ventilation and acid-base balance of the green turtle. Am J Physiol 239: R254–R258.

39. Labrada-Martagón V, Méndez-Rodriguez LC, Gardner SC, López-Castro M, Zenteno-Savin T (2010a) Health indices of the green turtle (*Chelonia mydas*) along the Pacific coast of Baja California Sur, Mexico. I. Blood biochemistry values. Chel Conserv Biol 9(2): 162–172. doi: 10.2744/ccb-0806.1

40. Labrada-Martagón V, Méndez-Rodriguez LC, Gardner SC, Cruz-Escalona VH, Zenteno-Savin T (2010b) Health indices of the green turtle (*Chelonia mydas*) along the Pacific coast of Baja California Sur, Mexico. II. Body condition index. Chel Conserv Biol 9(2): 173–183. doi: 10.2744/ccb-0807.1

41. Lutcavage ME, Lutz PL (1997) Diving Physiology. In: Lutz PL and Musick JA (eds.) The Biology of Sea Turtles. Mar Sci Ser CRC Press Boca Raton, FL.

42. .Mancini A, Koch V, Seminoff JA, Madon B (2011) Small-scale gillnet fisheries provoke massive green turtle (*Chelonia mydas*) mortality: a case study from Baja California Sur, Mexico. ORYX 46 (1), 69e77.

43. Parra M, Dêem SL, Espinoza E (2011) Green turtle (*Chelonia mydas*) mortality in the Galapagos Islands, Ecuador during the 2009–2010 nesting season. Mar Turtle News 130: 10–15.

44. Reisser J, Proietti M, Kinas P, Sazima I (2008) Photographic identification of sea turtles: method description and validation, with an estimation of tag loss. End Sp Res 5: 73–82. doi: 10.3354/esr00113

45. Samour JH, Hewlett JC, Silvanose C, Hasbún CR, AlpGhais SM (1998) Normal haematology of free-living green sea turtles (*Chelonia mydas*) from the United Arab Emirates. Comp Haematol Int 8(2): 102–107. doi: 10.1007/bf02642499

46. Santoro M, Meneses A (2007) Haematology and plasma chemistry of breeding Olive Ridley Sea Turtles (*Lepidochelys olivacea*). Vet Rec 161: 818.

47. Schofield G, Katselidis KA, Dimopoulos P, Pants JD (2008) Investigating the viability of photo-identification as an objective tool to study endangered sea turtle populations. J Exp Mar Bio Ecol 360: 103–108. doi: 10.1016/j.jembe.2008.04.005

48. Seminoff JA Jones TT, Marshall GJ (2006) Underwater behaviour of green turtles monitored with video-time-depth recorders: what's missing from dive profiles? Mar Ecol Prog Ser 322: 269–280. doi: 10.3354/meps322269

49. Seminoff JA, Zarate P, Coyne M, Foley DG (2008) Post-nesting migrations of Galapagos green turtles *Chelonia mydas* in relation to oceanographic conditions: integrating satellite telemetry with remotely sensed ocean data. Endan Spec Res 4: 57–72. doi: 10.3354/esr00066

50. Snoddy JE, Landon M, Slanvillain G, Southwood A (2009) Blood biochemistry of sea turtles captured in gillnets in the lower Cape Fear River, North Carolina, USA. J Wildlife Management 73(8): 1394–1401. doi: 10.2193/2008-472

51. **51.**Stabenau EK, Heming TA (1993) Determination of the constants of the Henderson-Hasselbalch equation, (alpha) CO2 and pKa, in sea turtle plasma. J Exp Bio 180(1): 311–314.

52. .Stacy NI, Innis CJ, Hernandez JA (2013) Development and evaluation of three mortality prediction indices for cold-stunned Kemp's ridley sea turtles (*Lepidochelys kempii*). Conserv Physiol. doi: 10.1093/conphys/cot003.

53. Wabnitz C, Pauly D (2008) Length-weight relationships and additional growth parameters for sea turtles. In: Palomares, MLD, Pauly D (eds.) Fish Cent Res Rep 16(10): 92–101.

54. Wibbels T (1999) Diagnosing the sex of sea turtles in foraging habitats. *In:* Eckert, KL, Bjorndal KA, Abeu-Grobois FA, Donnelly M (eds.). Research and Management Techniques for the Conservation of Sea Turtles. IUCN/SSC Marine Turtle Specialist Group Publication No. 4 1–5.

55. Wolf KN, Harms CA, Beasley JF (2008) Evaluation of five clinical chemistry analyzers for use in health assessment in sea turtles. J Am Vet Med Assoc 233: 470–475. doi: 10.2460/javma.233.3.470

56. Wood FE, Ebanks GK (1984) Blood cytology and hematology of the green sea turtle,*Chelonia mydas*. Herpetologica 40(3): 331–336.

57. Wyneken J, Lohmann KJ, Musick JA, eds. (2013) *The Biology of Sea Turtles Vol. III.* CRC Press, Inc., Boca Raton, FL. 379–397.

58. Yilmaz N, Tosunoglu M (2010) Hematology and some plasma biochemistry values of free–living freshwater turtles (*Emys orbicularis* and *Mauremys rivulata*) from Turkey. North-West J Zool 6: 109–117.

59. Zarate P (2002) Evaluación de la actividad de anidación de la Tortuga verde *Chelonia mydas*, en las islas Galápagos druante la temporada 2001. Fundación Charles Darwin. Presentado al Parque Nacional Galápagos. Ecuador. 35 pp.

60. Zhang F, Gu H, Li P (2011) A review of chelonian hematology. Asian Herp Res 2(1): 12–20. doi: 10.3724/sp.j.1245.2011.00012

Chapter 3

BIOENGINEERING FUNCTIONAL COPOLYMERS. XII. INTERACTION OF BORON-CONTAINING AND PEO BRANCHED DERIVATIVES OF POLY (MA-ALT-MVE) WITH HELA CELLS

Mustafa Türk[1], Zakir M. O. Rzayev[2], Gülcihan Kurucu[2]

[1] Department of Biology, Faculty of Science and Art, Kırıkkale University, Yahsihan, 71450 Kirikkale, Turkey

[2] Department of Chemical Engineering, Faculty Engineering, Hacettepe University, Beytepe, 06800 Ankara, Turkey

ABSTRACT

Novel boron-containing bioengineering copolymer and its α-hydoxy-ω-methoxy-poly(ethylene oxide (PEO) macrobranched derivatives were synthesized by (1) partially amidolysis of poly(maleic anhydride-alt-methyl vinyl ether) with ethanolamine ester of diphenylboronic acid and (2) esterification of synthesized B-containing copolymers with PEO. They had a combination of hydrophilic/hydrophobic linkages, free carboxylic groups, positive charges and an ionized organoboron linkage as antitumor sites, along with an ability to interact with HeLa cells. The structure, composition and properties (cytotoxicity and antitumor activity) of synthesized copolymers were investigated. In vitro cytotoxicity results, obtained by the fluore scence microscopy measurements indicate that unlike the virgin copolymer, boron-containing and PEO macrobranched derivatives exhibit higher antitumor activity. It was found that organoboron copolymer exhibits the most apoptotic and necrotic effects against HeLa cells whereas a minor effect relative to cancer cells was observed on L929 Fibroblast cells.

INTRODUCTION

The bioengineering functional polymers exhibit the characteristics of 1) alternating and random copolymers of maleic anhydride (MA) and 2) poly(ethylene oxide) (PEO), as well as 3) PEO grafted functional macromolecules. They are of great interest for many researchers due to their nontoxic, cell-compatible, biodegradable, stimuli-responsive properties, and therefore, a wide range of biomedical and bioengineering applications exist as drug or enzyme carriers and biomacromolecular conjugates both in diagnostics and chemotherapy as effective antitumor agents [1-8]. It is known that these copolymers can be regarded as pre-activated polymers due to the presence of anhydride moieties susceptible to the reaction with a primary amine of a biomolecule [9]. The alternating copolymers of maleic anhydride (MA) with methyl vinyl ether (MVE) or divinyl ether (DVE) were utilized in various applications in diagnostics [10,11] and in chemotherapy as effective antitumor agents [8]. Poly (MA-alt-DVE), known as pyran copolymer is one of the well known bioengineering polymers having a wide range of biological activity. It processes antitumor, antiviral, antibacterial and antifungal activities, induces interferon formation, and acts as an anticoagulant and anti-inflammatory agent [8,12-17]. Hirano et al. [18,19] reported that the poly (MA-alt-DVE) conjugated with bovine erythrocyte superoxide dismutase (SOD) is resistant against the proteolytic enzymes in serum, and shows a prolonged half-life in vivo. They established an increase in half-life after intravenous injection, as well as its decreased immunogenicity [19]. It was demonstrated that the copolymer-SOD conjugate shows anti-inflammatory effect against rat re-expansion pulmonary edema at the first step of leukocyte adhesion [15]. Maeda [20] discussed the development and therapeutic potential of prototype macromolecular drugs for use in cancer chemotherapy an artificial bioconjugate of neocarzinostatin (NCS) and poly(maleic acid-alt-styrene) copolymer. The biological response-modifiying effects, the mechanism of a tumor "enhanced permeability and retention" effect and the tumor-targeting mechanism of NCS-copolymer conjugate were also discussed. According to the author, a principal advantage in the use of this bioconjugate is the potential for a reduction or elimination of toxicity.

The copolymers of fumaric, citraconic and itaconic acid and their derivatives as isostructural analogues of MA, as well as copolymers of some N-substituted maleimides can also included to class of bioengineering polymer systems. Cam et al. [21] evaluated the in vitro cytotoxicities of glycinylmaleimide (GMI) copolymers using K-562 human leukemia cells and HeLa cells. They also evaluated the in vitro antitumor activities of copolymers against mice bearing sarcome 180. Monomeric GMI and its copolymers showed higher antitumor activity than well known 5-fluorouracil (5-FU) at any dosage tested.

One the other hand, growing interest and much effort have been also focused on the synthesis of organoboron low molecular-weight functional compounds, biopolymer and drugs with boron ligands and evaluation of their suitability for the bioengineering applications. Aromatic boronic acid and its functional derivatives, and some functionalized carboranes have become an very important class organic compounds, which are utilized in a variety of biological and medical applications, such as carbohydrate recognition [22], neutron capture therapy for cancer treatment as effective tumour-targeting agents [23,24], especially for brain tumours [25,26], and protease enzyme inhibition [27]. Kataoka et al. [28-32] synthesized a novel water-soluble polymer with lectin-like function by introducing phenylboronates, as sugar -recognizing moieties, into the side-chain of poly (N,N-dimethylacrylamide) [28,29]. According to the authors, at physiological pH medium, phenyl-boronates form an appreciably stable complex with sialic acid (Neu5Ac), a chacteristic anionic carbohydrate on the surface of the plasma membranes [30,31]. Authors suggested that boronate-containing polymer may be an effective immune-adjuvant for the induction of lymphokine-activatd killer (LAK) cell [31]. They also demonstrated that the copolymers of 3-acry- lamidophenylboronic acid and dimethylacrylamide with different compositions coated onto solid substrates support function as synthetic mitogens for mouse lymphocytes [32].

However, a wide range of functional polymer synthesis techniques can be utilized for the design of more effective synthetic routes to prepare new B-containing bioengineering polymers, especially copolymerization of organoboron monomer and chemical modification of biocompatible polymers with organoboron reactive compounds and monomers. Several researchers synthesized some bioengineering copolymers containing phenylboronic acid linkages by radical copolymerization and chemical modification methods, which are exhibit glugose-, RNA- and DNA-sensitive behavior [33-36]. Recently, we report the synthesis and chracterization of organoboron copolymers by complex-radical copolymerization of p-vinylphenylboronic acid with N-isopropylacrylamide (NIPA), maleic and citraconic anhydrides, maleimide and chemical modification of poly(NIPA rand-MA)s with organoboron amine, as well as synthesis of supramacromolecular poly(ethylene imine) macrocomplexes and PEO long branched derivatives of organoboron copolymers having stimuli-responsive and high HeLa cell transfection behavior [37-40].

The objective of this work is to develop novel boroncontaining functional copolymers with antitumor activity. In the present article, results of synthesis and characterization of a new generation of biocompatible boron-containing functional macromolecules having a combination of hydrophilic and hydrophobic segments, free carboxylic groups, positive charges and an ionized linkage as antitumor sities, along with an ability to conjugate with cancer cells were described and discussed. These organoboron copolymers were synthesized by 1) partially amidolysis of bioengineering alternating copolymer of maleic anhydride (MA) and methyl vinyl ether (MVE) with ethanolamine ester of diphenylboronic acid (EAPB) and 2) chemical modification (esterification) of synthesized organoboron copolymer with α-hydoxy-ω-methoxy-poly (ethylene oxide (PEO). Special attention was paid to the role of structural effects, especially to the influence of organoboron linkage, for the interaction of organoboron functional copolymers with HeLa cells and to the evaluation of citotoxisity and antitumour activity by using a combination of various methods such as statistical, hematoxylen/eosin staining, apoptotic and necrotic cell indexes, and M30 immunostaining analyses.

MATERIALS AND METHODS

Materials

Ethanolamine ester of diphenyl boronic acid (EAPB) (Sigma-Aldrich, Germany) was purified by recrystallization from anhydrous ethanol: m.p. 193.5° C (by DSC). 1 H NMR spectra (δ, ppm) in $CHCl_3$-d_1: CH_2-O 1.49, CH_2-NH_2 2.96, and 7.38-7.40 (1H), 7.19-7.24 (2H) and 7.13-7.16 (2H) for protons of p-, o-and m-positions in benzene ring, respectively. Poly(maleic anhydride-altmethyl vinyl ether), poly(MA-alt-MVE) (C1) (SigmaAldrich, Germany): M_n 80,000 g.mol^{-1}, T_g 148° C (by DSC); 1 H NMR spectra (δ, ppm) in DMSO-d6: CH_2 1.23, CH-O 2.11, O-CH_3 2.08 and CH-CH 3.38. α-Hydoxy-ω- methoxy-poly(ethylene oxide) (Fluka; PEO, M_n 2000 g.mol^{-1}):1 H NMR spectra (δ, ppm) in $CHCl_3$-d_1: CH_2-O 3.75-3.45, OH end group 2.61 and O-CH_3 end group 2.16.

Human cervix epithelioid carcinoma cell line (HeLa) was obtained from the tissue culture collection of the SAP Institute (Turkey). Cell culture flasks and other plastic material were purchased from Corning (USA). The growth medium, which is Dulbecco Modified Medium (DMEM) without L-glutamine supplemented fetal calf serum (FCS), and Trypsin-EDTA were purchased from Biological Industries (Israel). M30 CytoDEATH antibody (Roche).

Synthesis

Boron-contaning copolymer (C1-B) was synthesized by the partially amidolysis of succinic anhydride units of alternating copolymer C1 with EAPB, containing a primary amine group, in the 1,4-dioxane solution at 40° C for 3 h under nitrogen atmosphere at molar ratio of C1:EAPB = 2:1. Approbriate quantities of C1 and EAPB, solvent were placed in a standard Pyrex-glass tube and flushed with dried nitrogen gas for at leeast 3 min, then placed in a carousel type microreactor with a thermostated heater and magnetic mixer. The resulting copolymer C1-B was isolated from reaction mixture by precipitating with diethyl ether. Purification of copolymers was done by dissoving in dioxane and reprecipitating in diethyl ether, extraction with hexane and draying under vacuum at 50° C until constant weight.

PEO macrobranched copolymer (C1-B-PEO) was synthesized by the esterification of anhydride units of partially amidolysed C1-B copolymer with PEO, containing an end hydoxyl group, in the same conditions using in our previous publications 5,37..

Characterization

FTIR spectra of the organoboron copolymers (KBr pellet) were recorded with FT-IR Nicolet 510 spectrometer in the 4000-400 cm^{-1} range, where 30 scans were taken at 4 cm^{-1} resolution. ^{1}H $\{^{13}C\}$ NMR spectra were performed on a JEOL 6X-400 (400 MHz) spectrometer with DMSO-d$_6$ as a solvent at 25° C.

The differential scanning calorimetry (DSC) analysis was performed on a Shimadzu calorimeter (Japan) at a heating rate of 5o C/min, under nitrogen atmosphere. The X-ray diffraction (XRD) patterns were obtained from a Rigaku D-Max 2200 powder diffractometer. The XRD diffractograms were measured at 2θ, in the range 1-50° , using a Cu-K$_\alpha$ incident beam (λ = 1.5406 Å), monochromated by a Ni-filter. The scanning speed was 1 o /min, and the voltage and current of the X-ray tubes were 40 kV and 30 mA, respectively.

The number of living and dead cells were counted with a haemacytometer (C.A. Hausse & Son Phluila, USA) at X200 magnification. The number of apoptotic and necrotic cells were determined by Fluorescence Inverted Microscope (Olympus IX70, Japan). The cell images were also recorded using the both above mentions microscopes. Statistical analyses were performed using Student's t-test for unpaired data and P values of less than 0.05 were considered significant. Data are presented as means ± SEM (standard errors of the mean).

Cytotoxicity

For cytotoxicity experiments, HeLa cells and L929 Fibroblast cells respectively. (25x10^3 cells per well) were placed in DMEM by using 24-well plates. Different amounts of copolymers (C1, C1-B and C1-B-PEO) (about 50-500 µg.mL^{-1} in aqueous solutions) were put into wells containing cells, respectively. The plates were kept in the CO$_2$ incubator (37 °C in 5% CO2) for 2-24 h; the medium was replaced with fresh medium, and incubated at the same conditions for 24 h. Following of this incubation, HeLa cells and L929 Fibroblast cells were harvested with trypsin-EDTA, and then were dyed with trypan blue 41.. The viable cells were counted with a haemacytometer (C.A. Hausse & Son Phluila, USA), using light microscope.

Hematoxylen/Eosin Staining

HeLa cells and L929 Fibroblast cells (25×10^3 cells per well) were placed in DMEM by using 24-well plates. After treating with different amount functional copolymers (C1, C1-B and C1-B-PEO) (about 50-500 $\mu g.mL^{-1}$ in aqueous solutions) for 2-24 hours period, the medium was removed, the cells washed with distilled water and fixed in ethanol, and stained with Hematoxylen/Eosin. After staining, the cells were observed by light microscopy. By this way, cellular and nuclear morphology have been shown in cultured cells stained with Hematoxylen/Eosin.

Analysis of Apoptotic and Necrotic Cells

Double staining were performed to quantify the number of apoptotic cells in culture on basis of scoring of apoptotic cell nuclei. HeLa cells and L929 Fibroblast cells (25×10^3 cells per well) were placed in DMEM by using 24-well plates. After treating with different amount functional copolymers (C1, C1-B and, C1-B-PEO) (about 50-500 $\mu g.mL^{-1}$ in aqueous solutions) for 2-24 hours period, both attached and detached cells were collected, then washed with PBS and stained with Hoechst dye 3342 (2 $\mu g.mL^{-1}$), propodium iodide (PI) (1 $\mu g.mL^{-1}$) and DNAse free-RNAse (100 $\mu g.mL^{-1}$) for 15 min at room temperature. After that 10-50 mL of cell supension was smeared on slide and coverslip for examination by fluorescence microscopy 42,43.. The nuclei of normal cells were stained light blue but apoptotic cells were stained dark blue by the hoechst dye. The apoptotic cells were identified by their nuclear morphology as a nuclear fragmentation or chromatin condensation. Necrotic cells were staining red by PI. Necrotic cells lacking plasma membrane integrity and PI dye cross cell membrane, but PI dye don't cross non necrotic cell membrane. The number of apoptotic and necrotic cells in 10 randomly chosen microscopic fields were counted and the result expressed as a ratio of apoptotic and necrotic to normal cells.

M30 Immunostaining

The percentage of apoptotic cells was determined by M30 CytoDEATH antibody 44.. This is a monoclonal mouse immunoglobulin (Ig) G2b antibody (clone M30; Roche, Mannheim, Germany) that binds to a caspase-cleaved, formalin-resistant epitope of cytokeratin 18 cytoskeletal protein. The immunoreactivity of the M30 antibody is confined to the cytoplasm of apoptotic cells. HeLa cells (25×10^3 cells per well), treated with C1, C1-B and, C1-B-PEO copolymers (about 50-500 $\mu g. mL^{-1}$ in aqueous solutions) for about 2-24 h, were fixed in 10% neutral-buffered formalin for 15 min, treated with 0.3% hydrogen peroxide in methanol for 10 min to block the endogenous peroxidase activity, washed in the standard phosphate buffer solution, and then incubated with M30 antibody at room temperature for 1 h. In negative controls, preimmune mouse serum instead of primary antibody was used. Immunoreactions were revealed by the avidin-biotin complex technique using diaminobenzidine (DAB) as substrate. We counted the number of M30- positive cytoplasmic staining cells in all fields found at x400 final magnification. For each image, three randomly selected microscopic fields were observed, and at least 100 cells/field were evaluated. M30 CytoDEATH antibody was not sensetive to L929 Fibroblast. On account of this reason, M30 CytoDEATH antibody did not applied to L929 Fibroblast cells.

RESULTS AND DISCUSSION

Synthesis and Characterization of Organoboron Functional Copolymers

Boron-containing bioengineering functional copolymer (C1-B) and its α-hydoxy-ω-methoxy-poly(ethylene oxide) (PEO) long branched derivatives were synthesized by (1) amidolysis of succinic anhydride units of biocompatible poly(MA-alt-MVE) alternating copolymer (C1) with EAPB containing a primary amine group, and (2) esterification (grafting) of free anhydride units of partially amidolysed C1-B copolymer with PEO, containing an end hydoxyl group, respectively. General scheme of synthesis of the organoboron functional copolymer and its PEO branched derivative can be represented as follows (Scheme 1).

The synthesized boron-containing copolymers contain a combination of hydrophilic/ hydrophobic linkages, free carboxylic groups, positive charges and ionized organoboron linkage as antitumor sities, along with an ability to interact with canser biomacromolecules, especially with HeLa cells. The chemical and physical structure, composition and properties (temperature-responsiveness, glass-transition, melting and degradation temperatures, andantitumor activity and cytotoxicity) of synthesized copolymers were characterized by spectroscopy (FTIR, [1] H and ^{13}C NMR), viscometry, DSC, X-ray diffraction and Fluorescence microscopy analyses.

The results of chemical structural analysis of the syn thesized organoboron copolymers FTIR (KBr pellet) and ([1] H and ^{13}C) NMR spectroscopy (in DMSO-d$_6$ solution) were summarized in Table 1 (FTIR analysis data for C1-B) and illustrated in Figure 1 (NMR spectra of C1-B) and Figure 2 (FTIR spectra of C1-B-PEO). The formation of amide, carboxyl and organoboron groups in the structure of C1-B copolymer as results of amidolysis reaction was confirmed by apearance of the corresponded characteristic absorption bands for each monomer unit and diphenylboronic fragment in the spectra. Absorption bands at 1864 and 1781 cm⁻¹, relating to C=O groups of free anhydride units, indicated the partially amidolysis of these units as shown in Scheme 1.

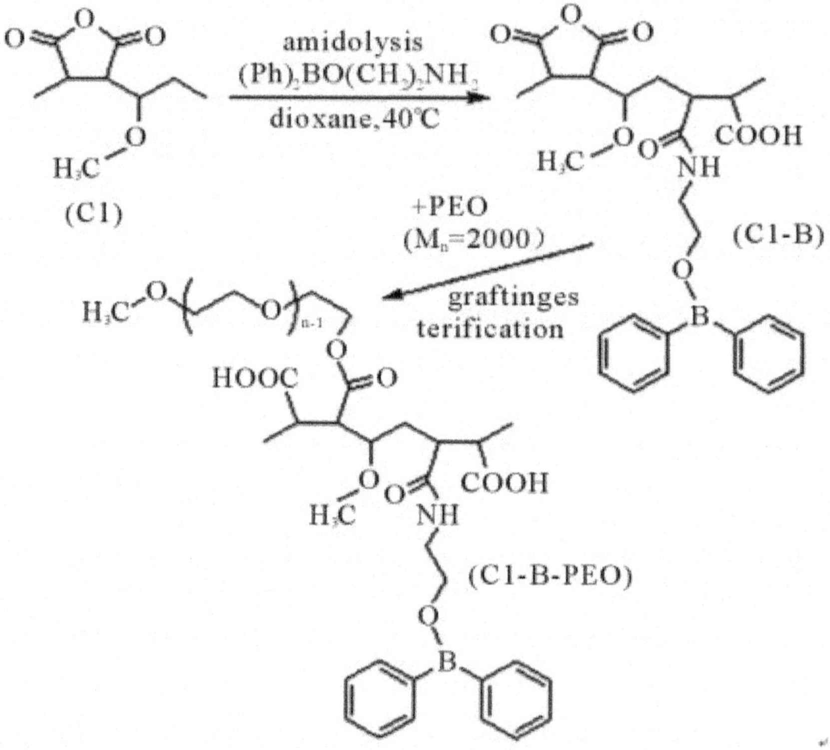

Scheme 1: Schematic representation of the synthesis routes of organoboron functional copolymers (C1-B and C1-B-PEO) by the amidolysis of poly(MA-alt-MVE) (C1) with organoboron amine (AEPB) and esterification of poly (MA-alt-MVE)-g -AEPB) (C1-B) with PEO, respectively.

Table 1: The results of FTIR analysis organoboron functional copolymer: Poly(MA-alt-MVE)-g-AEPBA) (C1-B)

Absorption bands (cm^{-1})	Band assignments
MA unit	
1980-1925 (w)	C=O (overtones)
1864 (m-s), 1781 (vs)	C=O stretching (anhydride)
1227 (s, broad), 1094 (s)	C–O and C–O–C bands
650 (w)	CH (in chain backbone)
MVE unit	
2942 (m-s)	CH$_3$ C–H stretching
2854 (m)	CH$_2$ C–H (chain backbone)
1475-1416 (m)	CH$_2$ and CH$_3$ deformation
1372 (m)	CH$_3$ deformaton (in O–CH$_3$)
975 (m)	CH$_3$ rocking
926 (vs)	C–O deformation
735 (m-w), 720 (w)	CH$_2$ and CH$_3$ deformation
Maleamide unit	
1736 (m-s)	C=O strtchıng (in -COOH)
1575-1510 (m-w)	COO$^-$ stretching (H-bonding)
1650 (m),1720 (m-w)	NH–C=O amide I band
1315 (w)	amide III band
Organoboron linkage	
3240 (w), 3100,1600 (m)	CH= (in aromatic ring)
1545 (m-w)	B–O stretching
1443 (m), 1420 (w)	B–Ph aromatic ring
1180 (m)	CH in-put-bending
770 (w)	CH out-put-bending
702 (m)	O-B-Ph aromatic ring

From the comparative analysis of [1] H and ^{13}C NMR spectra of virgin alternating copolymer (C1) and its organoboron derivative (C1–B) (Figure 2a, 2b), the following changes of the characteristic signals were observed: unlike the spectra of C1 copolymer having the peaks from chemical shifts of the CH and CH$_2$ backbone and CH$_3$ (in methoxy group) protons new signals from protons of amide NH, COOH and phenyl groups (in or ganoboron linkage) were appeared in the spectra of organoboron copolymer (C1–B). More detailed informations about micristructure of C1–B copolymer were pre-pared by analysis of ^{13}C NMR spectra (Figure 2c).

The following chemical shifts (d, ppm) of carbon atoms were observed in the spectra: 174.4 (–C=O of the maleamide and anhydride units), 128-136 (–B–C$_6$H$_5$ mono-substitued benzene ring), 77.2 (–CH–NH in organoboron linkage), 58.01 (–CH–O), 49.08 (–CH– CH-chain backbone), 30.3 (–CH$_3$-O), and 30.15 (–CH$_2$).

Chemical structure of C1–B–PEO long branched copolymer was confirmed by the appearance in the FTIR spectra (Figure 3) the following characteristic absorption bands (cm^{-1}): 3400 (strong broad peak for OH in Hbonded carboxyl groups), 2933-2735 for C–H stretching in CH$_2$ and CH$_3$), 2667 and 2600 (C-H stretching in CH$_2$–O of PEO branched segments), 2280 and 2135 (Fermi doublet for C–N band), 1986 and 1966 (overtones of C=O), 1746 (C=O of ester groups), 1710 (C=O of carboxyl groups), 1630 (NH–C=O amide I band), 1592 (phenyl groups), 1558 (H-bonded COO– stretching), 1545 (weak peak for B–O stretching), 1490 (C-H deformation for CH$_2$–O in PEO branches), 1480 and 1466 (CH$_2$ deformation), 1450 (B–Ph aromatic ring), 1405 (amide III band), 1372 and 1352 (CH3 deformation in O–CH$_3$), 1115 (broad peak for C–O band in CH$_2$–O and CH$_3$–O of PEO and MVE units, respectively), 948 (strong peak for C–O deformation in PEO branchs), and etc.

The comparative analysis of the XRD patterns of alternating copolymer and its organoboron derivative show a significant difference between physical structures of these copolymers (Figure 3). C1 copolymer has an amorphous structure, while C1-B copolymer exhibits pseudo-crystallinity behavior (without re-crystallization process due to macromolecular physical interactions via H-bonding, hydrophobic-hydrophilic interactions, etc.) with degree of pseudo-crystallinity χ_c = 26.2 % (by XRD analysis), glass-transition T$_g$ and pseudo-melt phase transition T$_m$ at 84.2° C and 136.3 ° C, respectively (by DSC analysis). It can be proposed that the producing the amphiphilic organoboron linkages in side chain of copolymer causes a formation of hydrophilic/hydrophobic balance, more polar amide and carboxyl groups, which are able to form strong H-bonded segments, and therefore, self-assembled suramacromolecular structure of C1-B copolymer as in other organoboron polymer systems.

Cytotoxicity of the Copolymer and its B-Containing and PEO Branched Derivatives

In this study, the comparative analysis of HeLa cells (cancel cells) and L929 Fibroblast cells (normal cells) has been investigated. The cytotoxicities of C1 copolymer and corresponding C1-B, C1-B/PEO derivatives were inquired about the utility for antitumor drugs. Figures 4 and 5 give the number of viable cancer and normal cells in each group after incubation of the cells with copolymer and organoboron copolymers at their different concentrations for 24 h incubating time in cell culture media, respectively. Under the same conditions, the wells containing cells without copolymers were also studied as a control. The following important results can be drawn from this graph which is illustrated in these figures. The C1 copolymer does not exhibit any observable toxicity in the chosen range of copolymer concentration. The toxicity of polymers containing boron (C1-B and C1-B-PEO) was significant, most probably due to hydrogen bonding supramacromolecular structure of these copolymers containing a combination of hydrophilic/hydrophobic linkages, free carboxylic groups, which are formed after partial amidolysis of anhydride containing copolymer C1 and full hydrolysis of free anhydride units in the chosen physiological medium where positive charges and ionized organoboronoxy groups also exist as antitumor sites along with an ability to interact with cells.

It was observed that an increase of C1-B and C1- B-PEO concentrations in each well caused higher degree of dying cells as compared to virgin C1 copolymer tested under the same conditions. C1-B copolymer exhibits relatively higher in vitro cytotoxicity than C1-B-PEO branched copolymer which can be explained by the higher content of organoboron linkages in C1-B copolymer. It is important to note that the boron containing side chain linkages, rather than the individual copolymers, increase the cytotoxicity more profoundly; an important feature which has a significant role in leading us to the present study. C1 copolymer had less toxicity compared to cultured cells at various quantities and different incubation times. On the contrary, the toxicity of C1-B and C1-B-PEO organoboron copolymers towards the HeLa cells increased by increasing their quantity from 50 to 500 $\mu g.mL^{-1}$, whereas, no significant change was observed with varying time. According to Figure 4, C1 did not show high toxicity at all although the copolymer amount was increased from 50 to 500 $\mu g.mL^{-1}$ whilst, a significant toxicity of C1-B andC1-B-PEO (100 $\mu g.mL^{-1}$ and above) started to be observed when cancer and normal cells (Figure 5) were incubated for about 4 h. As the amount of boron con taining polymers and their incubation time increased, toxicity to cultured cells was increased. C1-B-PEO and especially C1-B showed higher toxicity at 500 $\mu g.mL^{-1}$. Thus, it can be concluded that virgin C1 alternating copolymer does not exhibit any toxic effect on cultured HeLa cells, whereas, its organoboron and PEO branched derivatives are definitely toxic to cells. In particular, C1-B copolymer containing relatively high amount of organoboron linkages exhibits high toxicity toward cancer cells compared to normal cells at 500 $\mu g.mL^{-1}$ for 24 h.

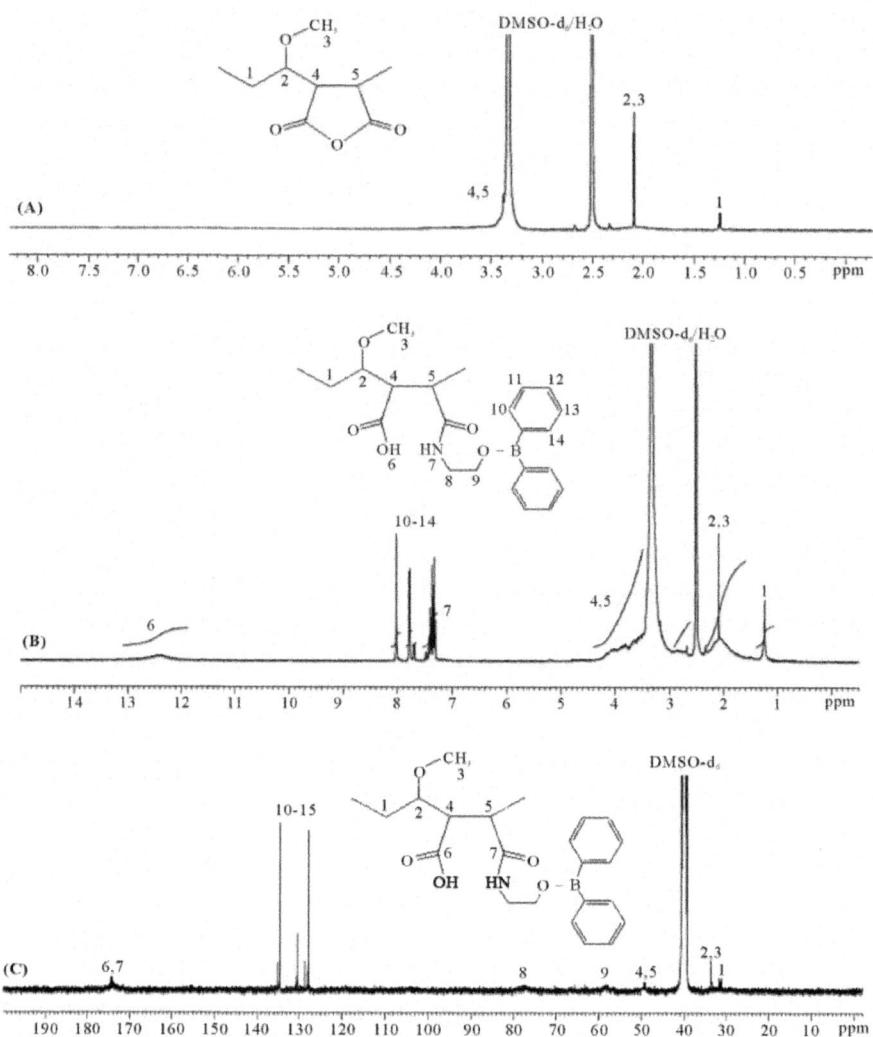

Figure 1: [1] H NMR spectra of (A) C1 copolymer and (B) C1-B organoboron copolymer; (C) ^{13}C NMR spectra of C1-B copolymer.

Staining Results

The important observations can be summarized as follows: we checked for apoptosis or necrosis with double staining (Hoescht 3342 and PI), M30 immunostaining for cancer cells. For the morphological observations, cancer and normal cells were stained by hematoxylen-eosin.

Hematoxylen-Eosin Staining Results

In this study, C-1 copolymers treated cancer and normal cells have intact nucleus of about 50-200 μg.mL^{-1} concentration during 2-14 h incubation. Cell morphology has not been changed at the same concentration for 2-14 h (Figure 6b). While C1-B and C1-B-PEO copolymers treated HeLa cells has no morphological changes at 50-200 μg.mL-1 concentration for about 2-4 h, they have vacuole formation in their cytoplasms with C1-B copolymer between 6-12 hours (Figure 6c). Vacuole formations determined rarely in normal cells (Figure 6e, f). In addition, cell membranes have lysed with C1-B copolymer around 12-24 h but, there was no change in their nuclei of cancer and normal cells. Moreover, some of the cells (30% and 15% for HeLa and fibroblast, respectively) have been detached from the well. Unaffected cells displayed similar morphological characteristics as with untreated (control) cells.

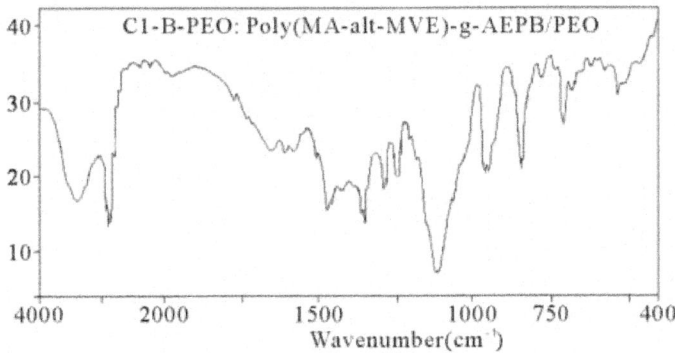

Figure 2: FTIR spectra of PEO macrobranched organoboron copolymer (C1-B-PEO).

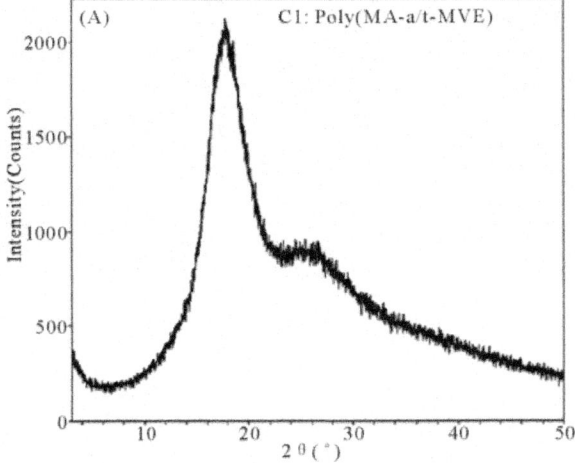

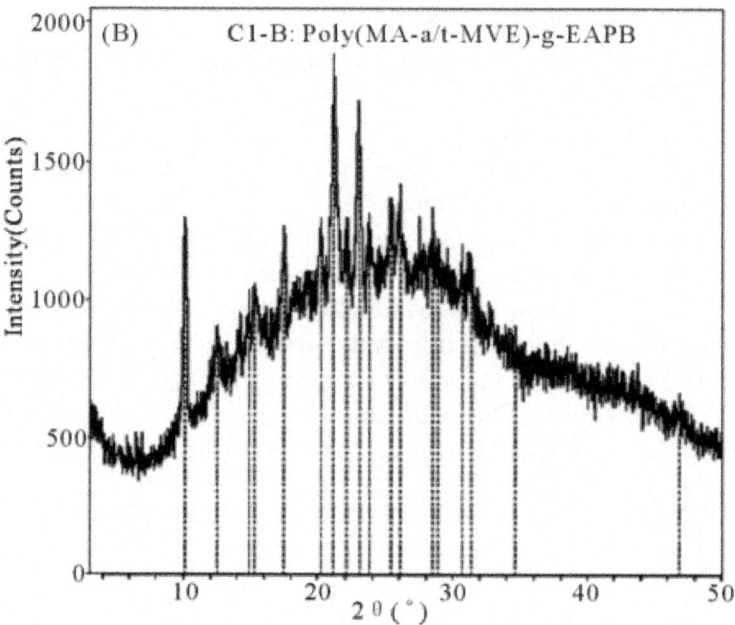

Figure 3: XRD patterns of (A) alternating copolymer (C1) and (B) its branched organoboron derivative.

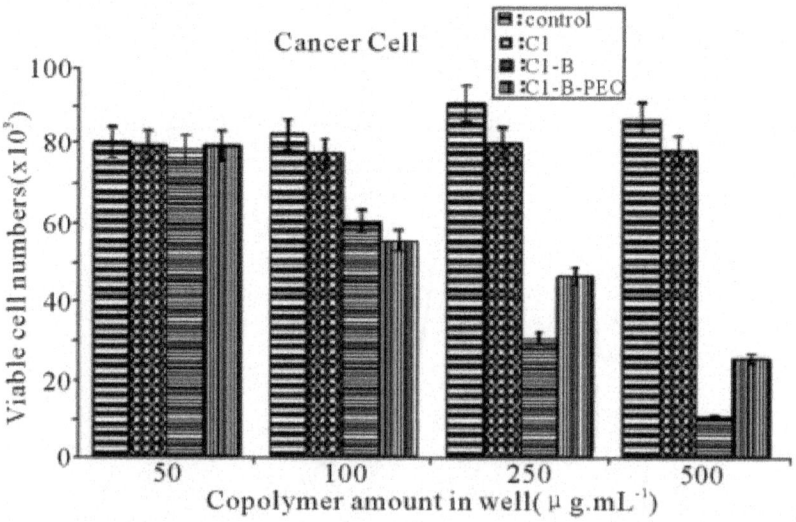

Figure 4: In vitro cytotoxicity of C1, C1-B and C1-B-PEO copolymers with different amount at 24 h incubation. Number of viable HeLa cells in wells. Results are presented as means ± SEM. * Significant difference from control ($p < 0.05$).

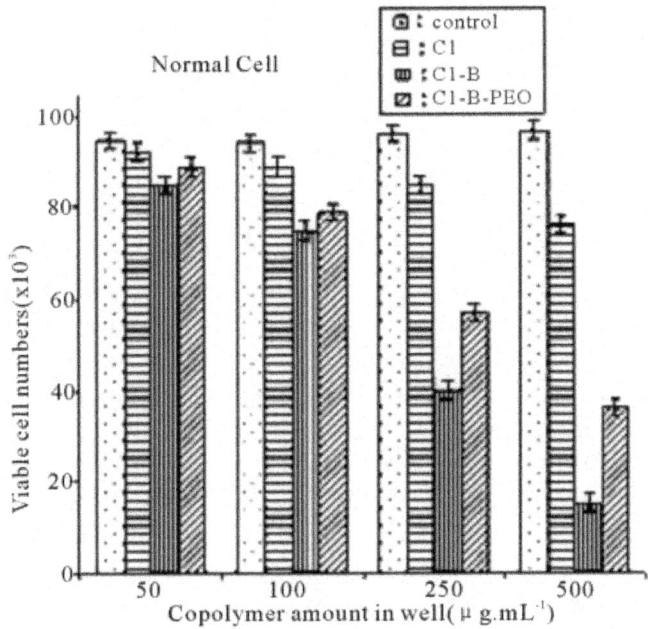

Figure 5: In vitro cytotoxicity of C1, C1-B and C1-B-PEO copolymers with different amount at 24 h incubation. Number of viable L929 Fibroblast cells in wells. Results are presented as means ± SEM. * Significant difference from control (p < 0.05).

Double Staining and M30 Immunostaining Results

In this study, if the HeLa and L929 Fibroblast cells treated by C1, C1-B, C1-B-PEO copolymers at low conce ntration for a short time, the number of apoptotic and necrotic cells was not high (Tables 2 and 3). However, if the polymer concentration and incubation time were increased, the number of apoptotic and necrotic cells was increased as well. Especially, the number of apoptotic and necrotic cells was increased when they were treated by C1-B copolymer at 500 μg.mL^{-1} concentration in cancer cell culture for 24 h (Table 2). The number of apoptotic and necrotic Fibroblast cells was not increased according to HeLa cells at the same concentration (Table 3). If cells were treated by the other copolymer under similar conditions, their apoptotic index was below 30 %. The results obtained at 500 μg.mL^{-1} concentration for 24 h are shown in Table 2. Meanwhile, apoptotic HeLa cells were immunostained by M30 antibody (Figure 7a, 7b). The double staining and M30 immunostaining results were similar to each other in HeLa cells. Apoptotic indexes of HeLa cells for M30 immunostaining were 12% for C1, 45 % for C1-B, 23 % for C1-B-PEO at 500 μg.mL^{-1} and 24 h incubation. Apoptotic L929 Fibroblast cells were stained only double staining method (Table 3). In addition to these polymers, especially boron containing polymers had toxic effects towards cancer and normal cells. But toxic effect of boron containing polymers was lower to normal cells than cancer cells. After an incubation at 50–500 $\mu g.mL^{-1}$ for 24 h period, C1 resulted in less apoptosis, while incubation with C1-B and C1-B-PEO at the same concentration and incubation time led to high apoptosis of HeLa cells compared to L929 Fibroblast cells. Both C1-B and C1-B-PEO may well inhibit cell growth and viability in HeLa (Figure 7b, c, d) and L929 Fibroblast cells (Figure 7e, f). One the other hand, around 50-500 $\mu g.mL^{-1}$ of C1-B and C1-B-PEO copolymer contents for 24 h gave rise to an increase in necrosis stained with PI dye (Figure 7c, e and Tables 2 and 3). It is important to note that incubation for 24 h with 500 $\mu g. mL^{-1}$ C1-B produced apoptosis supporting its high toxicity and necrotic effect. Furthermore, incubation without polymers as control cells resulted in a few PI-positive cells. Whereas, cells exposed to C1-B and C1-B-PEObecame highly PI-positive, suggesting that they were in necrosis. HeLa cells incubated with a high dose of boron containing copolymers resulted in rupture of cell membrane at around 12-24 h incubation period. Cell cytoplasm was discharged out of HeLa cells. On the otherhand, great of number vacuole originated in most of HeLa cells cytoplasma. It may have given rise to metabolicchanges of cells, affected by boron containing copolymers. C1-B copolymer was more toxic than virgin counder the testing conditions determined by us.

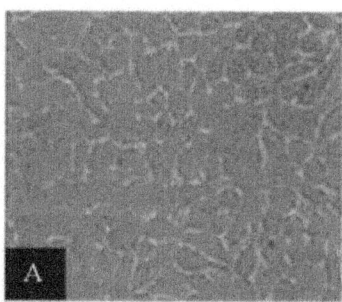

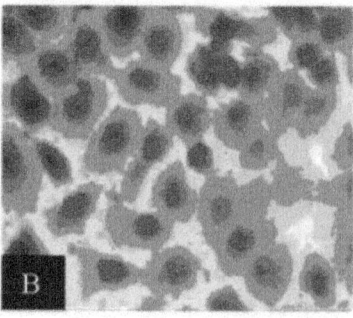

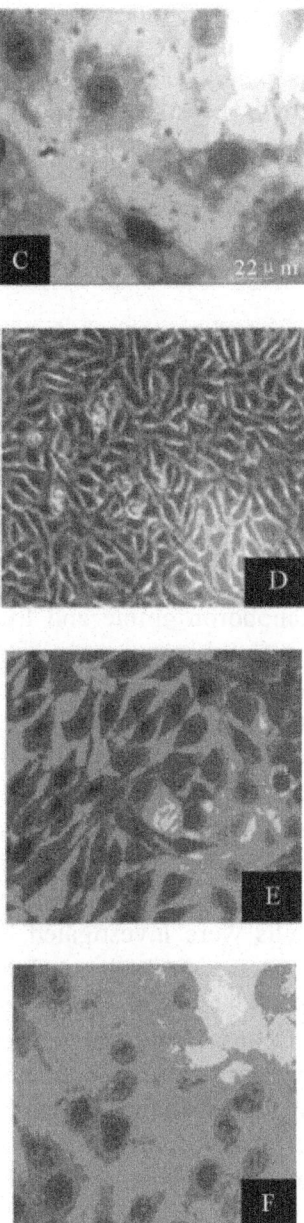

Figure 6: Light microscope image of (A) non stained HeLa cell culture as a control, (B) C1-B-PEO copolymer/HeLa cells conjugate (stained with hematoxilen-eosin dye); dense spots were showed nucleus of cells, and distinct violet were indicated cytoplasma of cells as a control, (C) Light microscope image of vacuole of HeLa cells cytoplasma; dense spots were showed nucleus of cells in C1-B copolymers (500 μg.mL⁻¹

consantration) at 24 h incubation. Light microscope image of (D) non stained L929 Fibroblast cell culture as a control, (B) C1-B-PEO copolymer/L929 Fibroblast cells conjugate (stained with hematoxilen-eosin dye); dense spots were showed nucleus of cells, and distinct violet were indicated cytoplasma of cells as a control, (C) Light microscope image of vacuole of L929 Fibroblast cells cytoplasma; dense spots were showed nucleus of cells in C1-B copolymers (500 µg.mL^{-1} consantration) at 24 h incubation. Images (A) and (D) taken under X200 magnification, others images taken under x400 magnification.

CONCLUSIONS

This work has attempted to develop novel bioengineering functional organoboron copolymers (C1-B and C1-BPEO), namely, amphiphilic macromolecules of which contained hydrophilic/hydrophobic fragments, ethylene amidodiphenylborinate linkages, long branched PEO segments and free carboxylic groups with an ability to conjugate with cancer HeLa cells. These copolymers were synthesized by amidolysis and esterification of anhydride units of poly(MA-alt-MVE) (C1) as a bio- compatible and non toxic polymer matrix with organoboron amine and PEO, respectively. Chemical and physical structure of organoboron copolymers were confirmed by FTIR and 1 H (13C) NMR spectroscopy and X-ray powder diffraction methods. The comparative analysis of novel organoboron functional copolymers with antitumor acivity towards cancer and normal cells was achieved. It was found that unlike the virgin amorphous C1 copolymer, organoboron copolymer (C1-B) exhibited semi-crystalline phase transition behaviour due to the formation of self-assembled supramacromolecular structures through strong intra-and intermolecular hy drogen bonding. The interactions of these copolymers with HeLa cells were investigated by using a combination of different methods such as cytotoxicity, statistical, hematoxylen/eosin staining, apoptotic and necrotic cell indexes, M30 immunostaining, double staining and M30 immunostaining, light and fluorescence microscopy analyses. In vitro cytotoxicities and antitumor activities of organoboron copolymers (C1-B and C1-B-PEO) against human cervix epithelioid carcinoma cell line (HeLa) was as well evaluated. It was observed that organoboron copolymers exhibited the most apoptotic and necrotic effects against HeLa cells whereas a minor effect relative to cancer cells was observed on L929 Fibroblast cells. Thus the obtained results allow us to propose that synthesized organoboron copolymers containing a combination of non toxic and biocompatibile polymer matrix and long branched PEO segments with functional groups as antitumor sities, can be utilized as therapeutic potential functional copolymer drugs, which are able to form an artificial bioconjugate with HeLa cells, in cancer chemotherapy.

Table 2: The comparative analysis of apoptotic and necrotic HeLa cell index for co-polymer (C1), organo-boron (C1-B) and organoboron PEO branched (C1-B-PEO) co-polymers at 24 h incubation

Polymer amount ($\mu g.mL^{-1}$)	Apoptotic cells (%)	Necrotic cells (%)
C1		
control	3 ± 2	4 ± 2
50	3 ± 1	5 ± 1.5
100	6 ± 1.5	9 ± 3
250	8.5 ± 1	13.8 ± 1.5
500	15 ± 2	27 ± 3
C1-B		
control	5 ± 3	2 ± 1
50	13 ± 2	20 ± 3
100	18 ± 2.5	31 ± 4
250	32 ± 1	41 ± 2
500	43 ± 2	53 ± 3
C1-B-PEO		
control	3.5 ± 1.5	5 ± 1.5
50	9 ± 1	11 ± 1.5
100	12 ± 1	16 ± 2
250	15 ± 2	25 ± 3
500	28 ± 2.5	43.5 ± 5

Table 3: The comparative analysis of apoptotic L 929 Fibroblast cells index for copo-lymers (C1), organoboron (C1-B) and organoboron PEO branched (C1-B-PEO) copo-lymers at 24 h incubation

Polymer amount ($\mu g.mL^{-1}$)	Apoptotic cells (%)	Necrotic cells (%)
C1		
control	2 ± 1	3 ± 1
50	3 ± 1	3 ± 1.5
100	4 ± 1	6 ± 2
250	7 ± 1	11 ± 1.5
500	9 ± 1.5	17 ± 2
C1-B		
control	2 ± 1	2 ± 1
50	7 ± 2	12 ± 2
100	12 ± 2.5	17 ± 3
250	16 ± 1	25 ± 2
500	27 ± 2	38 ± 2
C1-B-PEO		
control	3 ± 2	3 ± 1
50	4 ± 1	8 ± 1.5
100	6 ± 1	11 ± 2
250	9 ± 2	16 ± 3
500	17 ± 2.5	30 ± 1

Cancer cells   Normal cells

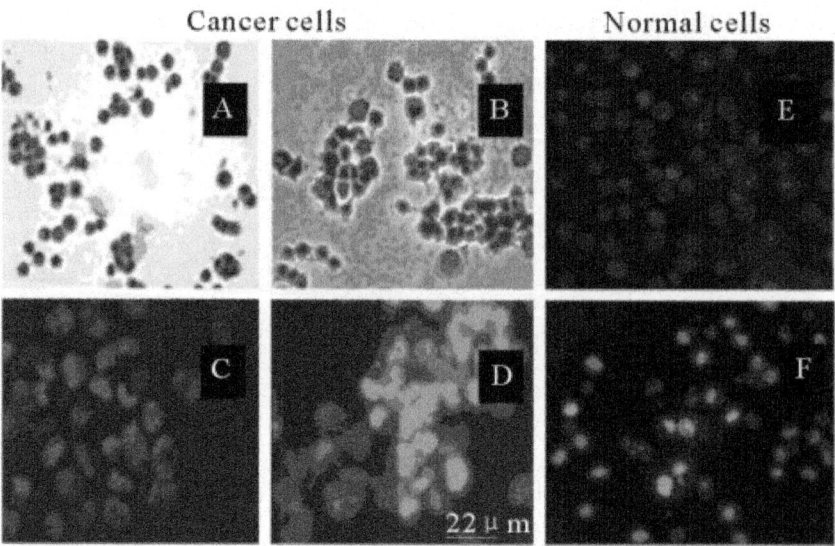

Figure 7: Light microscopy images of (A) virgin (nonapoptotic) HeLa cells as a control group (stained with M30 immunostaining kit), and (B) organoboron copolymer C1-B copolymer/HeLa cells conjugate (stained with M30 immunostaining kit), where brown cytoplasma of cells image indicates the formation of apoptotic cells; Fluorescence microscopy image of (C) nucleus of HeLa cells (stained with PI), where formation of red spots demostrates nucleus of necrotic cells, and (D) nucleus of HeLa cells (stained with Hoescht 3342), where dense spots indicates nucleus of apoptotic cells. Fluorescence microscopy image of (E) nucleus of L929 Fibroblast cells (stained with PI), where formation of red spots demostrates nucleus of necrotic cells and green spots demostrates nucleus of living cells, and (F) nucleus of L929 Fibroblast cells (stained with Hoescht 3342), where dense spots indicates nucleus of apoptotic cells. İmages of Cand D were recorded with x400 magnification, others image were recorded with x200 magnification.

ACKNOWLEDGEMENTS

The supports of the Turkish National Scientific and Technical Council (TÜBİTAK) through project TBAG-2486 and HU Scientific Research Foundation (BAB) through the BAB-2601006 project are kindly acknowledged.

REFERENCES

1.   Albertsson, P.A. (1986) Partition of Cell Particles and Macromolecules. New York, Wiley.

2.   Herold, D.A., Keil, K., Bruns, D.E. (1989) Oxidation of polyethylene glycols by alcohol dehydrogenase.Biochem Pharmacol, 38(1), 73.

3.   Sinha, V.R., Aggarwal, A., Trehan, A. (2004) Biodegradable PEGylated microspheres and nanospheres. Amer J Drug Deliv, 2(3), 157.

4.   Köşeli, V., Rzaev, Z.M.O., Pişkin, E. (2003) Bioengineering functional copolymers. III. Synthesis of biocompatible poly(NIPA-co-MA)-g-PEO/PEI macrocomplexes and their thermostabilization effect on the activity of the enzyme penicillin G acylase. J Polym Sci Part A: Polym Chem, 41(11), 1580.

5.   Mazi, H., Kibarer, G., Emregül, E., Rzaev, Z.M.O. (2006) Bioengineering functional copolymers. IX. Poly (maleic anhydride-co-hexene-1)-g-poly(ethylene oxide).. Macromol Biosci, 6(4), 311.

6.   Mazi, H., Emregül, E., Rzaev, Z.M.O., Kibarer, G. (2006) Preparation and properties of invertase immobilized on a poly(maleic anhydride-hexen-1) membrane. J Biomater Sci, Polym Ed, 17, 821.

7.   Rzaev, Z.M.O., Dinçer, S., Pişkin, E. (2007) Functional copolymers of N-isopropyl-acrylamide for bioengineering applications. Prog Polym Sci, 32(5), 534.

8.   Butler, G.B. (1992) Cyclopolymerization and cyclopolymerization. New York, Marcel Dekker.

9.   Veron, L., Bignicount, M.C.D., Delair, T., Pichot, C., Mandrand, B. (1996) Syntheses of poly N-(2,2-dimethoxyethyl)-N-methyl acrylamide. for the immobilization of oligo-nucleotides. J Appl Polym Sci, 60(2), 235.

10. Ladavière, C., Delair, T., Domard, A., Pichot, C., Man-drand, B. (1999) Covalent immobilization of biological molecules to maleic anhydride and methyl vinyl ether copolymers-A physico-chemical approach. J Appl Polym Sci, 71(6), 927.

11. Chaix, C., Minard-Basquin, C., Delair, T., Pichot, C., and Mandrand, B. (1998) Oligonucleotide synthesis maleic anhydride copolymers covalently bound to silica spherical support and characterization of the on obtained conjugates. J Appl Polym Sci 70(12), 2487.

12. Volkova, F., Gorshkova, M.Yu., Ivanov, P. E., Stotskaya, L. L. (2002) New scope for synthesis of divinyl ether and maleic anhydride copolymer with narrow molecular mass distribution. Polym Adv Technol, 13(10-12), 1067.

13. Izumrudov, V.A., Gorshkova, M.Yu., Volkova, F. (2005) Controlled phase separations in solution of soluble poly-electrolyte complex of DIVEMA (copolymer of divinyl ether and maleic anhydride. Eur Polym J, 41(6), 1251.

14. Ladavière, C., Delair, T., Domard, A., Pichot, C., Mandrand, B. (1999) Covalent immobilization of bovine serum albumin onto (maleic anhydride-alt-methyl vinyl ether) copolymers. J Appl Polym Sci, 72(12), 1565.

15. Delair, T., Badey, B., Domard, A., Pichot, C., Mandrand, B. (1997) Polym Adv Technol, 8(5), 297.

16. Patel, H., Raval, D.A., Madamwar, D., Sinha, T.J.M. (1997) Polymeric prodrugs. Synthesis, release study and antim-icrobial properties of polymer-bound acriflavine. Angew Makromol Chem, 245(1), 1.

17. Patel, H., Raval, D.A., Madamwar, D., Patel, S.R. (1998) Polymeric prodrug: Synthesis, release study and antimicrobial property of poly(styrene-co-maleicanhydride)-bound acriflavine. Angew Makromol Chem, 263(1), 25.

18. Hirano, T., Todorski, T., Kato, S., Yamamoto, H., Caliceti, P. (1994) Synthesis of the conjugate of superoxide dismutase with the copolymer of divinyl ether and maleic anhydride retaining enzymatic activity. J Control Release, 28(1-3), 203.

19. Hirano, T., Todorski, T., Morita, R., Kato, S., Ito, Y., Kim, K., Shukla, G., Veronese, F., Maeda, H., Ohashi, S. (1997) Anti-inflammatory effect of the conjugate of superoxide dismutase with the copolymer of divinyl ether and maleic anhydride against rat re-expansion pulmonary edema. J Control Release, 48(2-3), 131.

20. Maeda, H.H. (1991) SMANCS and polymer-conjugated macromolecular drugs: advantages in cancer chemotherapy. Adv Drug Delivery Rev, 6(2), 181.

21. Gam, G.-T., Jeong, J.-G., Lee, N.-J., Lee, W., Ha, C.-S., Cho, W.-J. (1995) Synthesis and biological activities of copolymers of N-glycinyl maleimide with methacrylic acid and vinyl acetate. J Appl Polym Sci, 57(2), 219.

22. Claracq, J., Santos, S., Duhamel, J., Dumousseaux, C., Corpart, J.M. (2002) Rigid interior of styrene-maleic anhydride copolymer aggregates probes by fluorescence spectroscopy. Langmuir, 18(10), 3829.

23. James, T.D., Sandanayake, S., Shinkay, S. (1996) Saccharide sensing with molecular receptors based on boronic acid. Angew Chem Inter Ed Eng, 35(17), 1910.

24. Barth, R.F., Yang, W., Rotaru, J.H., Moeschberger, M.L., Boesel, C.P., Soloway, A.H., Joel, D.D., Nawrocky, M.M., Ono, K., Goodman, J.H. (2000) Boron neutron cupture therapy of brain tumors: enchanced survival and cure following blood-brain barrier disruption and intracarotid

injection of sodium borocaptate and boronophenyl aniline. Int J Radiat Oncol Biol Phys, 47(1), 209.

25. Siebert, W. (Ed.) (1887): Advances in Boron Chemistry. Cambridge, Royal Society Chemistry.

26. Mishima Y. (Ed.) (1996) Cancer Neutron Capture Therapy. New York, Plenum Press.

27. Kettner, C.A. and Shenvi, A.B. (1984) Inhibition of the serine proteases leukocyte elastase, pancreatic elastase, cathepsin G, and chymotrypsin by peptide boronic acids. J Biol Chem, 259(24), 15106.

28. Miyazaki, H., Kikuchi, A., Kitano, S., Koyama, Y., Okano, T., Sakurai, Y., Kataoka, K. (1993) Boronate-containing polymer as novel mitogen for lymphocytes. Biochem Biophys Res Commun, 195(2), 829.

29. Aoki, T., Nagao, Y., Terada, E., Sanui, K., Ogata N., Yamada, N., Sakurai, Y., Kataoka, K., Okano, T. (1995) Endothelial cell differentiation into capillary structures by copolymer surfaces with phenylboronic acid group. J Biomater Sci Polym Ed, 7(7), 539.

30. Otsuka, H., Uchimura, E., Koshino, H., Okano, T., Kataoka, K. (2003) Anomalous binding profile of pheny-lboronic acid with N-acetylneuraminic acid (Neu5Ac) in aqueous solution with varying pH. J Am Chem Soc, 125(12), 3493.

31. Uchimura, E., Otsuka, H., Okano, T., Sakurai, S., Kataoka, K. (2001) Totally synthetic polymer with lectin-like function: Induction of killer cells by the copolymer of 3-acryl-amidophenylboronic acid with N,N-dimethylacrylamide. Biotech Bioeng, 72(3), 307.

32. Otsuka,H., Ikeya, T., Okano, T., Kataoka, K. (2006) Activation of lymphocyte proliferation by boronatecontaining polymer immobilised on substrate: The effect of boron content on lymphocyte proliferation. Eur Cells Mater, 12(1), 36.

33. Kataoka, K., Miyazaki, N., Okano, T., Sakurai, Y. (1994) Sensitive glucose-induced change of the lower critical solution temperature of poly N,N-dimethylacrylamideco-3-(acrylamido) phenylboronic acid. in physiological saline. Macromolecules, 27(4), 1061.

34. Uguzdoğan, E., Denktaş, E.B., Tuncel, A. (2002) RNAsensitive N-isopropylacrylamide/vinylphenyl boronic acid random copolymer. Macromol Biosci, 2(5), 214.

35. Uguzdoğan, E., Kayi, H., Denktaş, E.B., Patir, S., Tuncel, A. (2003) Stimuli-responsive properties of aminophenylboronic acid carrying thermosensitive copolymers. Polym Int, 52(5), 649.

36. Shiomori, K., Ivanov, A.E., Galaev, I.Yu., Kawano, Y., Mattiasson, B. (2004) Thermo-responsive properties of sugar sensitive copolymer of N-isopropylacrylamide and 3-(acrylamido)phenylboronic acid Macromol Chem Phys, 205(1), 27.

37. Rzayev, Z.M.O., Beşkardeş, O. (2007) Boron-containing functional copolymers for bioengineering applications. Collect Czech Chem Commun, 72(12), 1591.

38. Kahraman, G., Beşkardeş, O., Rzayev, Z.M.O., Pişkin, E. (2004) Bioengineering functional copolymers. VII. Synthesis and characterization of boron-containing self-assembled supramolecular architectures. Polymer, 45(17), 5813.

39. Çimen, E.K., Rzayev, Z.M.O., Pişkin, E. (2005) Bioengi-neering functional copolymers. V. Synthesis LCST, and thermal behavior of poly(N-isopropylacrylamide-co-p-vinyl-phenylboronic acid). J Appl Polym Sci, 95(3), 573.

40. Rzayev, Z.M.O., Erdoğan, D., Türk, M., Pişkin, E. (2008) Bioengineering functional copolymers. VIII. Stimuli-responsive boron-containing graft copolymers and their poly(ethylene imine) macrocomplexes and DNA conjugates. J Biol Chem, 36(2), 83.

41. Türk, M., Dincer, S., Yulug, I.G., Piskin, E. (2004) In vitro transfection of HeLa cells with temperature sensitive polycationic copolymers. J Control Release, 96(2), 325.

42. Choi, S.-J., Oh, J.-M., Choy, J.-H. (2009) Toxicological effects of inorganic nanoparticle nanoparticles on human lung cancer A549 cells. J Inorg Biochem, 103(3), 463.

43. Ulukaya, E., Kurt, A., Wood, E.J. (2001) 4-(N-hydroxyphenyl)retinamide can selectively induce apoptosis in human epidermoid carcinoma cells but not in normal dermal fibroblasts.Cancer Invest, 19(2), 145.

44. McPartland, J.L., Guzail, M.A., Kendall, C.H., Pringle, J.H. (2005) Apoptosis in chronic viral hepatitis parallels histological activity: an immunohistochemical investigation using antiactivated caspase-3 and M30 cytodeath antibody. Int J Exp Pathol, 86(1), 19.

Chapter 4

BIOENGINEERING FUNCTIONAL COPOLYMERS. XVII. INTERACTION OF ORGANOBORON AMIDE-ESTER BRANCHED DERIVATIVES OF POLY (ACRYLIC ACID) WITH CANCER CELLS

Mustafa Türk[1], Gülten Kahraman[2], Sevda A. Khalilova[3], Zakir M. O. Rzayev[4], Serpil Oguztüzün[1]

[1]Department of Biology, Faculty of Arts and Sciences, Kırıkkale University, Yahşihan, Turkey

[2]Sarayköy Nuclear Research and Training Center, Turkish Atomic Energy Authority, Ankara, Turkey

[3]Scientific Research Institute of Medicinal Prophylaxis, Ministry of Public Health, Baku, Azerbaijan

[4]Institute of Science & Engineering, Division of Nanoscience and Nanomedicine, Hacettepe University Ankara, Turkey

ABSTRACT

Novel bioengineering functional organoboron polymers were synthesized by 1) amidolysis of poly(acrcylic acid) (PAA) with 2-aminoethyldiphenyl borinate (2-AEPB), 2) esterification of organoboron PAA polymer (PAA-B) with a-hydroxy-w-methoxypoly(ethylene oxide) (PEO) as a compatibilizer and 3) conjugation of organoboron PEO branches (PAA-B-PEO) with folic acid (FA) as a targeting agent. Structure and composition of the synthesized polymers were characterized by FTIR-ATR and ^{1}H (^{13}C) NMR spectroscopy, chemical and physical analysis methods. Antitumor activity of organoboron functional polymer and its complex with FA (PAA-B-PEO-F) against cancer and normal cells were evaluated by using different biochemical methods such as cytotoxicity, statistical, apoptotic and necrotic cell indexes, double staining and caspase-3 immune staining, light and fluorescence inverted microscope analyses. It was found that citotoxicity and apoptotic/necrotic effects of polymers significantly depend on the structure and composition of studied polymers, and increase the following raw: PAA << PAA-B < PAA-B-PEO < PAA-B-PEO-F. Among them, PAA-B-PEO-F complex at 400 mg·

mL^{-1} **concentration as a therapeutic drug exhibits minimal toxicity toward the normal cells, but influential for** HeLa cancer cells.

INTRODUCTION

Many natural polymers such as polylysine, polyarginine, dextran derivatives, heparin and chitosan, and synthetic bioengineering polymers such as poly(acrylic acid) (PAA), copolymers of maleic anhydride, have now been reported to have direct or indirect antitumor activity via stimulation of the immune system [1-3]. In recent years, the PAA and its copolymers have been often used as carriers in drug release systems, because of their multifunctional nature, unique properties and good biocompatibility [4,5]. Dimitrov et al. [5] studied the biopharmaceutical characterization of hydrogels based on crosslinked PAA and showed that this studied systems provide retarded drug release and appear to be potential candidates for use in the pharmaceutical practice. PAA is grafted to the poly(ethylene glycol) hydro-gel by photo-induced graft polymerization. Due to carboxyl functionality of PAA, collagen and cell adhesion protein, they could be covalently immobilized on to the poly(ethylene glycol) hydrogel [6]. Most of the hydrogels utilized as adhesives for dermatological patches were composed using PAA and its salts as matrix polymers. The ionic interactions between the carboxyl groups of the polymer and polyvalent cations such as calcium, copper, and aluminum cause the formation of the chemical crosslinking used to increase their mechanical strengths [7].

Acrylate-based polymers, containing carboxylic groups, exhibit a swelling behaviour depending on pH and ionic strength of solution [8,9]. Argentiere et al. [10] investigated PAA nanogels as pH-sensitive carriers for biomedical applications. They prepared PAA–biopolymer nanogels by loading and release of an oligothiophene fluorophore and its albumin conjugate onto the PAA macromolecules. On the other hand, several synthetic boron-containing compounds exhibiting important biological properties were investigated as potential therapeutics [11,12]. The mild electrophilic nature of the boronic acid moiety has led to its use at the 'warhead' site of enzyme inhibitors, particularly for inhibiting proteases. For this purposes, several researchers developed some α-aminoboronic acid derivatives [11,13]. One such compound, the novel proteasome inhibitor bortezomib (Velcade) has been recently approved for clinical use as an anticancer agent for the treatment of myeloma [14].

Ban et al. [15] synthesized a series of o-carboranyl phenoxy derivatives as potent inducers for the activation of the 20S proteasome and as chemical probes for the investigation of proteasome-dependent degradation pathways. Other types of bioactive boron-containing compounds have been investigation as therapeutic agents. These include certain boron analogues of biomolecules

[16], diazaborine as an antibacterial and antimalarial agent [17], various antibacterial oxazaborolidines [18], the antibacterial diphenyl borinic esters to inhibit bacterial cell wall growth [19], the antifungal agent benzoxaborole (AN2690) [20], and an oestrogen receptor modulator containing a B–N bond [21]. Some organoboron compounds, including boronic acids and its functional derivatives, and carboranes, were also investigated as agents for boron-neutron capture therapy (BCNT) for the treatment of brain tumors [12,22, 23]. The goal of this work is synthesis and characterization of novel organoboron amide-ester derivatives by amidolysis of PAA with 2-aminoethyldiphenyl borinate (2-AEPB) and their a,ω-hydroxy-methoxypoly(ethylene oxide) (PEO) macrobranched derivative by grafting of synthesized organoboron polymer with PEO to improve the biocompatibility and degree of conjugation with cancer biomacromolecules. An important aspect of this work is comparative investigations of the interactions of these novel functionalized organoboron polymers with HeLa (human cervix carcinoma cell) cancer cells and L929 Fibroblast normal cells, and evaluation of their antitumor activity (cytotoxicity, apoptotic and necrotic effects) using various biochemical methods such as hematoxylen/eosin and immune cytochemical staining, light and fluorescence inverted microscopy analyses. Synthetic partway of the side-chain amide-, esterand carboxyl-functionalized organoboron polymers can be represented as follows (Scheme 1).

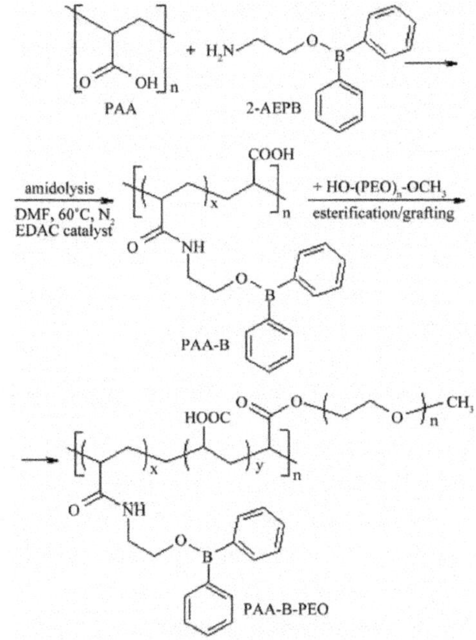

Scheme 1: Synthetic partway of the organoboron amideester-carboxyl functionalized polymers via amidolysis and esterification/grafting reactions.

EXPERIMENTAL

Materials

PAA (BDH) was used as 25% aqueous solution with M_w 230.000 g/mol and density 1.09 g/ml. 2-Aminoethyl diphenylborinate (2-AEPB) (Sigma-Aldrich, Germany) was purified by recrystallization from anhydrous ethanol: m.p. 193.5°C (by DSC); FTIR-ATR spectra of 2-AEPB, cm^{-1}: 3284 (vs) and 3220 (s) N-H stretching in NH_2, 3066 (vs)-2870(s) C-H stretching, 1611(vs) NH_2 bending and C=C stretching in phenyl groups, 1491(m) and 1334 (m) B-O band, 1432 (vs) fairly strong, sharp band due to benzene ring vibration in phenyl-boronic acid linkage, 1263-1154 (s) fairly strong, sharp bands due to C-N stretching in C-NH_2, 1061(vs) N-H bending in NH_2 and 750-710(s) sharp bands due to boron-phenyl linkage; ^1H NMR spectra (δ, ppm) in CHCl$_3$-d$_1$: CH$_2$-O 1.49, CH$_2$-NH$_2$2.96, and 7.38-7.40 (1H), 7.19-7.24 (2H) and 7.13-7.16 (2H) for protons of p-, oand m-positions in benzene ring, respectively.

a-Hydroxy-ω-methoxy-PEO (M_n 2000 g · mol^{-1}) (Fluka). ^1H NMR spectra (δ, ppm) in CHCl$_3$-d$_1$: CH$_2$-O3.75-3.45, OH end group 2.61 and O-CH$_3$ end group 2.16.

N-Ethyl-N-(3-dimethylaminopropyl)carbodiimide hydrochloride (EDAC) as a catalyst and folic acid (FA) as a targeting agent were supported from Aldrich-Sigma (Germany). All solvents and reagents were of analytical grade and used without purification HeLa (human cervix carcinoma cell) cancer cells and L929 Fibroblast cells were obtained from the tissue culture collection of the SAP Institute (Ankara, Turkey). Cell culture flasks and other plastic material were purchased from Corning (NY, USA). The growth medium, which is Dulbecco Modified Medium (DMEM) without L-glutamine supplemented fetal calf serum (FCS), and Trypsin-EDTA were purchased from Biological Industries (Kibbutz Beit Haemek, Israel). The primary antibody, caspase-3 was purchased from Lab Vision (Germany).

Synthesis of 2-Amidoethyldiphenylborinate-poly(Acrylic Acid)

Amidolysis of PAA with 2-AEPB using various [PAA]/[2-AEPB] mole ratios was carried out in N,N'-dimethylformamide (DMF) at 60°C with EDAC catalyst under the nitrogen atmosphere using a standard Pyrex-glass reactor supplied by a mixer, temperature control unit and condenser. Reaction conditions: [2-AEPB] = 0.066 mol · L^{-1}, mole ratios of [PAA]/ [2-AEPB] = 1: 1, 3: 1, 5: 1 and EDAC = 1.0 wt %. Appropriate quantities of PAA, 2-AEPB, DMF and EDAC were placed in a reactor and the reaction mixture was flushed with dried nitrogen gas for at least 2 min, then sealed and placed in a thermo stated silicon oil bath at 60°C to intensive mixing for 5 h. The organoboron amide polymer was isolated from reaction mixture by precipitation with diethyl ether and dried under vacuum. Synthesized organoboron polymer has the following average parameters: T_g (by DSC) 190.6°C, $[h]_{in}$ in deionized water at 25°C 2.72 dL $\cdot$ g^{-1}; FTIR-ATR spectra of PAA-g-2-AEPB (KBr pellet), cm^{-1}: 1702 (m) C=O stretching (amide I band), 1642 (w) and 1542 (m) N-H deformation (amide II band), 1446(m) and 1407(w) C-N stretching (amide III band); ^{1}H NMR spectra (in DMSO-d$_6$ at 25°C) δ ppm: protons of phenyl groups 7.9, 2H from CH$_2$ in -CH$_2$-CO-NHfragment 7.3, 2H from B-O-CH$_2$ group 3.6, 2H from NH-CH$_2$ group 3.0 and 2H from backbone -CH-CH- 3.3, 2.2 and 1.2-1.7; ^{13}C NMR spectra (δ ppm): C=O of PAA unit/amide linkage 177, CH= in phenyl groups 162-158, backbone CH$_2$ and CH128-127 and 57, NH-CH$_2$ 41-42, CH$_2$-O 31-36.

Synthesis of PEO Branched Organboron Copolymer

The esterification (grafting) of organoboron amide of PAA, containing 19.24 mol % of organoboron groups, with PEO (M$_n$ 2000 g $\cdot$ mol^{-1}) at organoboron polymer/PEO feed molar ratio 1: 0.01 was carried out in DMF at 60°C for 1 h. PEO branched polymer was isolated from reaction mixture by precipitation with diethyl ether and dried 40°C under vacuum. Prepared PEO ester of organoboron polymer has the following average characteristics: T_g 175.8°C (by DSC); $[h]_{in}$ in deionized water at 25°C 0.9 dL $\cdot$ g^{-1}; ^{1}H NMR spectra (in DMSO-d$_6$ at 25°C) δ ppm: protons of phenyl groups 7.8, CH$_2$ in CH$_2$-CO-NHamide group 7.2, 2H in B-O-CH$_2$ 3.5, CH$_2$CH$_2$ in PEO branch 3.35, 3H in OCH$_3$end group 3.1 and 2H in NH-CH$_2$ 2.9; ^{13}C NMR spectra (δ ppm): C=O in –COOH 176, CH= in phenyl groups 162, backbone CH$_2$ and CH 128 and 127, respectively, O-CH$_2$ in PEO 69, end OCH$_3$group of PEO 57, NH-CH$_2$ 41-42, and CH$_2$-O-B 31-36.

Characterization

Fourier transform infrared (FTIR-ATR) spectra were recorded with FTIR Nicolet 8700 spectrometer in the 3700 - 600 cm^{-1} range. ^{1}H and ^{13}C NMR spectra were performed on a Bruker Avance (300 MHz) spectrometer with DMSO-d$_6$ as a solvent at 25°C. Thermo gravimetric (TGA) and differential scanning calorimetric (DSC) analyses were performed in a TGA-DTA (Perkin Elmer TG/DTA6300) and a DSC2010 Thermal Analyzers, respectively, under nitrogen atmosphere at a heating rate of 10°C/min. The intrinsic viscosity [h]$_{in}$ values of the organoboron polymers were determined in deionized water at 25°C ± 0.1°C in the concentration range 0.003 - 0.06 g $\cdot$ dL^{-1} using an Ubbelohde viscometer.

Analyses of cytotoxicity, apoptotic and necrotic cells with double staining and immune cytochemical stains of the synthesized novel organoboron functional polymers were performed according to the modified methods using in our recent published work [25,26]. For cytotoxicity experiments, HeLa and L 929 Fibroblast cells (50×10^3 cells per well) were placed in DMEM by using 24-well plates. Different amounts of pristine PPA and organoboron polymers (PAA-B, PPA-PEO-B and PAA-PEO-B-F) (about 0-650 mg $^\bullet$ mL^{-1} in aqueous solutions) were put into wells containing cells, respectively. The plates were kept in the CO_2 incubator (37°C in 5% CO_2) for 24 h; the medium was replaced with fresh medium, and incubated at the same conditions for 24 h. Following of this incubation, HeLa cells were harvested with trypsin–EDTA, and then were dyed with trypan blue. The number of living and dead cells were counted with a haemacytometer (C.A. Hausse & Son Phluila, USA), using light microscope at ×200 magnification. Analysis of apoptotic and necrotic cells with double staining were performed to quantify the number of apoptotic cells in culture on basis of scoring of apoptotic cell nuclei. HeLa and L929 fibroblast cells (20×10^3 cells per well) were placed in DMEM by using 24-well plates. After treating with different amount functional oligomers (about 0 - 650 mg $^\bullet$ mL^{-1} in aqueous solutions) for 24 hours period, both attached and detached cells were collected, then washed with PBS and stained with Hoechst dye 33342 (2 mg $^\bullet$ mL^{-1}), propodium iodide (PI) (1.0 mg $^\bullet$ mL^{-1}) and DNAse free-RNAse (100 mg $^\bullet$ mL^{-1}) for 15 min at room temperature. After that 10 - 50 mL of cell suspension was smeared on slide and cover slip for examination by fluorescence microscopy. The nuclei of normal cells were stained light blue but apoptotic cells were stained dark blue by the Hoechst dye.

The apoptotic cells were identified by their nuclear morphology as a nuclear fragmentation or chromatin condensation. Necrotic cells were staining red by PI. Necrotic cells lacking plasma membrane integrity and PI dye cross cell membrane, but PI dye don't cross non necrotic cell membrane. The number of apoptotic and necrotic cells in 10 randomly chosen microscopic fields were counted and the result expressed as a ratio of apoptotic and necrotic to normal cells. The number of apoptotic and necrotic cells were determined by light and fluorescence inverted microscope (Leica, Germany). The cell images were also recorded using the both above mentions microscopes with DAPI and FITC filters, respectively.

For immunocytochemical stains, about 2 ml HeLa (20×10^3 cells per well), treated with functional oligomers (about 0 - 650 mg $^\bullet$ mL^{-1} in aqueous solutions) suspension was centrifuged for 5 min in a Hettich centrifuge. Cytospin preparations were fixed in 70 % ethanol for immunocytochemistry.

For an indirect immunocytochemical procedure, cytology specimens were treated with 3% H_2O_2 for 10 min, taken to water, and then rinsed in PBS (pH 7.4) for 5 min. Nonspecific protein binding was blocked on specimen by incubating with blocking solution for 10 min. The primary antibody, caspase-3 (Lab Vision) used at 1:300 dilution, incubated for 1 h at room temperature. Specimens were washed with PBS buffer (pH 7.4) and incubated in biotinylated secondary antibody solution for 10 min. Diaminobenzidine (Dako) served as the chromagen and Mayer's hematoxylin as the counter stain. For the negative control the primary antibody was omitted in one of the slides.

The immunocytochemical staining results were controlled independent and blindly by observers without the knowledge of treatment. The immunoreactivity of the caspase-3 antibody is confined to the cytoplasm of apoptotic cells. We counted the number of caspase-3-positive cytoplasmic staining cells in all fields found at ×400 final magnification. For each image, three randomly selected microscopic fields were observed, and at least of 100 cells/field were evaluated.

RESULTS AND DISCUSSION

Structure of Organoboron Polymers

The structures of synthesized organoboron polymers and their PEO branches were confirmed by FTIR-ATR and ^1H (^{13}C) NMR analysis. Comparative analysis of FTIR-ATR spectra (**Figure 1**) of 2-AEPB, PAA and its organoboron derivative indicates that the characteristic bands of acid C=O groups disappearance in the spectra of PAA-B-1 polymer prepared from 1:1 molar feed ratio of PAA: 2-AEPB.

The formation of amide bound in this organoboron polymer is confirmed by the appearance of new bands such as 1702 (amide-I band), 1642 and 1542 (amide-II band), 1446 cm^{-1} and 1407 (amide-III band). Simultaneously a very broad band between 3500 and 2500 cm^{-1} appearances in spectra due to increase hydrogen bonded fragments in organoboron polymer (PAA-B-1). As the intensities of amide bands significantly decrease in spectra of PAA-B-2 and PAA-B-3 polymers prepared from

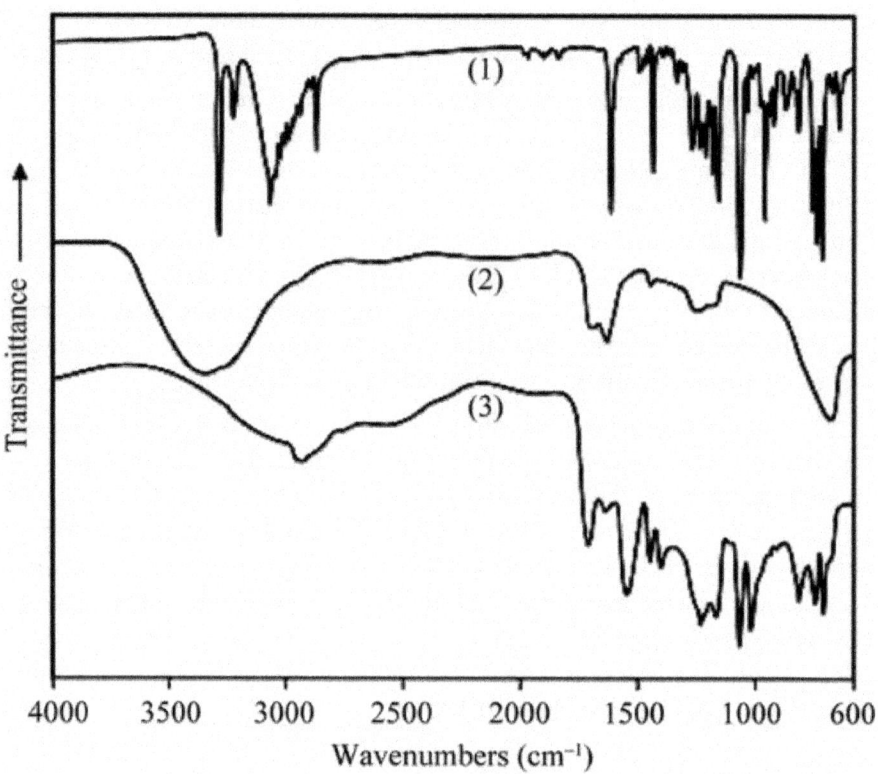

Figure 1: FTIR spectra: (1) 2-AEPB, (2) PAA and (3) PAAg-2-AEPB-1 organoboron amide polymer.

PAA/ 2-AEPB mixtures enchasing with PAA (PAA>> 2-AEPB), the intensities of free acid group bands increase. Similar effect has been observed from comparative analysis of the ^{1}H and ^{13}C NMR spectra of A-B-2and its PEO branch (PAA-B-PEO). The results of this analysis are illustrated in Figures 2 and 3. The formation of H-bonded amide linkages is confirmed by a presence of characteristic broad peaks at 7.3 and 177 ppm in the ^{1}H and ^{13}C NMR spectra of PAA-B-2, respectively (**Figure 2**). In addition, the presence of characteristic proton peaks of organoboron linkages such as quarter phenyl peak at 7.9 ppm, triplet B-O-CH$_2$ peak at 3.6 ppm andquarter NH-CH$_2$ peak at 3.0 ppm (**Figure 2**(a)) also confirmed that 2-AEPB is covalently bound to anhydride units. In the ^{13}C NMR spectra of PAA-B-2 polymer (**Figure 2**(b)), the characteristic carbon resonances (162, 158, 41, 42, 31 and 36 ppm) from organoboron fragment are also observed.

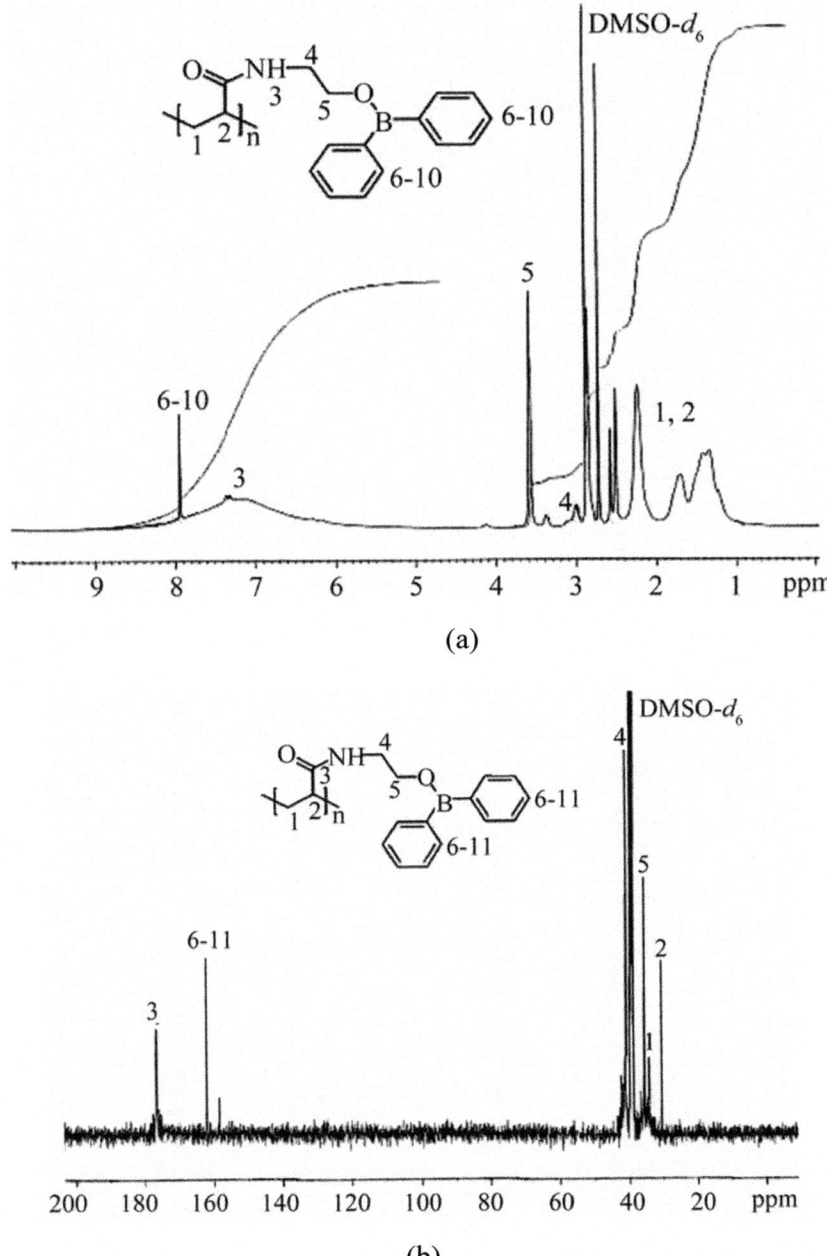

Figure 2: (a) ¹H NMR and (b)¹³C NMR spectra of organoboron polymer (PAA-B) in DMSO-d₆.

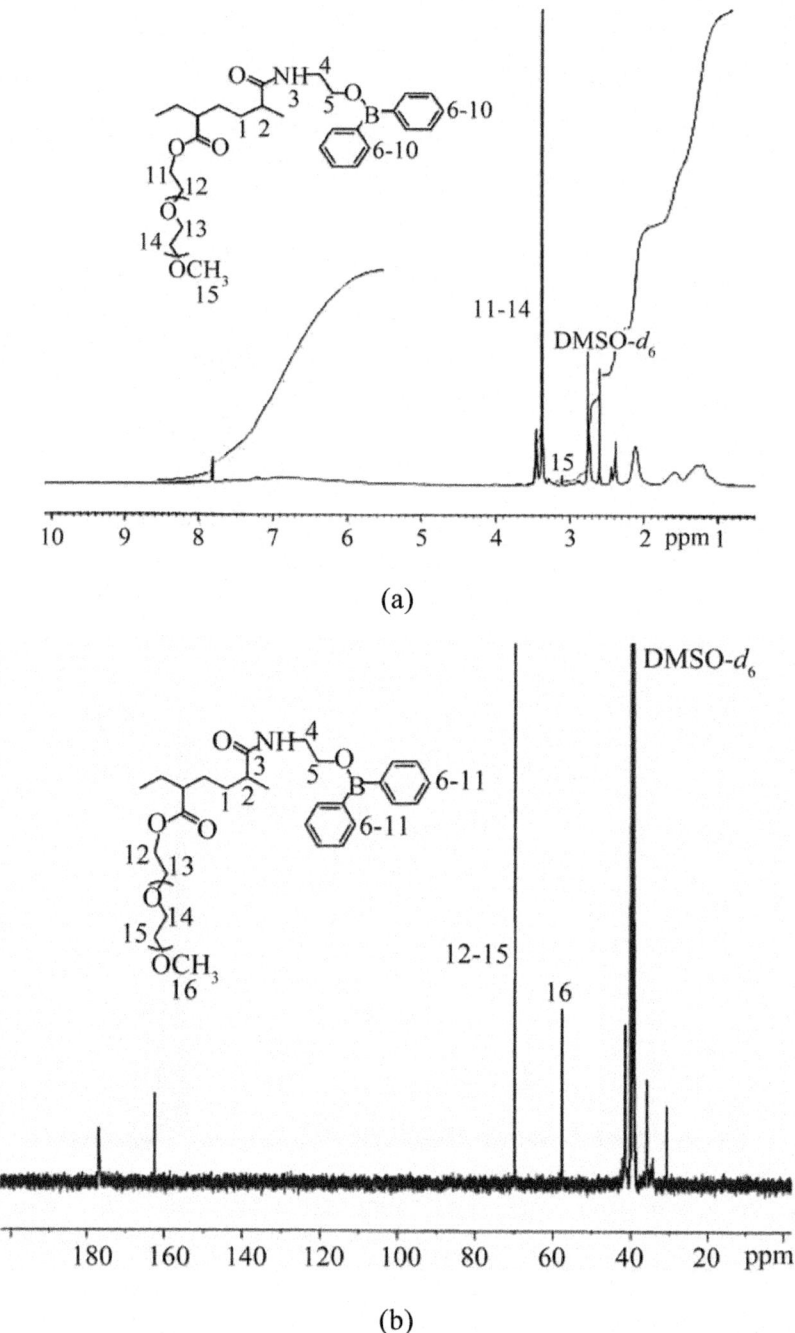

Figure 3: (a) ^{1}H NMR and (b) ^{13}C NMR spectra of PAA-gAEPB-2-g-PEO in DMSO-d$_6$.

^{1}H (^{13}C) NMR spectra of PEO grafted organoboron PAA polymer (PAA-B-PEO) were illustrated in**Figure 3**. The observed proton signals from of side-chain PEO branches at 3.4 and 3.1 ppm for $(CH_2\text{-}CH_2\text{-}O)_n$ units (**Figure 3**(a)) and carbon atom resonances (69 ppm for O-CH$_2$ and 57 ppm for OCH$_3$ end group) (**Figure 3**(b)) may be served an additional fact to confirm the formation of side-chain macrobranched PEO linkages.

Functional Polymer Composition–Property Relationship

The results of intrinsic viscosity measurements from the plots of h$_{sp}$c (specific viscosity) vs. c (polymer concentration in deionized water) for the organoboron PAA and PEO derivatives having different compositions are illustrated in **Figure 4**. A visible decrease of h$_{in}$ value with

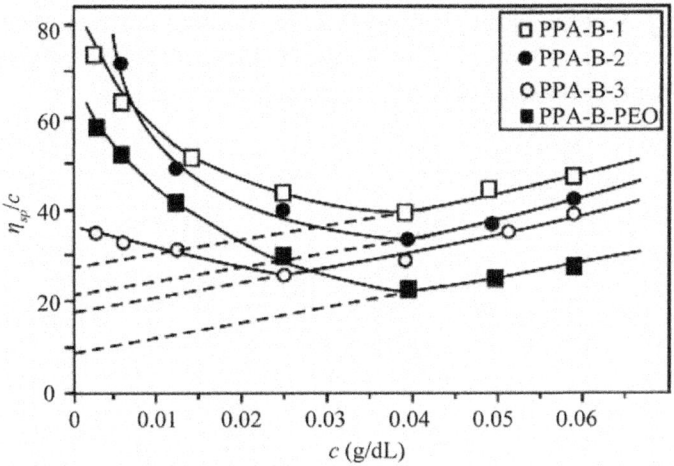

Figure 4: The plots of η$_{sp}$/c (specific viscoity) versus c (polymer concentrations in deionized water) for the determination of intrinsic viscosity and evaluation of polymer composition-viscosity relationships (dilution effect and polyelectrolyte behaviour): –□– PPA-B-1, –·– PPA-B-2, –o– PAA-B-3 and –■– PAA-B-PEO.

Table 1. Some characteristic parameters of organoboron amide (PAA-B) and PEO branched (PAA-B-PEO) derivatives of PAA

Functional polymers	B content (%)	$[\eta]_{in}$ (dL · g^{-1}) in water at 25°C	T_g (°C)
PAA-B-1	3.64	2.75	190.6
PAA-B-2	2.45	2.16	185.0
PAA-B-3	1.85	1.74	181.2
PAA-B-2-PEO	2.16	0.85	175.8

increasing the organoboron fragment in PAA polymer was observed (**Table 1**). Unlike the PEO branched derivative the organoboron polymers exhibit typical polyelectrolyte behavior, i.e., increase in viscosity with a dilution of polymer water solution, which can be explained by specific behavior of complexed macromolecules and their conformational changes resulting in the expansion of polymer coil in the dilution solution. Similar effect was observed for the other carboxyl-containing polymers [27,28].

Thermal behavior and phase transitons of synthesized organoboron polymers were investigated by differential scanning calorimetric (DSC) and thermal gravimetrical analysis (TGA) methods. The obtained results were summarized in **Figure 5**. It was found that PAA and its organoboron and PEO branched derivatives exhibit amorphous structure with characteristically broad endo-peaks, which are associated with the glass-transition temperatures (T_g), significantly depend on the composition and content of organoboron linkages in the functional polymers. The higher values of T_g are observed for the polymers containing relatively high organoboron linkages.

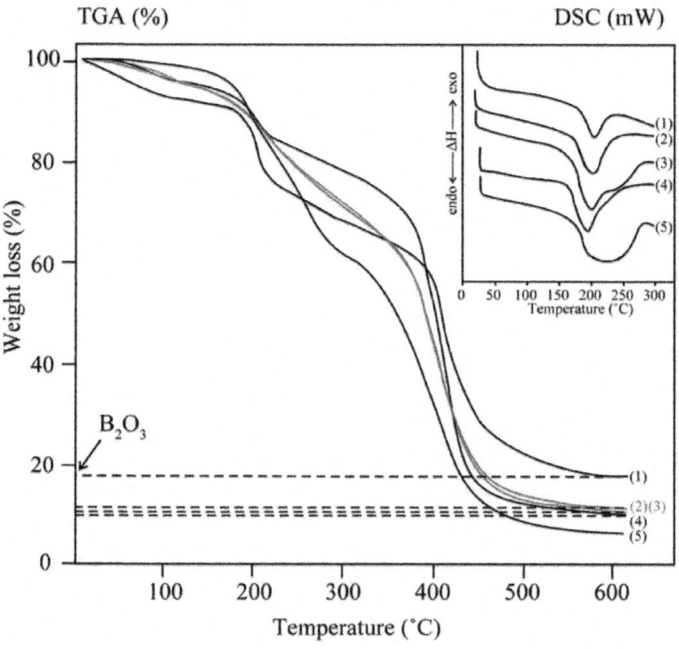

Figure 5. DSC and TGA curves of functional organoboron polymers: (1) PAA-B-1, (2) PAA-B-2, (3) PAA-B-3, (4) PAA-B-PEO and (5) pristine PAA. Heating rate 10°C / min under a nitrogen atmosphere.

Therefore, rigid H-bonded structure provides high T_g in the organoboron polymers (curves 1-3). The results of TGA analyses (**Figure 5**) indicate that the organoboron polymer and PEO branched derivative of PAA show higher thermal stability which increases with increasing degree of grafted organoboron linkage in the polymer. The observed two step degradation of the PAA and its functionalized derivatives indicates occurrence of some macromolecular reactions. TGA analyses also allow us to determine the content of boron in studied functionalized polymers, results of which are summarized in **Table 1**.

Cytotoxicity

The obtained cytotoxicity results of the pristine PAA and its organoboron amide (PAA-B) and organoboron amide-ester (PAA-B-PEO) branches, and PAA-BPEO/ folic acid complex (PAA-B-PEO-F) on cancer cells using a trypan blue staining were illustrated in **Figure 6**. As seen from plots of concentration of polymers versus percent of cell viability, the toxicity of pristine PAA against cancer and normal cells decreased with an increasing in polymer concentration from 100 to 200mg $\cdot$ mL^{-1}for 24 h incubation at 37°C. If polymer concentration was higher than 20 mg $\cdot$ mL^{-1}, its toxicity increased, especially higher toxicity exhibits for 24 h incubation. The toxicity of PAA-B (organoboron amide polymer) was more significant than other polymer systems.

Figure 6 shows that the number of viable cells is above 80% for normal and cancer cells after incubation of the cells with PAA-B at concentrations around 100 - 200 mg $\cdot$ mL^{-1} for 24 h incubating time in cell culture media.

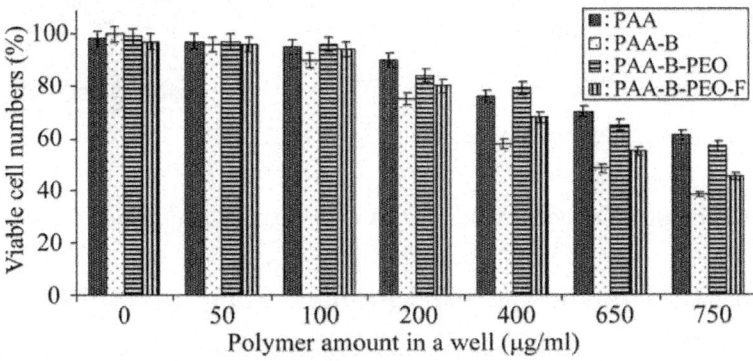

Figure 6: In vitro cytotoxicity of pristine PAA and functionalized organoboron.polymers with different amounts at 24 h incubation. Results are presented as means ± SEM.

The number of viable cells was over 50 % or normal cells in the range of 400 - 650 mg $\cdot$ mL^{-1} concentration. However, the toxicity of cancer cells was increased beginning from 400 mg $\cdot$ mL^{-1}.

The PAA-B and PAA-B-PEO-F polymers had higher toxicity for cancer cells (55% alive cells) than normal cells (68% alive cells) in 650 mg $\cdot$ mL^{-1} concentration of polymer...FA complex. It was observed that the cytotoxicity of PAA-B-PEO decreases as compared with organoboron polymers, which can be explained by compatibilizing effect of PEO branched linkages. The cytotoxicity of PEO containing polymer was lower than those without PEO at 400 - 650 mg $\cdot$ mL^{-1} concentration (**Figure 6**). To improve the targeting of polymer macromolecules to cancer cells, folic acid (FA) was inserted to the structure through complex-formation.

The formation of organoboron polymer...FA complex through interaction of amide or carboxylic groups with pseudo-aromatic amine of FA and its conjugation with HeLa cells may be schematically represented as follows (Scheme 2).

When the polymer...FA complex was incubated, the toxicity of the HeLa cells was higher than that of the normal cells, because cancer cells had more FA receptors than normal cells. Therefore PAA-B-PEO-F complex can be utilized as a therapeutic drug at 200 - 400 mg $\cdot$ mL^{-1} concentration, where its toxicity was minimal for normal cells, but influential for cancer cells.

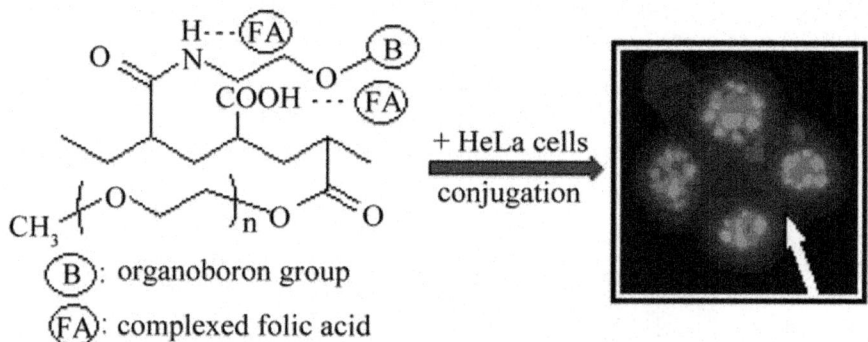

Scheme 2: Proposed structure of PAA...FA complex and conjugation with HeLa cells.

Double Staining and Capase 3 Immune Staining results

Apoptotic index was obtained by both double staining and caspase 3 immune staining methods. The results were presented in **Table 2**. If the cells treated by PAA-B at 400 - 650 mg $\cdot$ mL^{-1} concentration, the number of apoptotic cells was not high. While PPA-B-PEO-F complex at 400 mg $\cdot$ mL^{-1} concentration exhibits the highest apoptotic ratios on cancer cells. In addition, the number of apoptotic cells was high as well for the organoboron polymer/folic acid complex at 400 mg $\cdot$ mL^{-1}concentration (**Table 2**). The results of light and fluorescent microscope investigation of the interaction of organoboron polymers with cancer cells were illustrated in **Figure 7**.

The cytoplasm's of apoptotic cells treated with complex were stained brown (**Figure 7**(b)) but, the cytoplasm of non apoptotic cells were not stained brown (**Figure 7**(a)). According to double staining results, apoptotic cells' nucleus stained bright blue and compartmentalized **Figure 7**(d)), but non-apoptotic cells' nuclei stained lifeless blue (**Figure 7**(c)). When the PEO-containing branched polymers applied, the number of apoptotic cells was decreased. However, the number of apoptotic cells was increased as 7 % when they treated by PAA-PEO-BF complex in 400 mg $\cdot$ mL^{-1}concentration. Apoptotic index for Fibroblast cells was 18 % at 400 mg $\cdot$ mL^{-1} concentration of PAA-PEO-B-F. Moreover, there was no significant change on apoptotic index (18%) of Fibroblast cell targeting by folic acid.

Apoptotic indexes in cancer and normal cells was estimated of caspase-3 and double staining result. The important observations can be summarized as follows: we checked for apoptosis or necrosis with double staining (Hoechst 33342 and PI) and caspase 3 immune staining. It was observed that both the cytotoxicity and necrotic indexes of synthesized functional organoboron polymers show approximately same values

Table 2: The comparative analysis of apoptotic and necrotic HeLa cells index for (I) PAA, and its (II) organoboron amide (PAA-B), (III) PEO branched (PAA-B-PEO) and (IV) FA complexed (PAA-B-PEO-F) derivatives at 24 h incubation. Results are presented as means ± SEM

[Polymer] ($\mu g \cdot mL^{-1}$)	Apoptotic index (%)				Necrotic index (%)			
	I	II	III	IV	I	II	III	IV
0	1 ± 1	1 ± 1	1 ± 1	1 ± 1	1 ± 1	1 ± 1	1 ± 1	1 ± 1
100	5 ± 1	10 ± 2	12 ± 2	16 ± 1	2 ± 1	8 ± 2	2 ± 1	7 ± 1
200	2 ± 2	16 ± 2	15 ± 1	20 ± 1	10 ± 2	25 ± 3	7 ± 1	18 ± 2
400	5 ± 1	21 ± 1	18 ± 2	5 ± 2	24 ± 2	46 ± 3	20 ± 1	30 ± 3
600	6 ± 1	6 ± 1	15 ± 2	20 ± 1	34 ± 2	53 ± 3	33 ± 2	45 ± 2

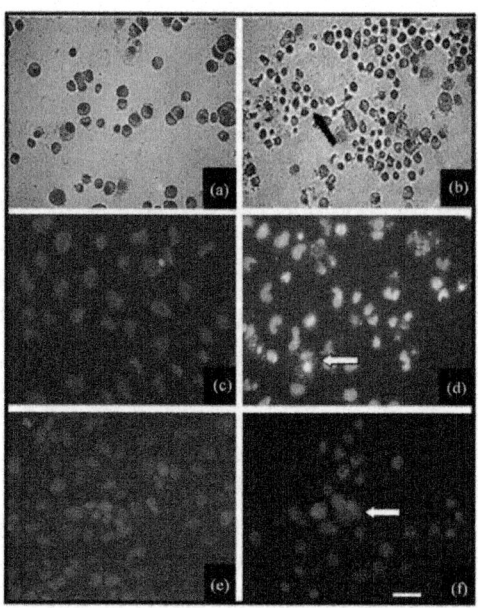

Figure 7: Light microscope images of (a) virgin (non-apoptotic) HeLa cells as a control group (stained with caspas-3 immune staining kit), and (b) 400 mg · mL⁻¹ concentration of organoboron polymer (PolyAC-B-2-PEO-F)/HeLa cells conjugate (stained with caspas-3 immune staining kit); (c) Fluorescent microscope image of nucleus of untreated HeLa cells (stained with Hoechst 33342 dye) as a control, (d) nucleus of HeLa cells (stained with Hoechst 33342); (e) Fluorescent microscope image of nucleus of untreated HeLa cells (stained with PI dye) as a control; (f) nucleus of HeLa cells (stained with PI dye); Photos (c) and (d) taken under DAPI filter, photos (e) and (f) taken under FITC filter. Figure (a) and (b) were recorded with ×200 magnification. Figure (d), (e) and (f) were recorded with ×400 magnification. Scale bar is 20 μm.

The polymers with lower concentrations (100-200 mg · mL⁻¹) decrease in necrosis stained with PI dye. While the necrotic indexes of normal and cancer cells increase at relatively higher concentration (400 mg · mL⁻¹) of polymers, especially PAA-B (**Table 2**and Figures 7(e) and (f)). However, when the polymer containing PEO was incubated to cancer cells, the necrotic index was decreased in cancer and normal cells.

Fluorescent microscope image of nucleus of untreated HeLa cells (stained with PI dye) as a control was presented in **Figure** 7(e), where formation of green spots indicates the nucleus of non-necrotic cells. Cancer cells exposed to polymer...FA complex became highly PI-positive. This observed fact indicated that the cells were undergoing to necrosis. HeLa and Fibroblast cells incorporated with PPA-B polymer provide a lysing the cellmembrane (necrosis) and relatively higher necrotic indexes 53% and 43% for the cancer and normal cells, respectively. When both the cells treated with PAAPEO-B-F copolymer, necrotic indexes decrease for HeLa (49%) and Fibroblast cells (41%).

CONCLUSIONS

This work presents the synthesis and characterization of organoboron, PEO branched and FA complexed derivatives of PPA and investigation of their antitumor activity (cytotoxicity, apoptotic and necrotic effects) toward HeLa and Fibroblast cells by using a combination of various biochemical, statistical and microscopy methods. It was observed that antitumor activity significantly depends on the structure, amount of ionizable free carboxylic groups, organoboron linkages and complexed fragments in the functionalized polymers, and changes in the following row: PAA << (PAA-B)s < PPA-B-PEO < PAA-B-PEO-F. Among them, PAA-B-PEO-F copolymer system showed promising antitumor activity against cancer cells through apoptosis and necrosis induced caspase-3-dependent partway. Apoptotic indexes in cancer and normal cells were estimated of caspase-3 immune staining and double staining (Hoechst 33342 and PI) results. These observations are confirmed the realization of apoptosis and necrosis processes in the interaction of functionalized polymers with normal and cancer cells. Apoptotic index of cancer cells were obtained higher than normal cells. Especially, apoptotic effect of FA containing copolymer was increased compared with non-targeted copolymers. HeLa and Fibroblast cells incorporated with organobron polymer provide a lysing the cell-membrane (necrosis) and relatively higher necrotic indexes 53% and 43% for the cancer and normal cells, respectively. It was found that interactions of both the cells with PPA-PEO-B-F copolymer were decreased the necrotic indexes for HeLa (45%) and Fibroblast cells (41%). Utilization of this novel organoboron polymer as precursors in boron neutron capture therapy (BNCT) will be a subject for our future investigations.

ACKNOWLEDGEMENTS

The supports of this work by the TAEK (Turkish Atomic Energy Authority) and TÜBİTAK (Turkish Scientific and Technology Research Council) through TAEK-A3.H2.P2.01 and TBAG-2386 projects, respectively, are gratefully acknowledged.

REFERENCES

1.   L. Seymour, "Synthetic Polymers with Intrinsic AntiCancer Activity," Journal of Bioactive and Compatible Polymers, Vol. 6, No. 2, 1991, pp.

178-216.doi:10.1177/088391159100600205

2. J. Liao and R. M. Ottenbrite, "Controlled Drug Delivery: Challenges and Strategies," ACS, Washington DC, 1997.

3. S. Akhtar, "Non-Viral Cancer Gene Therapy: Beyond Delivery," Gene Therapy, Vol. 13, No. 5, 2006, pp. 739- 740. doi:10.1038/sj.gt.3302692

4. M. Dittgen, M. Durrani and K. Lehmann, "Acrylic Polymers. A Review of Pharmaceutical Applications," S. T. P. Pharma Science, Vol. 7, No. 6, 1997, pp. 403-437.

5. M. Dimitrov, M. Lambovi, S. Shenkov, V. Dosseva and V. Y. Baranovski, "Hydrogels Based on the Chemically Crosslinked Polyacrylic Acid: Biopharmaceutical Characterization," Acta Pharmaceutica, Vol. 53, No. 1, 2003, pp. 25-31.

6. W. Lee, T. G. Lee and W.-G. Koh, "Grafting of Poly- (Acrylic Acid) on the Poly(Ethylene Glycol) Hydrogel Using Surface-Initiated Photopolymerization for Covalent Immobilization of Collagen," Industrial & Engineering Chemistry Research, Vol. 13, No. 7, 2007, pp. 1195- 1200.

7. Y. Onuki, M. Nishikawa, M. Morishita and K. Takayama, "Development of Photocrosslinked Polyacrylic Acid Hydrogel as an Adhesive for Dermotological Patches: Involvement of Formulation Factors in Physical Properties and Pharmacological Effects," International Journal of Pharmacology, Vol. 349, No. 1-2, 2008, pp. 47-52.doi:10.1016/j. ijpharm.2007.07.021

8. B. R. Saunders, H. M. Crowther and B. Vincent, "Poly- [(methyl methacrylate)-co-(methacrylic acid)] Microgel particles: Swelling Control Using pH, Cononsolvency, and Osmotic Deswelling," Macromolecules, Vol. 30, No. 3, 1997, pp. 482-487.doi:10.1021/ma961277f

9. T. Sawai, S. Yamazaki, Y. Ikariyama and M. Aizawa, "pH-Responsive Swelling of the Ultrafine Microsphere," Macromolecules, Vol. 24, No. 8, 1991, pp. 2117-2118.doi:10.1021/ma00008a067

10. S. Argentiere, L. Blasi, G. Ciccarella, G. Barbarella, R. Cingolani and G. Gigli, "Synthesis of Poly(Acrylic Acid) Nanogels and Application in Loading and Release of on Oligothiophene Fluorophore and Its Bovine Albumin Conjugate," Macromolecular Symposia, Vol. 281, No. 1, 2009, pp. 69-76. doi:10.1002/masy.200950709

11. N. A. Petasis, "Expanding Roles for Organoboron Compounds–Versatile and Valuable Molecules for Synthetic, BIOLOGICal and Medicinal Chemistry," Australian Journal of Chemistry, Vol. 60, No. 11, 2007, pp. 795-798. doi:10.1071/CH07360

12. W. Yang, S. Gao and B. Wang, "Boronic Acid Compounds as Potential Pharmaceutical Agents," Medicinal Research Reviews, Vol. 23, 2003, pp. 346-368.doi:10.1002/med.10043

13. V. M. Dembitsky and M. Srebnik, "Synthesis and Biological Activity of a-Aminoboronic Acid, Aminocarboranes and Their Derivatives," Tetrahedron, Vol. 59, No. 5, 2003, pp. 579-593. doi:10.1016/S0040-4020(02)01618-6

14. P. G. Richardson, C. Mitsiades, T. Hideshima and K. C. Anderson, "Bortezomib: Proteasome Inhibition as an Effective Anticancer Therapy," Annual Review of Medicine, Vol. 57, 2006, pp. 33-47. doi:10.1146/annurev.med.57.042905.122625

15. H. S. Ban, H. Minegishi, K. Shimizu, M. Maruyama, Y. Yasui and H. Nakamura, "Discovery of Carboranes as Inducers of 20S Proteasome Activity," Chemistry & Medicinal Chemistry, Vol. 5, No. 8, 2010, pp. 1236-1241. doi:10.1002/cmdc.201000112

16. C. Morin, "The Chemistry of Boron Analogues of Bio Molecules," Tetrahedron, Vol. 50, No. 44, 1994, pp. 12521-12569. doi:10.1016/S0040-4020(01)89389-3

17. C. Baldock, G.-J. de Boer, J. B. Rafferty, A. R. Stuitje and D. W. Rice, "Mechanism of Action of Diazaborines," Biochemical Pharmacology, Vol. 55, No. 10, 1998, pp. 1541- 1549.

18. Jabbour, D. Steinberg, V. M. Dembitsky, A. Moussaieff, B. Zaks and M. Srebnik, "Synthesis and Evaluation of Oxazaborolidines for Antibacterial Activity against Streptococcus Mutants," Journal of Medicinal Chemistry, Vol. 47, No. 10, 2004, pp. 2409-2410. doi:10.1021/jm049899b

19. S. J. Benkovic, S. J. Baker, M. R. K. Alley, Y.-H. Woo, Y.-K. Zhang, T. Akama, W. Mao, J. Baboval, P. T. Ravi Rajagopalan, W. Wall, L. S. Kahng, A.Tavassoli and L. Shapiro, "Identitication of Borinic Esters as Inhibitors of Bacterial Cell Growth and Bacterial Methyltransferases, CcrM and MenH," Journal of Medicinal Chemistry, Vol. 48, No. 23, 2005, pp. 7468-7476. doi:10.1021/jm050676a

20. S. J. Baker, Y.-K. Zhang, T. Akama, A. Lau, H. Zhou, V. Hernandez, W. Mao, M. R. K. Alley, V. Sanders and J. J. Plattner, "Discovery of a New Boron-Containing Antifungal Agent," Fluoro-1,3-dihydro-1-hydroxy-2,1-benzoxa-borole (AN2690), for the potential treatment of onychomycosis," Journal of Medicinal Chemistry, Vol. 49, No. 15, 2006, pp. 4447-4450. doi:10.1021/jm0603724

21. H. B. Zhou, K. W. Nettles, J. B. Bruning, Y. Kim, A. Joachimiak, S. Sharma, K. E. Carlson, F. Stossi, B. S. Katzenellenbogen, G. L. Greene

and J. A. Katzenellenbogen, "Elemental Isomerism: A Boron-Nitrogen Surrogate for a Carbon-Carbon Double Bond Increases the Chemical Diversity of Estrogen Receptor Ligands," Chemistry & Biology, Vol. 14, No. 5, 2007, pp. 659-669. doi:10.1016/j.chembiol.2007.04.009

22. J. F. Valliant, K. J. Guenther, A. S. King, P. Morel, P. Schaffer, O. O. Sogbein and K. A. Stephenson, "The Medical Chemistry of Carborones," Coordination Chemistry Reviews, Vol. 232, No. 1-2, 2002, pp. 173-230. doi:10.1016/S0010-8545(02)00087-5

23. W. Chen, S. C. Mehta and D. R. Lu, "Selective Boron Drug Delivery to Brain Tumors for Boron Neutron Capture Therapy," Advanced Drug Delivery Reviews, Vol. 26, No. 2-3, 1997, pp. 231-247. doi:10.1016/S0169-409X(97)00037-9

24. F. Shosseler, F. Ilmain and S. J. Candau, "Structure and Properties of Partially Neutralized Poly(Acrylic Acid) Gels," Macromolecules, Vol. 24, No. 1, 1991, pp. 225- 234.doi:10.1021/ma00001a035

25. M. Türk, Z. M. O. Rzayev and S. A. Khalilova, "Bioengineering Functional Copolymers. XIV. Synthesis and Interaction of Poly(N-isopropyl Acrylamide-co-2,3-dihydro-2H-pyran-alt-maleic Anhydride)s with SCLC Cancer Cells," Bioorganic & Medicinal Chemistry, Vol. 18, No. 22, 2010, pp. 7975-7984. doi:10.1016/j.bmc.2010.09.031

26. Türk, Z. M. O. Rzayev and G. Kurucu, "Bioengineering Functional Copolymers. XII. Interaction of Boron-Containing and PEO Branched Derivatives of Poly(MA-alt-MVE) with HeLa Cells," Health, Vol. 2, No. 1, 2010, pp. 51-61. doi:10.4236/health.2010.21009

27. T. Shimisu and A. Minakata, "Effect of Divalent Cations on the Volume of a Maleic Acid Copolymer Gel Examined by Incorporating Lysozyme," European Polymer Journal, Vol. 38, No. 6, 2002, pp. 1113-1120. doi:10.1016/S0014-3057(01)00283-X

28. O. Nobumichi and S. Shintaro, "Conformational Characterization of a Maleic Acid Copolymer with an Inflexible Side Chain," Journal of Macromolecular Science: Pure and Applied Chemistry, Vol. 27, No. 7, 1990, pp. 861-873. doi:10.1080/10601329008544810

Chapter 5

AFLATOXINS BIOCHEMISTRY AND MOLECULAR BIOLOGY - BIOTECHNOLOGICAL APPROACHES FOR CONTROL IN CROPS

Laura Mejía-Teniente, Angel María Chapa-Oliver, Moises Alejandro Vazquez-Cruz, Irineo Torres-Pacheco and Ramón Gerardo Guevara-González

Facultad de Ingeniería, CA Ingeniería de Biosistemas, Universidad Autónoma de Querétaro Centro Universitario Cerro de las Campanas s/n, Querétaro, Qro, México

INTRODUCTION

Fungi play a very important, but yet mostly unexplored role. Their widespread occurrence on land and in marine life makes them a challenge and a risk for humans   (Bräse et al., 2009). Fungi are ingenious producers of complex natural products which   show a broad range of biological activities (Bohnert et al., 2010). However, a specific   characteristic is the production of toxins. Mycotoxins (from "myco" fungus and toxin), are   nonvolatile, relatively low-molecular weight, fungal secondary metabolic products (Bräse   et al., 2009). The most agriculturally important micotoxins are aflatoxins (AF) which are a group of highly toxic metabolites, studied primarly because of their negative effects on   human health. Aflatoxins belong to a group of difuranocumarinic derivatives structurally   related, and are produced meanly by fungi of genus Aspergillus spp. Its production   depends on many factors such as substrate, temperature, pH, relative humidity and the   presence of other fungi. It has been identified 18 types of aflatoxins; the most frequent in   foods are B1, B2, G1, G2, M1, and M2 (Bhatnagar et al., 2002). These secondary metabolites contaminate a number of oilseed crops during growth of the fungus and this can result in   severe negative economic and health impacts (Cary et al., 2009). The higher levels of   aflatoxins have been found in cotton and maize seeds, peanuts, and nuts. In grains like   wheat, rice, rye or barley the presence of aflatoxins is less frequent. Mycotoxins may also occur in conjugated form, either soluble (masked mycotoxins) or incorporated into/ associated with/attached to macromolecules (bound mycotoxins). These conjugated mycotoxins can emerge after metabolization by living plants, fungi and mammals or after food processing. Awareness of such altered forms of mycotoxins is increasing, but reliable analytical methods, measurement standards, occurrence, and toxicity data are still lacking (Berthiller et al., 2009). A variety of studies has been conducted in order to understand the process of crop contamination by aflatoxins. Mycotoxins are dangerous metabolites that are often carcinogenic, and they represent a serious threat to both animal and human health (Reverberi et al., 2010).

Mycotoxins are considered secondary metabolites because they are not necessary for fungal growth and are simply a product of primary metabolic processes. The functions of mycotoxins have not clearly established, but they are believed to play a role in eliminating other microorganisms competing in the same environment (Bräse et al., 2009). The biosynthesis and regulation of these toxins represent one of the most studied areas of all the fungal secondary metabolites. Much of the information obtained on the AF biosynthetic genes and regulation of AF biosynthesis was obtained through studies using A. flavus and A. parasiticus and also the model fungus Aspergillus nidulans that produces sterigmatocystin (ST), the penultimate precursor to AF. Further studies in A. nidulans and A. flavus and also of the fungus-host plant interaction have identified a number of genetic factors that link secondary metabolism and morphological differentiation processes in A. flavus as well as filamentous fungi in general (Cary et al.,   2009). Recent investigations of the molecular mechanism of AF biosynthesis showed that the genes required for biosynthesis are in a 70 kb gene cluster. These genes encode for the   proteins required in the oxidative and regulatory steps in the aflatoxins byosinthesis. A   positive regulatory gene, aflR, coding for a sequence-specific, zincfinger DNA-binding protein is located in the cluster and is required for transcriptional activation of most, if not all, of the aflatoxin structural genes. Some of the genes in the cluster also encode other   enzymes such as cytochrome P450-type monooxygenases, dehydrogenases, methyltransferases, and polyketide and fatty acid synthases (Bhatnagar et al., 2003). The   application of genomic DNA sequencing and functional genomics, powerful technologies   that allow scientists to study a whole set of genes in an organism, is one of the most   exciting developments in aflatoxin research (Yu et al., 2004; Bennett et al., 2007). Moreover,   the rapid development of high throughput sequencing made it possible in genetic research to advance from single gene cloning to whole genome sequencing. Tremendous   advances have also been made in understanding the genetics of four non-aflatoxigenic Aspergillus species, A. oryzae, A. sojae, A. niger and A. fumigatus. Currently, the whole   genome sequencing and/or Expressed Sequence Tag (EST) projects for A. flavus have been   completed (Bhatnagar et al., 2006). The characterization of genes involved in aflatoxin   formation affords the opportunity to examine the mechanism of molecular regulation of the aflatoxin biosynthetic pathway, particularly during the interaction between aflatoxin- producing fungi and plants (Bhatnagar et al., 2003).Aflatoxin contamination in crops is a   worldwide food safety concern due that are compound carcinogenic highly and   mutagenic in animals and human (Yin et al., 2008). Therefore their management in   agricultural (pre-harvest, harvest and post-harvest) is of importance vital, so quantity in   food and feed is closely monitored and regulated in most countries for example, in the   European Union has a maximum level of 2 ng/g for B1 and 4 ng/g for total aflatoxins in crops (van Egmond and Jonker, 2004).

OCCURRENCE OF MYCOTOXINS

Mycotoxins occur in many varieties of fungi. Several mycotoxins are unique to one species, but most mycotoxins are produced by more than one species. The most important mycotoxins are aflatoxins, ochratoxins, deoxynivalenol (DON), searalenone, fumonisin, T- 2 toxin, and T-2 like toxins. However, food borne mycotoxins likely to be of greatest significance in tropical developing countries are the fumonisins and aflatoxins (Kumar et al., 2008; Muthomi et al., 2009). Aflatoxins are carcinogenic secondary metabolites produced by several species of Aspergillus section Flavi, including Aspergillus flavus Link,

Aspergillus parasiticus Speare, and Aspergillius nominus Kutzman, Horn, and Hesseltime. The fungus forms sclerotia which allow it to survive in soil for extended periods of time (Schneiddeger & Payne, 2003). Conditions such as high temperatures and moisture, unseasonal rains during harvest and flash floods lead to fungal proliferation and production of mycotoxins (Bhat & Vasanthi, 2003). About 4.5 billion people in developing countries are chronically exposed to aflatoxin and the CODEX recommended sanitary and phytosanitary standards set for aflatoxins adversely affect grain trade in developing countries (Gebrehiwet et al., 2007). Concerns for human and livestock health have led several countries to constantly monitor and regulate aflatoxin contamination of agricultural commodities (Wang & Tang, 2005). Since the discovery of aflatoxins in the early 1960s, many studies have been conducted to assess the occurrence and to describe the ecology of aflatoxin-producing fungi in natural and agricultural environments. Aspergillus flavus is the most abundant aflatoxin-producing species associated with corn (Abbas et al., 2004a). While aflatoxins occur mostly in maize and groundnuts, the prevalence of fumonisins in maize is 100% (Wagacha & Muthomi, 2008). Mycotoxins have negative impact on human health, animal productivity and trade (Wagacha & Muthomi, 2008; Wu, 2006). Aflatoxin B1 is the most toxic and is associated with liver cancer and immune suppression (Sheppard, 2008). Exposure to large doses (> 6000 mg) of aflatoxin may cause acute toxicity with lethal effect, whereas exposure to small doses for prolonged periods is carcinogenic (Groopmann & Kensler, 1999). There may be an interaction between chronic mycotoxins exposure and malnutrition, immune-suppression, impaired growth, and diseases such as malaria and HIV/AIDS (Williams et al., 2004). Mycotoxin poisoning may be compounded by the co-ocurrence of aflatoxins with other mycotoxins such as fumonisins, zearalenone and deoxynivalenol (Kimanya et al., 2008; Pietri et al., 2009). However, the presence of mycotoxins in food is often overlooked due to public ignorance about their existence, lack of regulatory mechanisms, dumping of food products and the introduction of contaminated commodities into the human food chain during chronic food shortage due to drought, wars, political, and economic instability. The largest mycotoxinpoisoning epidemic in the last decade occurred in Kenya in 2004. Aflatoxin poisoning was associated with eating home grown maize stored under damp conditions (Lewis et al., 2005). Acute aflatoxin poisoning has continued to occur severally in Eastern and Central provinces of Kenya (CDC, 2004). In the 2004 aflatoxin-poisoning outbreak, the concentrations of aflatoxin B1 in maize was high as 4,400 ppb, which is 220 times greater than the 20 ppb regulatory limit. The outbreak covered more than seven districts and resulted in 317 case-patients and 125 deaths (Lewis et al., 2005). The association of mycotoxins with human and animal health is not a recent phenomenon; for example, in the past, ergotism was suspected of being a toxicosis resulting from these toxic fungal metabolites. Nowadays, more is known regarding this family of compounds. Mycotoxins were considered as a storage phenomenon whereby grains becoming moldy during storage allowed for the production of these secondary metabolites proven to be toxic when consumed by man and other animals. Subsequently, aflatoxins and mycotoxins of several kinds were found to be formed during development of crop plants in the field. The determination of which of the many known mycotoxins are significant can be based upon their frequency of occurrence and/or the severity of the disease that they produce, especially if they are known to be carcinogenic. The diseases (mycotoxicoses) caused by these mycotoxins are quite varied and involve a wide range of susceptible animal species including humans. Most of these diseases occur after consumption of mycotoxin contaminated grain or products made from such grains but other routes of exposure exist. The diagnosis of mycotoxicoses may prove to be difficult because of the similarity of signs of disease to those caused by other agents. Therefore, diagnosis of a mycotoxicoses is dependent upon adequate testing for mycotoxins involving sampling, sample preparation and analysis (Richard, 2007).

Toxicology Of Mycotoxins

Mycotoxins primarly occur in the mycelium of the toxigenic moulds and may also be found in the spores of these organisms and cause a toxic response, termed a mycotoxicoses, when ingested by higher vertebrates and other animals (Bennett & Klich, 2003). These secondary metabolites are synthesized during the end of the exponential phase of growth and appear to have no biological significance with respect to mould growth/development or competitiveness. All moulds are not toxigenic and while some mycotoxins are produced by only a limited number of species, others may be produced a relatively large range from several genera (Hussein & Brasel, 2001). The toxic effect of mycotoxin ingestion in both humans and animals depends on a number of factors including intake levels, duration of exposure, toxin species, mechanisms of action, metabolism, and defense mechanisms (Galvano et al., 2001). Consumption of mycotoxin-contaminated food or feed does however lead to the induction of teratogenic, carcinogenic, oestrogenic, neurotoxic, and immunosuppressive effect in humans and/or animals (Atroshi et al., 2002). The mycotoxins of most significance from both a public health and agronomic perspective include the aflatoxins, trichotecenes, fumonisins, ocharotoxin A (OTA), patulin, tremorgenic toxins, and ergot alkaloids (Papp et al., 2002).

AFLATOXINS

Aflatoxin was initially identified as toxic after investigations of the death of 100,000 turkeys in the United Kingdom in 1960 (Blout, 1961). This prompted a major revolution in mycotoxin research resulting in intensive testing of mycotoxins in any moldy products. Since then several Aspergilli have been identified as capable of producing aflatoxins. The two most agriculturally important species are Aspergillus flavus and A. parasiticus, which are found throughout the world, being present in both the soil and the air (Abbas et al., 2005). When conidia (spores) encounter a suitable nutrient source and favorable environmental conditions (hot and dry conditions) the fungus rapidly colonizes and produces aflatoxin (Payne, 1992). Contamination of agricultural commodities by aflatoxin is a serious problem due to the substantial health effect it has on humans and animals. The use of agrochemicals (fungicides), timely irrigation, and alternate cropping systems have independently shown limited success in preventing aflatoxin contamination. Integration of these tactics will be required to manage such a difficult problem (Cleveland et al., 2003). A more recent and promising technology is the use of non toxigenic strains of Aspergillus as biocontrol agents. However, to maximize this methodology and to prevent the colonization of multiple crops by A. flavus and related species (A. parasiticus and A. nominus), it is critical that a complete understanding of the ecology of these unique fungi be developed (Abbas et al., 2009). Aflatoxins are toxic compounds chemically related to bisfuranocoumarin that are produced by A. flavus and A. parasiticus strains (Abbas et al., 2004b). These two aflatoxigenic species have been frequently studied due to their impact on agricultural commodities and their devastating effects on livestock. The name aflatoxin comes from the genus Aspergillus which is where the letter "a" in aflatoxin is derived and "fla" from the species name flavus. In agricultural grains the fungi A. flavus and A. parasiticus are capable of producing four major aflatoxins (AfB1, AfB2, AfG1, and AfG2). A. flavus tipically produces only the B toxins (Abbas et al., 2004b). Corn and cottonseed are typically contaminated with the aflatoxin B1, produced after colonization by A. flavus (Klich, 1986). A. parasiticus is more prevalent in peanuts than any other crop; however, it is typically outcompeted by A. flavus when the two fungi are both present (Horn et al., 1995). These fungi are ubiquitous in the environment, being readily isolated from plants, air, soil, and insects (Wicklow et al., 2003). Soil populations of A. flavus in soils under maize cultivation can range from 200 to >300,000 colony-forming units (CFU) g-1soil (Zablotowicz et al., 2007) and can constitute from ≤0.2% to ≤8% of the culturable soil fungi population. The major soil property associated with maintaining soil populations of A. flavus is soil organic matter. Higher populations of A. flavus are maintained in the soil surface of no-till compared to conventional-till soils (Zablotowicz et al., 2007). The presence of Aspergillus species in dust can compromise individuals with elevated allergies to the fungus or its products (Benndorf et al., 2008). Of more concern is the colonization of certain food and feed crops (corn, cottonseed, peanuts, and some tree nuts) by the fungus, where it may produce a high concentration of these chemical compounds, specifically aflatoxin, to cause them to be considered contaminated and unfit for their intended use (Abbas et al., 2009). When suitable environmental conditions arise, sclerotia and conidia germinate into mycelia that produce numerous conidiophores and release conidia into the air that can be available for colonizing plants. Although A. flavus colonizes a plant structure, it doesn´t necessarily produce aflatoxin to excessive levels. In this manner, A. flavus is an opportunistic pathogen in a similar context to the opportunistic human pathogens Pseudomonas fluorescens and Burkholderia capacia. These bacteria may colonize in low levels in compromised individuals, such as burn patients or the immunocompromised, and become pathogenic. In the same context, healthy plant tissues are less prone to be extensively colonized by A. flavus. However, under heat stress and moisture deficit, corn reproductive structures are readily susceptible to high levels of aflatoxin contamination, (O'Brian et al., 2007). Therefore, inoculum potential modified by life cycle of the fungus is as critical as the environment and the host. The A. flavus life cycle can be divided into two major phases:

the colonization of plant residues in the soil, and the infection of crop tissues, including grain and seeds of actively growing plant tissues. At the beginning of the growing season, usually in spring and sometimes at the end of winter, when sclerotia are exposed to the soil surface, they quickly germinate and form new conidial inoculum. This new inoculum will be vectored by insects and carried by the wind to begin the colonization and infection of the freshly planted crops (Horn, 2007). During the growing season, infected plant tissues can serve as sources of secondary conidial inoculum, which colonize new non-infected plant tissues (Fig.1). Despite our understanding of how the initial and secondary inocula occur for plant infection, little information is available about the saprotrophic activities of these fungi in soil. Recently, Accinelli et al. (2008) confirmed the presence of A. flavus in the soil actively synthesizing aflatoxins. However, not all A. flavus and A. parasiticus isolates produce aflatoxins (Abbas et al., 2004b).

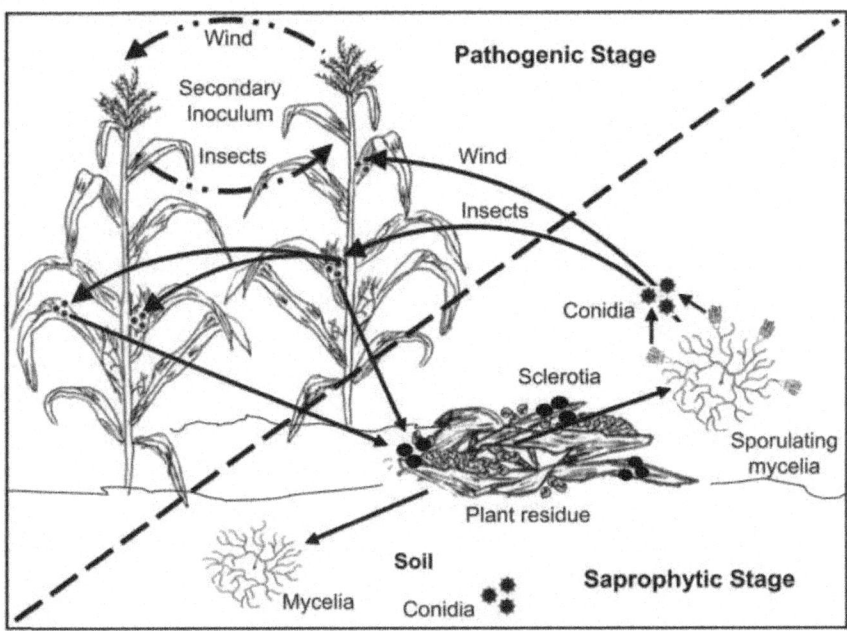

Figure.1: Life cycle of A. flavus in a corn cropping system (Abbas et al., 2009).

Fungi are classified as nonaflatoxigenic if they do not produce aflatoxins but produce other toxins. If fungi produce no toxins at all, they are classified as nontoxigenic. Generally, in any environment, the frequency of aflatoxigenic isolates can range from 50% to 80% (Abbas et al., 2004a). The relative distribution of aflatoxigenic versus nonaflatoxigenic isolates is modulated by many factors including plant species present, soil composition, cropping history, crop management, and environment conditions, including rain fall and temperature (Abbas et al., 2004b). Each of these factors can reduce the levels of A. flavus, for example, noncultivated fields near cultivated land are observed to have very low populations of A. flavus (Horn, 2007). Similarly, the frequency of drought is a factor in populations of fungi, with significant drops in soil populations of A. flavus after several years without drought. The conidia remain dormant in soil and only germinate when nutrient sources are present (Zablotowicz et al., 2007). The behavior of Aspergilli structures in soil needs to be investigated and evaluated thoroughly, especially in agricultural soils, due to the fungal structures serving as the primary inoculum resulting in aflatoxin contamination in agricultural commodities (Abbas et al., 2009)

Biosynthesis

Aflatoxins the most carcinogenic substances known to date have gained much interest among organic chemists since the elucidation of their structure by Buchi and co-workers in 1963. Even though numerous syntheses of racemic aflatoxins were reported in the following years , it took 40 years for the first enantioselective total synthesis of (-)-aflatoxin B1 and B2 to be published by Trost et al. (2003), their approach resembles in part (construction of the DE ring system) the first total synthesis of (±)-aflatoxin by Buchi et al. (1967). The biosynthetic pathway in A. flavus consists of approximately 23 enzymatic reactions and at least 15 intermediates (reviewed in Bhatnagar et al., 2006; Bräse et al., 2009) encoded by 25 identified genes clustered within a 70-kb DNA region on chromosome III (Bhatnagar et al., 2006a; Cary & Ehrlich, 2006; Smith et al., 2007; Cary & Calvo, 2008). The initial substrate acetate is used to generate polyketides with the first stable pathway intermediate being anthraquinone norsolorinic acid (NOR) (Bennett et al., 1997). This is followed by anthraquinones, xanthones, and ultimately aflatoxins synthesis (Yu et al., 2004). Few regulators of this process have been identified (Cary & Calvo, 2008), and a general model based on Aspergillus has recently been reviewed by Georgianna & Payne (2009) (Figure.2). In addition to pathway-specific regulators, production of aflatoxins is also under the control of a number of global regulatory networks that respond to environmental and nutritional cues. These include responses to nutritional factors such as carbon and nitrogen sources and environmental factors such as pH, light, oxidative stress, and temperature. Nitrogen source plays an important role in aflatoxin biosynthesis (Bhatnagar et al., 1986). In general, nitrate inhibits aflatoxin production, while ammonium salts are conducive (Cary & Calvo, 2008). Ammonium acetate does not have any significant impact on the level of OTArelated pks (the gene encoding for a polyketide synthase) expression. Nevertheless, this compound does lead to an increase in OTA production (Abbas et al., 2009).

Some aminoacids as proline, asparagines, and tryptophane significantly increase the biosynthesis of aflatoxins B1 and G1 in A. parasiticus (Payne and Hagler, 1983). Tryptophane acts by up-regulating aflatoxin gene expression in A. parasiticus and down-regulating it in A. flavus. Some nitrogen sources can also be non-conducive for OTA production in A. ochraceus, and their inhibitory effect is probably exerted at the transcriptional level (O'Callaghan et al., 2006). The influence of carbon sources on aflatoxins and OTA biosynthesis has been studied for decades and it has produce contradictory results (Abbas et al., 2009). Aflatoxin biosynthesis is induced by simple sugars such as glucose and sucrose that are present or generated by fungal hydrolytic enzymes during invasion of seed tissues (Cary & Calvo, 2008). A key factor determining whether a carbon source can support aflatoxin production and fungal growth is its availability to both hexose monophosphate and glycolitic pathways. This finding was confirmed by the identification of a set of genes including enoA and pbcA genes, both these genes are upregulated in response to sucrose supplementation (Price et al., 2006). The addition of different simple sugars may have opposite effects on OTA synthesis depending on the culture media used. Nevertheless, lactose exhibited a significant enhancing effect on OTA biosynthesis both in restrictive and conducive media, whilst glucose can show a repressive effect on OTA synthesis (Abbas et al., 2009). This negative effect may be partially explained by the involvement of CreA, the regulator of the carbon repression system which also acts as a controller of the secondary metabolism in many fungal species (Roze et al., 2004). Other environmental factors, such as temperature, water activity and pH, strongly influence mycotoxin biosynthesis. Some examples have been provided for OTA and aflatoxins biosynthesis (Ramirez et al., 2006; Ribeiro et al., 2006). The optimal temperature for production of aflatoxins is approximately 30°C (Boller & Schroeder, 1974). The establishment of temperature as an important component of infection by A. flavus and subsequent aflatoxin contamination has been clearly demonstrated under controlled greenhouse conditions (Payne et al., 1988). Some efforts to illustrate a relationship between temperature and aflatoxin contamination were unsuccesfull (Stoloff and Lillehoj, 1981). The reason for this phenomenon can be traced to the finding that a detectable relationship exists only during years when amounts of contamination are high (McMillian et al., 1985). Conclusions of this work were that high temperatures do significantly contribute to the contamination process and the ultimate amount of aflatoxin which is produced. Naturally, nothing can be done to control ambient temperatures, but it is possible to avoid their full impact during the later stages of kernel filling by early planting (Abbas et al., 2009). Relative humidity above 86% also promotes colonization and aflatoxin production in the field (Plasencia, 2004).

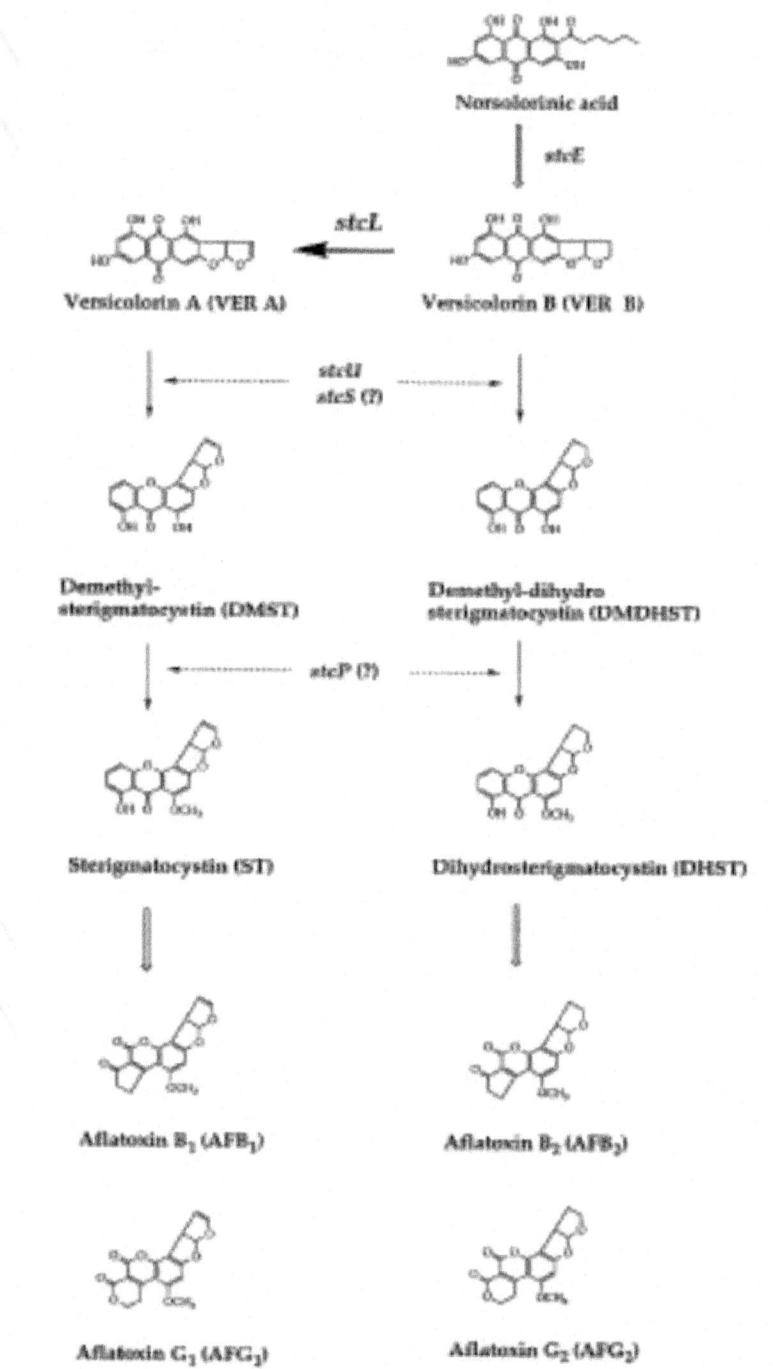

Figure.2: AF/ST biosynthetic pathway in Aspergillus spp. (Kelkar et al., 1997).

Aflatoxin production, in general, is greatest in acidic medium and tends to decrease as the pH of the medium increases (Keller et al., 1997). Response to changes in pH is regulated by the globally acting transcription factor PacC, which is posttranslationally modified by a pHsensing protease (Tillburn et al., 1995). PacC binding sites indentified in the promoters of aflatoxins biosynthetic genes could be involved in negative regulation of aflatoxins biosynthesis during growth at alkaline pH (Ehrlich et al., 2002). Fungal development also appears to respond to changes in pH as sclerotial production was found to be reduced by 50% at pH 4.0 or less while aflatoxins production was at its maximal (Cotty, 1988). According to Georgianna and Payne (2009), only temperature has a greater influence on aflatoxin biosynthesis than pH. pH values lower than 4.0 are needed for aflatoxin production, and generally, the lower the pH value, the higher is the toxin synthesis (Klich, 2007).

In addition to temperature, water activity, and pH, the application of suboptimal concentrations of fungicides can boost mycotoxin biosynthesis (Schmidt-Heydt et al., 2007; D'Mello et al., 1998). A more appropriate general strategy is therefore to investigate natural products within the crop which confer resistance to Aspergillus colonization and growth, and/or aflatoxin biosynthesis. Two classes of protective natural factors exist in nature: phytoalexins, inducible metabolites, formed after invasion de novo, e.g. by activation of latent enzyme systems; phytoanticipins, constitutive metabolites, present in situ, either in the active form or easily generated from a precursor. Since phytoalexins are produced only in response to fungal attack, it is obvious that their presence would lag behind the infection and levels capable of suppressing aflatoxin would be difficult to regulate. In contrast, phytoanticipins are always present and such factors offer the potential for enhancement through breeding and selection of more resistant cultivar, or even genetic manipulation to introduce or enhance their levels. Once such compounds have been identified, it is only necessary to ensure that they are present in large enough quantities and in tissues from which fungal growth and aflatoxin deposition must be excluded (Campbell et al., 2003). Currently available methods of removing aflatoxins from tree nuts after contamination are impractical and expensive (Scott, 1998). There is a need to design new and environmentally safe methods of reducing infection by aflatoxigenic aspergilla and to inhibit aflatoxin biosynthesis.

Genetics of aflatoxin biosynthesis

Cloning of genes involved in aflatoxin biosynthesis is the key to understanding the molecular biology of the pathway (Trail et al., 1995). There are 21 enzymatic steps required for aflatoxin biosynthesis and the genes for these enzymes have been cloned (Bhatnagar et al., 2003). Molecular research has targeted the genetics, biosynthesis, and regulation of aflatoxin formation in A. flavus and A. parasiticus. Aflatoxins are biosynthesized by a type II polyketide synthase and it has been known for a long time that the first stable step in the biosynthetic pathway is the norsolorinic acid, an anthraquinone (Bennett et al., 1997). A complex series of post-polyketide synthase steps follow, yielding a series of increasingly toxigenic anthraquinone and difurocoumarin metabolites (Trail et al., 1995).

Sterigmatocystin (ST) is a late metabolite in the aflatoxin pathway and is also produced as a final biosynthetic product by a number of species. It is now known that ST and aflatoxins share almost identical biochemical pathways (Bhatnager et al., 2003). Aflatoxin (AF) was one of the first fungal secondary metabolites shown to have all its biosynthetic genes organized within a DNA cluster (Figure.3). These genes, along with the pathway specific regulatory genes aflR and aflS, reside within a 70 kb DNA cluster near the telomere of chromosome 3 (Sweeney et al., 1999; Georgianna and Payne, 2009). Research on A. flavus, A. parasiticus and A. nidulans has led to our current understanding of the enzymatic steps in the AF biosynthetic pathway, as well as the genetic organization of the biosynthetic cluster. A. nidulans does not produce AF but has all of the genes and enzymatic steps preceding the production of ST. The AF and ST pathways appear to have a common biosynthetic scheme up to the formation of ST, and thus information gained from both pathways has been used to study AF regulation (Georgianna & Payne, 2009). The biosynthetic and regulatory genes required for ST production in A. nidulans are homologous to those required for aflatoxin production in A. flavus and A. parasiticus and they also are clustered. The physical order of the genes in the cluster largely coincides with the sequential enzymatic steps of the pathway and both gene organization and structure are conserved within A. favus and A. parasiticus (Sweeney et al., 1999; Bhatnagar et al., 2006). Of the 25 genes identified in the pathway, only four (norA, norB, aflT, and ordB) have yet to have the function of their protein product determined experimentally. Only one of these genes, aflR, appears to encode a transcription factor (Bhatnagar et al., 2006, 2003). The expression of the structural genes in both aflatoxin and ST biosynthesis is regulated by a regulatory gene, aflR, which encodes a GAL4-type C6 zinc binuclear DNA-binding protein (Bhatnagar et al., 2003). This gene is located in the cluster and is required for transcriptional activation of most, if not all, of the aflatoxin structural genes. Adjacent to and divergently transcribed from the aflR gene is aflJ.

This gene is also involved in the regulation of the aflatoxin gene cluster because no aflatoxin pathway intermediates are produced when it is disrupted. The gene product of aflJ has no sequence homology with any other genes or proteins present in databases. It interacts with aflR but not with the structural genes of the pathway. It has been speculated that aflJ is an aflR coactivator (Yu et al., 2002; Bennett et al., 2007). The function of most of the aflatoxin gene products has been deduced either by genetic or biochemical means (Bhatnagar et al., 2006). Two of the genes of the ST gene cluster in A. nidulans, stcJ and stcK, encode the K- and Lsubunit of a fatty acid synthase (FAS) which is specific for the formation of the hexanoate starter of ST. Disrupted stcJ/stcK mutants do not synthesise ST, but retain the ability to do it when provided with hexanoic acid (Sweeney et al., 1999).

The protein set requested for ST/AF transduction regulatory pathways includes: FlbA, an RGS (regulator of G-protein signaling) protein; FluG, an early acting development regulator; FadA, the alpha subunit of a heterotrimeric G-protein; and PkaA, encoding the catalytic subunit of protein kinase A. When FadA is activated following the signal "perception" both directly and indirectly it is able to inhibit AflR activity. FlbA whose activation is dependent on FluG, suppresses FadA and triggers AflR activation (Reverberi et al., 2010)

The pathway specific regulator gene

Two genes, aflR and aflS, located divergently adjacent to each other within the AF cluster are involved in the regulation of AF/ST gene expression. The gene aflR encodes a sequencespecific DNA-binding binuclear zinc cluster (Zn(II)2Cys6) protein, required for transcriptional activation of most, if not all, of the structural genes (Georgianna and Payne, 2009). It was first cloned from an A. flavus cosmid library by showing that it could restore aflatoxin-producing ability to a mutant blocked in all steps of aflatoxin biosynthesis. An increase in the copy number of aflR somehow altered normal regulation of aflatoxin biosynthesis (Bhatnagar et al., 2003). The aflR locus has been compared among isolates of AF producers such as A. parasiticus and A. flavus. These comparisons revealed differences in

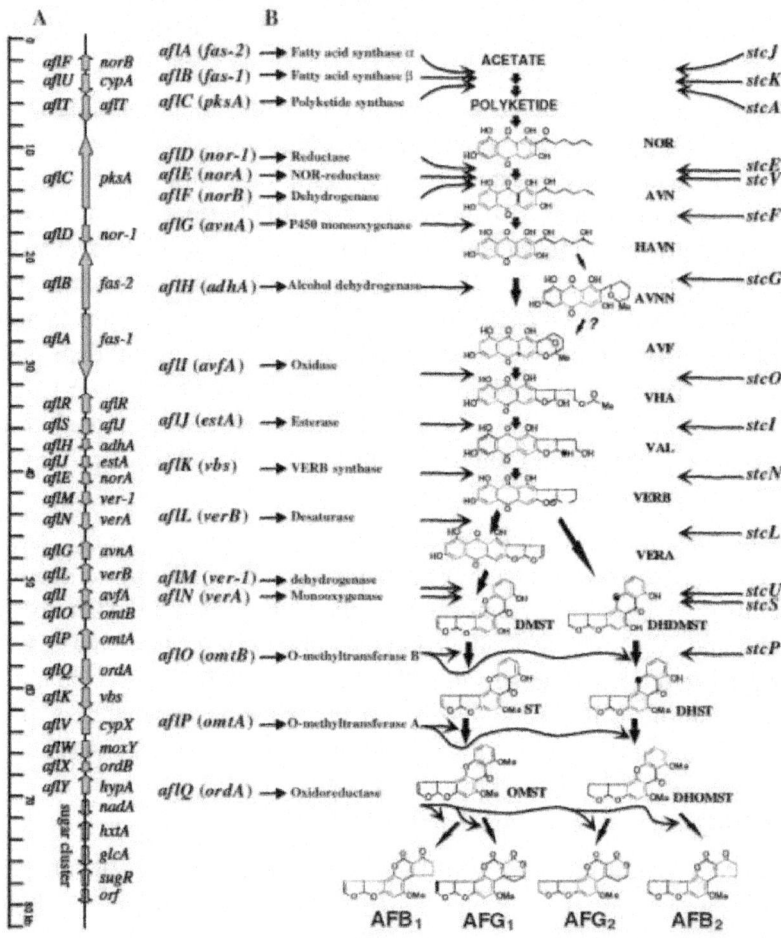

Figure.3: The gene cluster responsible for aflatoxins biosynthesis in A. flavus and A. parasiticus. A) Clustered genes (arrows indicate the direction of gene transcription) and B) the AF biosynthetic pathway (Bhatnagar et al., 2006).

Many promoter regulatory elements such as PacC and AreA binding sites. The aflR gene is also found in A. nidulans and A. fumigatus. Despite clear differences in the sequence of AflR between A. nidulans and A. flavus, function is conserved. AflR from A. flavus is able to drive expression of the ST cluster in an A. nidulans aflR deletion strain (Carbone et al., 2007; Georgianna and Payne, 2009). AflR binds to the palindromic motif 5'-TCGN5CGA-3' (also called AflR binding motif) in the promoter region of aflatoxin structural genes in A. parasiticus, A. flavus, and A. nidulans. The promoter regions of the majority of aflatoxin genes have at least one 5'-TCGN5CGA-3' binding site within 200 bp of the translation start site, though some putative binding sites have been identified further upstream. AflR probably binds to its recognition site as a dimer. The gene, aflR may be self-regulated, as well as, under the influence of negative regulators. Upstream elements may be involved in negative regulation of aflR promoter activity. When aflR is disrupted, no structural gene transcript can be detected. Introduction of an additional copy leads to overproduction of aflatoxin biosynthetic pathway intermediates (Fernandes et al., 1998; Bennett et al., 2007). Electrophoretic mobility shift assays (EMSA) have been used to thoroughly examine promoters for AflR binding in 11 different genes from the AF cluster, with three of these genes having sites that deviate from the predicted AflR binding motif, and an additional three AF genes for which AflR binding sites could not be demonstrated. Among these genes are aflE, aflC, aflJ, aflM, aflK, aflQ, aflP, aflR, and aflG. All of these genes have predicted sites and demonstrate some degree of AflR binding in EMSA assays. Moreover, they were differentially expressed between WT and the DaflR mutant, suggesting that AflR is required to activate their expression (Price et al., 2006; Georginna and Payne, 2009). Aflatoxins biosynthesis is also regulated by aflS (formerly aflJ), a gene that resides next to aflR. The genes aflS and aflR are divergently transcribed, but have independent promoters. The intergenic region between them, however, is short and it is possible that they share binding sites for transcription factors or other regulatory elements (Ehrlich and Cotty, 2002; Georgianna and Payne, 2009). The roles of AflR and AflS were examined by studying the expression of pathway genes in transformants of A. flavus strain 649-1 that received the respective genes individually. Strain 649-1 lacks the entire AF biosynthetic cluster but has the necessary upstream regulatory elements to drive the transcription of aflR (Du et al., 2007). These studies showed that AflR is sufficient to initiate gene transcription of early, mid, and late genes in the pathway, and that AflS enhances the transcription of early and mid aflatoxin pathway genes. Moreover, the induced expression of A. flavus aflR in A. nidulans, under conditions in which ST biosynthesis is normally suppressed, resulted in activation of genes in the ST biosynthetic pathway. These studies demonstrated that aflR function is conserved in widely different Aspergillus spp (Bhatnagar et al., 2003). Roles for AflS have been suggested to be as diverse as aiding in transport of pathway intermediates to the interaction of AflS with AflR for altered AF pathway transcription. The observation that AflS binds to AflR argues that AflS modulates aflatoxin expression through its interaction with AflR (Chang, 2003; Georgianna and Payne, 2009). Metabolite feeding studies showed that a functional aflR allele is required for accumulation of NOR, the first stable intermediate in the aflatoxin biosynthetic pathway. When this gene was disrupted, the fungi were incapable of aflatoxin metabolite production or transcription of nor- 1, but otherwise grew normally (Bhatnagar et al., 2003). In addition to the binding sites for AflR, there are binding sites within the cluster for other transcriptional factors that may play important roles in transcriptional regulation of the AF cluster. A novel cAMP-response element, CRE1, site has been studied specifically in the aflD (nor-1) promoter of A. parasiticus (Georgianna and Payne, 2009).

Aflatoxins and fungal development

The association between fungal morphological development and secondary metabolism, including aflatoxin production, has been observed for many years (Calvo et al., 2002). The environmental conditions required for secondary metabolism and sporulation are similar, and both processes occur at about the same time (Reiss, 1982; Bennett et al., 2007). A number of studies have identified a genetic connection between aflatoxin/sterigmatocystin biosynthesis and fungal development. In Aspergillus, several observations linked a fluffy phenotype to loss of AF/ST production. The available well characterized fluffy mutants in A. nidulans were instrumental in the discovery of a signal transduction pathway regulating both conidiation and ST/AF biosynthesis. These mutants are deficient in ST formation (Weiser et al., 1994). Proteins identified as belonging to this signal transduction pathway include FlbA, an RGS (Regulator of G-protein Signaling) protein, FluG, an early acting development regulator, FadA, the alpha subunit of a heterotrimeric G-protein and PkaA, encoding the catalytic subunit of protein kinase A (Gerogianna and Payne, 2009). Furthermore, a possible transcription regulatory gene, veA, has been identified in A. nidulans and A. parasiticus and this gene controls both toxin production and sexual development. Both A. nidulans and A. parasiticus veA mutants fail to produce ST or aflatoxin. Moreover, A. nidulans and A. parasiticus do not produce cleistothecia (sexual fruiting bodies harboring ascospores) and sclerotia (asexual overwintering structures) respectively. Finally, a number of genetic loci were identified in A. nidulans mutants that resulted in loss of ST production but had normal developmental processes. Complementation studies with one of these mutants identified a gene called laeA. This gene encodes an enzyme with sequence similarity to methyltransferases and appears to be required for expression of ST. LaeA homologs have been found in a number of filamentous fungi and in all species examined, disruption of laeA resulted in loss of secondary metabolite production while overexpression of laeA results in hyperproduction of the secondary metabolite (Bhatnagar et al., 2006; Reverberi et al., 2010).

Economic impact of aflatoxins

Aspergillus spp. is a fungal that grows and produces aflatoxins in climes ubiquitous but is commonly found in warm and humid climates (Dohlman, 2003). Hence most commodities from tropical countries, especially peanut and maize, are likely to be easily contaminated with aflatoxins (Bley, 2009). Aflatoxin contamination of human and animal feeds poses serious health and economic risks worldwide (Bley, 2009). The economic impact of aflatoxin contamination is difficult to measure, but the following losses have been documented. In United States (US) from 1990 to 1996, litigation costs of $34 million from aflatoxin contamination occurred. In 1998, corn farmers lost $40 million as a result of aflatoxin contaminated grain (AMCE, 2010). The FAO estimates that 25% of the world food crops are affected by mycotoxins each year and constitute a loss at post-harvest (FAO, 1997). According to Cardwell et al (2004) aflatoxin contamination of agricultural crops causes annual losses of more than $750 million in Africa. Dohlman (2003) defined mycotoxin as toxic by-products of mould infestations affecting about one-quarter of global food and feed crop output. Newly in the US, it was reported that income losses due to AF contamination cost an average of more than US$100 million per year to US producers (Coulibaly et al., 2008). As of this date, the average direct loss to the US is estimated at $200 million annually for corn. Indirect losses because of contaminated byproducts, such as distillers' grain, compound these losses. Ultimately, all contribute to increased costs to consumers (AMCE, 2010). Jolly et al. (2009) also reveal that post-harvest losses of crops are greater than the improvements made in primary production. In other hand, Otsuki et al (2001) has calculated that the European Union (EU) regulation on aflatoxins costs Africa $670 million each year in exports of cereals, dried fruit and nuts. But another study (World Bank, 2005) indicated that Otsuki et al. had overestimated the impact of the EU aflatoxin standard on Africa, and that the largest losses were incurred by Turkey,

Brazil, and Iran. However, several studies have indicated that these costs may increase not only for Africa but for other countries that are suppliers of grains of the EU (Otsuki et al., 2001; Wu, 2004). This due to that the regulation on aflatoxins is among the strictest in the world, at 4 ng/g total aflatoxins for all foods except peanuts (15 ng/g). The EU regulation standards on aflatoxins are base in the ALARA principle (As Low As Reasonably Achievable) which has a strong potential impact on nations attempting to export foods that are susceptible to aflatoxins contamination into the EU (Wu, 2008). In the study of 2004, Wu estimated a $450 million annual loss to the U.S., China, Argentina, and sub-Saharan African peanut markets if the EU aflatoxin standard were adopted worldwide. Nevertheless, in other study realized in 2008, Wu also mentions that under certain conditions, export markets may actually benefit from the strict EU standard. These conditions include a consistently high-quality product, and a global scene that allows market shifts. Even lower-quality export markets can benefit from the strict EU standard, primarily by technology forcing. Nevertheless, if the above conditions are not met, export markets suffer from the strict EU standard. Recent studies have linked aflatoxins production in foods to environmental conditions, poor processing and lack of proper storage facilities in developing countries (Farombi, 2006; Hell et al., 2000; Kaaya and Kyamuhangire, 2006).

Control of aflatoxin contamination in crops

Mycotoxin contamination often is an additive process, beginning in the field and increasing during harvest, drying, and storage (Wilson and Abramson, 1992). Environmental conditions are extremely important in pre-harvest mycotoxin contamination of grain and oilseed crops. Aflatoxin generation is favored in years with above average temperature and below average rainfall (Wilson and Abramson, 1992). Fungal contamination both at preharvest and post-harvest is determined by a range of factors which can be classified into four main groups including: intrinsic nutritional factors, extrinsic factors, processing factors and implicit microbial factors (Sinha, 1995). The Figure.4 summarises the factors which affect fungal colonization of stored grain (Megan and Aldred, 2008). Strategies to address the food safety and economic issues employ both pre-harvest and post harvest measures to reduce the risk of mycotoxin contamination in food and feed (Dorner, 2004). Pre-harvest control includes good cultural practices, biocontrol and development of resistant varieties of crops through new biotechnologies.

The good cultural practices consist in planting adapted varieties, proper fertilization, weed control, and necessary irrigation as well as crop rotation, cropping pattern, and use of biopesticides as protective actions that reduce mycotoxin contamination of field crops. Among the strategies of biotechnology in the pre-harvest control is the development of transgenic plants resistant to fungal infection as well as crops capable of catabolism/interference with toxin production. Pre-harvest prevention especially through host resistance is probably the best and widely explored strategy for control of mycotoxins (Kumar and Kumari, 2010; Bhatnagar, 2010). Post-harvest control is based mainly eliminate or inactivate mycotoxins in grains and other commodities. Among the methods used in this control, are physical separation, detoxification, biological inactivation, chemical inactivation, and decreasing the bioavailability of mycotoxins to the host animal

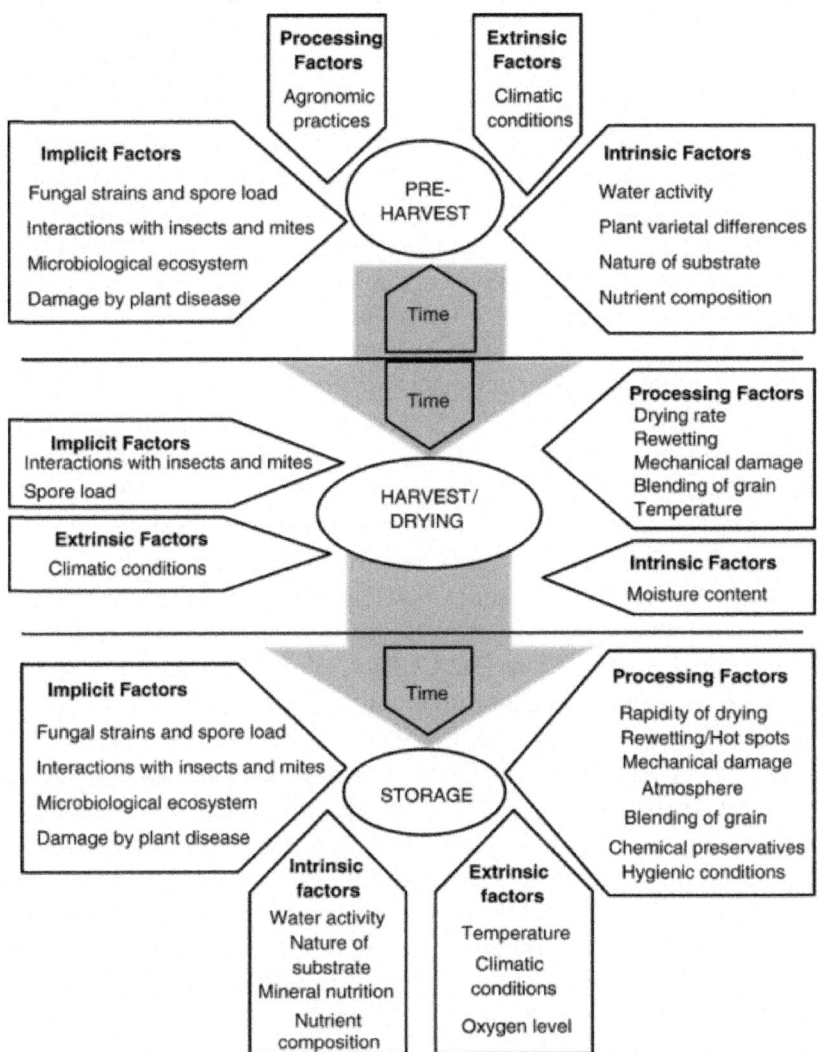

Figure.4: Interaction between intrinsic and extrinsic factors in the food chain which influences mould spoilage and mycotoxin production in stored commodities (Magan et al., 2004).

(Richard, J.L. et al., 2003). Because of the detrimental effects of mycotoxins, a number of strategies have been developed to help prevent the growth of mycotoxigenic fungi as well as to decontaminate and/or detoxify mycotoxin contaminated foods and animal feeds (Rustom, 1997). These strategies include: the prevention of mycotoxin contamination, detoxification of mycotoxins present in food and feed, as well as the inhibition of mycotoxin absorption in the gastrointestinal tract. Mycotoxin contamination may occur in the field before harvest, during harvesting, or during storage and processing. Thus methods can conveniently be divided into pre-harvest, harvesting and postharvest strategies (Heathcote & Hibbert, 1978). Whereas certain treatments have been found to reduce specific mycotoxin formation in different commodities, the complete elimination of mycotoxin contaminated commodities is currently not realistically achievable. Several codes of practice have been developed by Codex Alimentarius for the prevention and reduction of mycotoxins in cereals, peanuts, apple products, and raw materials. The elaboration and acceptance of a General Code of Practice by codex will provide uniform guidance for all countries to consider in attempting to control and manage contamination by various mycotoxins. In order for this practice to be effective, it will be necessary for the producers in each country to consider the general principles given in the Code, taking into account their local crops, climate, and agronomic practices, before attempting to implement provisions in the Code. The recommendations for the reduction of various mycotoxins in cereals are divided into two parts: recommended practices based on Good Agricultural Practice (GAP) and Good Manufacturing Practice (GMP); a complementary management system to consider in the future is the Hazard Analysis Critical Control Point (HACCP) (Codex Alimentarius Commission, 2002). Mycotoxins are secondary metabolites which are produced by several fungi mainly belonging to the genera: Aspergillus, Penicillium, Fusarium, and Alternaria. While Aspergillus and Penicillium species are generally found as contaminants in food during dry and storage, Fusarium and Alternaria spp. can produce mycotoxins before or immediately after harvesting (Sweeney & Dobson, 1999). Up until now, approximately 400 secondary metabolites with toxigenic potential produced by more than 100 moulds, have been reported, with the Food and Agricultural Organization (FAO) estimating that as much as 25% of the world's agricultural commodities are contaminated with mycotoxins leading to significant economic losses (Kabak et al., 2006). Although A. flavus is readily isolated from diverse environmental samples, soil and plant tissues or residues are considered the natural habitat of this fungus (Jaime-Garcia & Cotty, 2004). Soil serves as a reservoir for primary inoculums for the infection of susceptible crops. Information concerning the soil ecology of A. flavus is consequently considered a prerequisite for developing effective measures to prevent and to control aflatoxin contamination of crops (Zablotowics et al., 2007). Soil and crop management practices and a number of environmental factors can influence the population size and spatial distribution of A. flavus in cultivated soils (Abbas et al., 2004b). The population size of A. flavus has been correlated with soil organic matter and nutritional status, with the most fertile soils containing the greatest concentration of aspergilli (Zablotowics et al., 2007). Subsequently, as more soils are managed under no tillage systems, a higher inoculums of this fungus may result, which could contribute to increased pre-harvest aflatoxin contamination of susceptible crops. It should be noted that the post-harvest control is a corrective method, and in this Chapter be addressed essentially biotechnological approaches that serve as preventive methods from emergency and development of Aspergillus flavus and consequently; inhibition synthesis of aflatoxins –a pre-harvest level–. Such approaches include of biologic control methods and use of elicitors.

Pre-harvest control strategies

It is well established that mycotoxin contamination of agricultural product can occur in the field as well as during storage (Wilkinson, 1999). Since phytopathogenic fungi such as Fusarium and Alternaria spp can produce mycotoxins before or immediately post harvesting several strategies have been developed including biological and cultural control practices to help mycotoxin contamination occurring in this way.

Prevention strategies in cereals

The main mycotoxin hazards associated with wheat pre-harvest in Europe are the toxins that are produced by fungi belonging to the genus Fusarium in the growing crop. Mycotoxins produced by these fungi include zearalenone (ZEN) as well as trichothecenes and include nivalenol (NIV), deoxynivalenol (DON) and T-2 toxin. Fusarium species are also responsible for a serious disease called Fusarium Head Blight (FHB), which can result in significant losses in both crop yield and quality. It is important to note that although Fusarium infection is generally considered to be a pre-harvest problem, it is possible for poor drying practices to lead to an increased susceptibility for storage mycotoxin contamination (Aldred & Magan, 2004).

Resistant varieties and transgenics

Research has demonstrated that insecticides cannot be applied economically to control corn insects well enough to reduce aflatoxin to acceptable levels. The most successful approach has been the use of corn resistant to ear-feeding insects. Several authors have shown that Bacillus thuringensis (Bt)-transformed corn hybrids, which are resistant to ear-feeding insects, reduce aflatoxin contamination of the grain. The adoption of Bt corn hybrids has given producers crop with increased insect resistance, however these hybrids may only reduce aflatoxin contamination under certain circumstances. However, commercial production of these genetically modified hybrids is not allowed in some nations. Several sources of natural resistance to insects have been identified, and crosses between insect- and aflatoxin-resistant lines have shown potential to increase resistance to both insect damage and aflatoxin contamination (Williams et al., 2002). Ideally, management of aflatoxin contamination should begin with the employment of resistant genotypes as has been demonstrated by several U.S. breeding programs. In Mexico the wide genetic diversity of maize has not been fully exploited to identify resistance to aflatoxin contamination in breeding programs, thus impeding the reduction of aflatoxin levels in the field. Additional complications come from the fact that transgenic maize expressing insecticidal protein or any other trait to reduce aflatoxin is not viable in Mexico due to a government prohibition on the use of genetically modified maize (Plasencia, 2004). Four major genetically controlled components for which variability exist appear to be involved in determining the fate of A. flavus-grain interaction: 1) resistance to the infection process, 2) resistance to toxin production, 3) plant resistance to insect damage, and 4) tolerance to environmental stress (Widstrom, 1987). The latter two components have an indirect influence since their effects only reduce aflatoxin contamination but do not prevent it. Although differences among genotypes have been found, heritability of the trait appears to be low, and the genotype/environment interaction may often mask true differences among genotypes (Plasencia, 2004). There are many new and exciting pre-harvest prevention strategies being explored that involve new biotechnologies. These new approaches involve the design and production of plants that reduce the incidence of fungal infection, restrict the growth of toxigenic fungi, or prevent toxic accumulation. Biocontrols using non-toxigenic biocompetitive agents is also a potentially useful strategy in corn. However, the possibility of recombination with toxigenic strains is a concern (Abdel-Wahhab & Kholif, 2008).

The differences between crop species appear to differ between countries. This is probably due to the differences in genetic pool within each country's breeding program and the different environmental and agronomic conditions in which crops are cultivated (Edwards, 2004). With respect to genetic resistance to Aspergillus infection and subsequent aflatoxin production, since the early 1970s, much work has been done to identify genetically resistant crop genotypes in both laboratory and field based experiments to help control of aflatoxigenic mould growth and aflatoxin and aflatoxin biosynthesis (D'Mello et al., 1998). This has led to the identification of a number of well-characterized sources of both resistance of Aspergillus flavus infection and to aflatoxin production. These include kernel proteins such as a 14-kDa trypsin-inhibiting protein and others including globulin 1 and 2 and a 22-kDa zeamatin protein (Chen et al., 2001). Although the role of insects in fostering

Aspergillus colonization of maize kernels is well documented, there is little evidence that transgenic corn expressing insecticidal proteins has a significant effect on reducing aflatoxin contamination. In contrast, several studies have reported a protective effect of Cry-type proteins in maize to Fusarium kernel rot and fumonisin accumulation (Dowd, 2003). Cry-type proteins constitute a family of insecticidal proteins from Bacillus thuringensis, whose genes have been incorporated into several crops to confer protection against insect pests. In corn, several hybrids expressing distinct Cry-type proteins have been developed and widely used in the U.S., Canada, Argentina, and other maize-producing countries (Plasencia, 2004). The distribution of aflatoxin in agricultural commodities has been fairly well characterized because of its importance to food supply. However, little is known on the occurrence and fate of aflatoxin in soil. Radiological assays conducted to assess the fate of aflatoxin B1 (AFB1) in soil indicated that a low level of mineralization of AFB1 to CO2 was observed, with less than 1-8% mineralized in 120 days (Angle, 1986). Not surprisingly, several microorganisms have the potential to degrade aflatoxins, especially bacteria, e.g., Flavobacterium and Mycobacterium (Hormisch et al., 2004). In addition, A. flavus also is capable of degrading aflatoxins during later stages of mycelial growth in pure culture (Huyhn & Lloyd, 1984). In recent years, molecular techniques have increased the possibilities to characterize soil microbial ecology. While molecular methods have been extensively used for studying soil bacteria, these techniques have been applied to studying soil fungi, such as the biological control agents Colletotrichum coccodes (Dauch et al., 2003), Trichoderma (Weaver et al., 2005), and mycorrizal fungi (Ma et al., 2005). Amplification of specific DNA fragments using polymerase chain reaction (PCR) and specific gene probes is extremely sensitive and has the potential to detect the presence of A. flavus in agricultural commodities (Manonmani et al., 2005). Since all of the genes involved in the aflatoxin biosynthesis pathway have been identified and cloned (Yu et al., 2004a, 2004b), and the entire genome of A. flavus sequenced (Payne et al., 2006), molecular methods for the detection of Aspergillus should be fairly readily adapted by using biosynthetic pathway genes as probes, as evidenced by the recent work differentiating toxigenic and atoxigenic A. flavus-utilizing aflatoxin gene expression using the reverse transcriptionpolymerase chain reaction (RT-PCR) (Degola et al., 2007). Application of these molecular techniques to A. flavus soil ecology should greatly enhance our understanding of this fungus. Aspergillus flavus is commonly considered a saprophytic fungus; however, its ability to colonize growing crops and inflict economic damage clearly shows that it can and does function as an opportunistic pathogen. Despite the elucidation of many aspects influencing A. flavus ability to colonize crops and accumulate aflatoxins, its activity and potential to produce aflatoxins in soil and in crop residues has remained unexplored (Accinelli et al., 2008). One interesting approach is the engineering of cereal plants to catabolize fumonisins in situ. Typically, these approaches require considerable research and development but have the potential of ultimately producing low cost and effective solutions to the mycotoxin problem in corn and other cereals. Thus this level of prevention is the most important and effective plan for reducing fungal growth and mycotoxin production.

Field management

Appropriate field management practices including crop rotation, soil cultivation, irrigation, and fertilization approaches are known to influence mycotoxin formation in the field. Crop rotation is important and focuses on breaking the chain production of infectious material, for example by using wheat/legume rotations. The use of maize in a rotation is to be avoided however, as maize is also susceptible to Fusarium infection and can lead to carryover onto wheat via stubble/crop residues (Nicholson et al., 2003). Dill-Macky and Jones (2000) observed that FHB disease severity and DON contamination of grain was significantly different when the previous crop was maize, wheat, or soya bean; with the highest levels following maize and the lowest levels following soya bean.

Soil cultivation can be divided into ploughing, where the top 10-30 cm of soil is inverted; minimum tillage, where the crop debris is mixed with the top 10-20 cm of soil; and no till, where seeds are directly drilled into the previous crop stubble with minimum disturbance to the soil structure. Ay crop husbandry that results in the removal, destruction, or burial of infected crop residues is likely to reduce the Fusarium inoculum for the following crop. Dill-Mackey and Jones (2000) reported that no till (direct drilling) after wheat or maize significantly increase DON contamination of the following wheat crop compared to ploughing, but no till had no effect when the previous crop was soya bean. Irrigation is also a valuable method of reducing plant stress in some growing situations. It is first necessary that all plants in the field have an adequate supply of water if irrigation is used. It is known that excess precipitation during anthesis makes conditions favorable for dissemination and infection by Fusarium spp., so irrigation during anthesis and during ripening of the crops, specifically wheat, barley, and rye, should be avoided. The soil must be tested to determine if there is need to apply fertilizer and soil conditioners to assure adequate soil pH and plant nutrition to avoid plant stress, especially during seed development. Fertilizer regimes may affect FHB incidence and severity either by altering the rate of residue decomposition, by creating a physiological stress on the host plant or by altering the crop canopy structure. Martin et al., (1991) observed the increasing N from 70 to 170 kg/ha significantly increased the incidence of Fusarium infection grain in wheat, barley, and triticale. Recent work by Lemmens et al., (2004) has shown that a significant increase in FHB intensity and DON contamination in the grain was observed with increasing a mineral N fertilizer from 0 to 80 kg/ha. This group concluded that in practical crop husbandry, FHB cannot be sufficiently controlled by only manipulating the N input.

Environmental conditions

Environmental conditions such as relative humidity and temperature are known to have an important effect on the onset of FHB. For example, it has been shown that moisture conditions at anthesis are critical in Fusarium infection of the ears (Aldred & Magan, 2004); while Lacey et al. (1999) have shown that Fusarium infection in the UK is exacerbated by wet periods at a critical time in early flowering in the summer, which is the optimum window for susceptibility. Equally, there is evidence that droughed-damaged plants are

more susceptible to infection, so crop planting should be timed to avoid both high temperature and drought stress during the period of seed development and maturation. On the other hand, the planning of harvesting grain at low moisture content and fully maturity may be an important control point in the preventing of mycotoxin contamination, unless allowing the crop to continue to full maturity would subject it to extreme heat, rainfall or drought conditions. Delayed harvest of the grain already infected by Fusarium species is known to cause a significant increase in the mycotoxin content of the crop.

Biotechnological approaches

Biological control of aflatoxins

The first approach which we will discuss is the biological control, which is focuses in the use of living organisms to control pests (insects, weeds, diseases and disease vectors) in agriculture. The objective of the biologic control is to stimulate the colonization of antagonist organism on plant surfaces to reduce the inoculum of the pathogens (FAO, 2004). Different organisms, including bacteria, yeasts and nontoxigenic Aspergillus fungi, have been tested for their ability in the control of aflatoxin contamination (Yin et al., 2008). According to reported by Palumbo et al., (2006) several bacterial species as Bacillus spp., Lactobacilli spp., Pseudomonas spp., Ralstonia spp. and Burkholderia spp., have shown the ability to inhibit fungal growth and production of aflatoxins by Aspergillus spp. in laboratory experiments (Yin et al., 2008), the same effect was observed in strains of B. subtilis and P. solanacearum isolated from the nonrhizophere of maize soil were also able to inhibit aflatoxin accumulation (Nesci et al., 2005). In other experiments, is showed that Bacillus subtilis prevented aflatoxin contamination in corn in field tests when ears were inoculated with the bacterium 48 hours before inoculation with A. flavus (Cuero et al., 1991). However, no reduction in aflatoxin occurred when bacteria were inoculated 48 hours after inoculation with A. flavus. Bacillus subtilis (NK-330) did not inhibit aflatoxin contamination in peanuts when it was applied to pods prior to warehouse storage for 56 days (Smith et al., 1990). Saprophytic yeasts isolated from fruits of almond, pistachio, and walnut trees inhibited aflatoxin production by A. flavus in vitro (Hua et al., 1999; Masoud and Kaltoft, 2006). A strain of Candida krusei and a strain of Pichia anomala reduced aflatoxins production by 96% and 99%, respectively, in a Petri dish assay.

Efforts are underway to apply these yeasts to almond and pistachio orchards to determine their potential for aflatoxin reduction under crop production conditions (Hua, 2002). Although they were considered to be potential biocontrol agents for management of aflatoxins, further field experiments are necessary to test their efficacies in reducing aflatoxins contamination under field conditions (Yin et al., 2008). Alternatively, a limited number of biocompetitive microorganisms have been shown for the management of Fusarium infections. Antagonistic bacteria and yeasts may also lead to reductions in pre-harvest mycotoxin contamination. For instance, Bacillus subtilis has been shown to reduce mycotoxin contamination by F. verticilloides during the endophytic growth phase. Similarly antagonistic yeasts such as Cryptococcus nodaensis have also been shown to inhibit various Fusarium species (Cleveland et al., 2003). Recent glasshouse studies by Diamond and Coke (2003) involving the pre-inoculation of wheat ears at anthesis, with the two non-host pathogens, Phoma betae and Pytium ultimum showed a reduction in disease development and severity caused by F. culmorum, F. avenaceum, F. poae, and M. nivale. A. flavus is not considered to be an aggressive invader of pre-harvest corn ear tissue. However, developing grain when damaged is easily contaminated by the pathogen (Diener et al.,1987).

The association between insect damage and fungal infection of corn ears was first recognized by Riley (1882) reported molds appearing on corn-ear tips soon after being infested with insect larvae. Garman and Jewett (1914) reported that in years with high insect populations, the incidence of moldy ears in field corn increased. Efforts to determine the specific role of insects in the A. flavus infection process increased dramatically when aflatoxin was recognized as a health concern, leading to recognition that ear feeding insects (e.g., corn earworm, Helicoverpa zea; European corn borer, Ostrinia nubilalis; fall armyworm, Spodoptera frugiperda; western bean cutworm, Striacosta albicosta; and southwestern corn borer, Diatraea grandiosella) can increase aflatoxin levels in pre-harvest corn (Catangui & Berg, 2006). The difficulty in establishing the relationship between insect damage and aflatoxin incidence is in part due to A. flavus ability to colonize silks, infect kernels, and produce aflatoxins in developing ears under insect-free conditions (Jones et al., 1980), and in part due to unknown factors that result in conflicting information (Abbas et al., 2009). Because the relationship between insect damage to corn ears and aflatoxin is heavily influenced by environmental conditions, success in managing aflatoxin contamination via insect control has been highly variable. The greatest success to date regarding biological control of aflatoxins contamination in the field has been achieved through competitive exclusion by applying on aflatoxigenic strains of Aspergillus flavus and Aspergillus parasiticus to soil of developing crops.

These strains are typically referred to as atoxigenic or nontoxigenic, but those designations are often used with reference to production of aflatoxins only (Dorner, 2004). According to Yin et al., (2008) the use of non-toxigenic Aspergillus strains is a strategy based on the application of nontoxigenic strains to competitively exclude naturally toxigenic strains in the same niche and compete for crop substrates. Thus, for competitive exclusion to be effective, the biocontrol nontoxigenic strains must be predominant in the agricultural environments when the crops are susceptible to be infected by the toxigenic strains (Cole and Cotty, 1990; Cotty, 1994; Dorner, 2004). For this to work, the applied strains must occupy the same niche as the naturally occurring toxigenic strains and compete for crop substrates (Dorner, 2004). Two primary factors exist that determine the effectiveness of this strategy. First, the applied strain(s) must be truly competitive and dominant relative to the toxigenic strains that are already present. Second, the formulation used to apply the competing strain(s) must be effective in delivering the necessary quantity of conidia to achieve a competitive advantage. In addition, the timing of that application is crucial for ensuring that the necessary competitive level is present when the threat of crop infection is greatest (Cotty, 1989; Dorner, 2004). Should be noted, that not only species of Aspergillus used for biological control are capable of producing aflatoxins, but also a variety of other toxins and toxic precursors to aflatoxins including cyclopiazonic acid, sterigmatocystin and related compounds, and the versicolorins (Cole and Cox, 1981). In the research realized by Cotty (1990) in greenhouse, demonstrated the ability of seven non-aflatoxigenic strains of A. flavus to reduce aflatoxins contamination of cottonseed when were co-inoculated with toxigenic strains. Six of these strains show significantly reduced the amount of aflatoxins produced in cottonseed by the toxigenic strain. Strain 36 (AF36) produced the largest reduction in aflatoxin under these conditions and it was Biological Control of Aflatoxin Contamination of Crops

429 subsequently shown to reduce aflatoxin contamination of cottonseed in the field when applied on colonized wheat seed (Cotty, 1994). This strain has been registered on cotton for control of aflatoxin contamination of cottonseed in Arizona, USA. It is also on a schedule for registration on pistachio in California. Additionally, this biocontrol agent was also tested for control of aflatoxin in corn (Cotty, 1996). When corn ears were either co-inoculated with AF36 and a toxigenic strain of A. flavus or inoculated with AF36 at 24 h prior to inoculation with the toxigenic strain, subsequent aflatoxins concentrations were significantly reduced, compared to inoculation with the toxigenic strain alone (Brown et al., 1991). Also have been demonstrated that other strains of a. flavus and a. parasiticus are capable of reduce aflatoxin contamination in crops; as is case of A. flavus NRRL 21882, a naturally occurring strain isolated from a peanut in Georgia in 1991, that has been used in diverse studies where has been verified its efficacy for reducing contamination in the field. This strain is the active ingredient in an EPA-registered biopesticide called afla-guard1. A color mutant of this strain, NRRL 21368, was used in several early studies and also found to be effective when used in conjunction with a color mutant of A. parasiticus (NRRL 21369) (Dorner et al., 1998, 1999b). Atoxigenic strain technology based provides an opportunity to reduce the overall risk of contamina-tion during all phases of aflatoxin contamination including in the field during crop development, in storage or at any other time after harvest until the mature crop is eventually utilized. Atoxigenic strains are but one example of how improved knowledge of both the contamination process and the etiologic agents can result in improved methods for limiting human exposure to aflatoxins.

Chemical agents and use of elicitors to aflatoxin inhibition

Another factor which is known to increase the susceptibility of agricultural commodities to toxigenic mould is injury due to insect, bird, or rodent damage (Smith et al., 1994). Insect damage and fungal infection must be controlled in the vicinity of the crop by proper use of registered insecticides, fungicides, and other appropriate practices within an integrated pest management control. Part of the integrated control of FHB in wheat production involves the use of fungicides, but this introduces a complication as far as trichothecenes are concerned as there is evidence that under certain conditions, fungicide use may actually stimulate toxin production. This raises particular concerns, since circumstances may arise where the obvious manifestations of FHB are reduced or even eliminated and yet high levels of mycotoxins may be present. Clearly grain affected in this way cannot be identified by visual inspection for signs of FHB (e.g., pink grains) and, in fact, cannot be identified until a specific mycotoxin analysis is carried out (Simpson et al., 2001). Early investigation in vitro indicates that the fungicide chlobenthiazone is highly effective in inhibiting aflatoxin biosynthesis by cultures of A. flavus; however, aflatoxin synthesis by A. parasiticus was, in fact, stimulated by the fungicide (Wheeler, 1991). Various surfactants, including some used in pesticide formulations, reduced aflatoxin biosynthesis by >96% (Rodriguez & Mahoney, 1994). Use of natural oils from thyme (Kumar et al., 2008), and other herbs has also been studied and shown to repress aflatoxin in certain crops in Asia. The herbicide glufosinate has been reported as having antifungal activity against certain phytopathogenic fungi in vitro (Uchimiya et al., 1993) and has shown activity in reducing infection of corn kernels in vitro (Tubajika & Damann, 2002). Higher levels of aflatoxin were observed in glyphosateresistant corn compared with traditional corn hybrids. Thus, effects of glyphosate on in vitro growth of A. flavus in pure culture and on native soil populations were examined, finding that high levels of glyphosate (> 5mM) were required for inhibition. In addition, application of greater amounts was found to have no effect on A. flavus populations. Interestingly, A. flavus when grown on glyphosate water agar media, produced 20% of aflatoxin produced on water agar without glyphosate (Abbas et al., 2009). Research carried out on fungicide use in terms of FHB and mycotoxin development has produced very interesting results. In particular, fungicides in common use have been shown to have differential effects against toxin-forming Fusarium species and related non-toxing-forming pathogens such as Microdochium nivale on ears (Simpson et al., 2001). The outcome of the use of fungicides seems to depend on the fungal species present, and the effect that the particular fungicide has on these species. For example, in recent work commissioned by the Home Grown Cereal Authority, in an experimental situation where Fusarium culmorum and M. nivale where both present, the use of azoxystrobin showed a significant reduction in disease levels while increasing the levels of DON present in grain. This was believed to be the result of selective inhibition of M. nivale by azoxystrobin. M. nivale is a natural competitor of toxin-forming Fusarium species, particularly F. culmorum. Removal of M. nivale by the fungicide probably allowed development of the toxigenic species in its place with concomitant increase in toxin formation. This is an important finding as it indicates that the impact of the fungicide is not directly related to mycotoxin production. It follows from these findings that where FHB is caused by Fusarium species in the absence of Microdochium, disease development is associated with higher levels of toxin (Magan et al., 2002). Ioos et al. (2005) also carried out a screen on the efficacy of fungicides, azeoxystrobin, metconazole, and tebuconazole at anthesis against Fusarium spp., M. nivale and on years on naturally infected fields of soft wheat, durum wheat, and barley. The infection levels of F. graminearum, F. culmorum, and

M, nivale were significantly reduced by the application Fusarium mycotoxin concentration over three of fungicides, with tebuconazole and metconazole effectively controlling the Fusarium spp., but they had little effect on M. nivale. Although this conclusion concurs with Simpson et al. (2001) for tebuconazole, their benefits were apparently seasonal-with tebuconazole controlling these fungi in 2001, while having little effect in 200 and 2002. The second approach involves the application of elicitors in crops susceptible to A. flavus, with the aim of protecting the plant of subsequent aflatoxins contamination. This because that the elicitors are capable molecules from activating multiple reactions defense that are induced and agrouped both histological level of physical barrier as a biochemist with the de novo synthesis of proteins associated with pathogenicity (PR), in the absence of the pathogen. Besides serves as aguide of intracellular events that end in activation of signal transduction cascades and hormonal pathways, triggering the induced resistance (IR) and consequently activation of plant immunity to ivironmental stresses (Riveros, 2001; Odjacova and Hadjiivanova, 2001; Garcia-Brugger et al., 2006; Bent and Mackey, 2007; Holopainen et al., 2009; Mejía-Teniente et al., 2011). Between the elicitors that have been investigated for more control of aflatoxin contamination in crops of commercial interest is the jasmonic acid (JA) and related compounds, as well as ethylene (ET). One factor influencing the production of aflatoxin is the presence of high levels of oxidized fatty acids such as fatty acid hydroperoxides, which can form in plant material either preharvest under stress or postharvest under improper storage conditions, correlates with high levels of aflatoxin production (Goodrich-Tanrikulu et al., 1995). Fatty acid hydroperoxides can be formed by autooxidation, or enzymically by lipoxygenases acting on a-linoleic and a-linolenic acids (Vick, 1993). These hydroperoxides stimulate the formation of aflatoxins by A. flavus and A. parasiticus (Fabbri et al., 1983; Fanelli and Fabbri, 1989). Degradation of the hydroperoxides by later steps in the plant lipoxygenase pathway leads to multiple byproducts, depending on the polyunsaturated fatty acid substrate, the positional specificity of the lipoxygenase, and the activities of enzymes catalysing the subsequent steps. The jasmonic acid (JA) is α- linolenic acid metabolite, via lipoxygenase and hydroperoxide dehydratase, is jasmonic acid (JA) (Vick, 1993). JA and closely related compounds, such as its methyl ester, MeJA, are endogenous plant growth regulators both higher and lower plants (Staswick, 1992; Sembdner and Parthier, 1993). JA and MeJ are two well-characterized plant growth regulators that exert a vast variety of biological activities in plants as the activation of defense responses (for review see Sembdener and Parthier, 1993). Among the diverse plant defense mechanisms, recent findings have demonstrated that low-concentrations of JA or MeJ induce protein inhibitors (Farmer and Ryan, 1992), thionin (Andresen et al., 1992;

Epple et al., 1995) and several plant defense enzymes such as PAL (Gundlach et al., 1992), LOX (Bell and Mullet, 1991) and chalcone synthase (Creelman et al., 1992). MeJ is volatile suggesting its action could be exerted in gaseous form, similar to the plant hormone, ethylene. Goodrich-Tanrikulu et al., (1995) reporting the effect of MeJA on aflatoxins production and growth of Aspergillus flavus in vitro. They Found that at concentrations MeJA of 10-3-10-8 M in the growth medium was inhibited aflatoxin production, by as much as 96%. Besides that when cultures were exposed to MeJA vapour similarly was inhibited aflatoxin production, observing that the amount of aflatoxin produced depended on the timing of the exposure. MeJA treatment also delayed spore germination and was inhibited the production of a mycelial pigment. These fungal responses resemble plant jasmonate responses. In other hand, Zeringue (2002) carried out a series of experiments where artificially wounded 22–27-day old developing cotton bolls were initially inoculated with, (1) a cell-free, hot water-soluble mycelial extract (CFME) of an atoxigenic strain of Aspergillus flavus or with, (2) chitosan lactate (CHL) or with, (3) CFME or CHL and then exposed to gaseous methyl jasmonate (MJ) or, (4) exposed to MJ alone. The results indicated a two- or three-fold increase in the production of the phytoalexins when gaseous MJ was added in combination to the CFME or the CHL elicitors. While the effects of aflatoxin B1 production after the developing cotton bolls pretreated with CFME, CHL or with CFME–MJ, CHL–MJ or only with MJ, showed a lower aflatoxin (Table 2, taken of Zeringue, 2002). All pretreatments resulted in some degree of aflatoxin B1 inhibition in the seeds underlying the treatment. CFME pretreatment resulted in a 88% inhibition of aflatoxin B1 and CHL resulted in a 64% inhibition (Table 2). CFME–MJ boll treatment resulted in the maximum aflatoxin B1 inhibition (95%) compared to CHL–MJ (75%). These series of experiments demonstrate a correlation between increased phytoalexin induction with a decreased aflatoxin B1 formation under the influence of volatile MJ in combination with selected elicitors. Phytoalexins are synthesized and accumulated at the site of microbial infection or as shown in this study localized at the site of the placement of elicitors (carpel discs). Besides, these results demonstrate an added inducement of phytoalexins and aflatoxin B1 inhibition produced by MJ treatment in combination with elicitors. This inducement is perhaps produced by an added signal/signals that activates other secondary pathways that either enhance the concentrations of the demonstrated phytoalexins or inhibit aflatoxin B1 biosynthesis or both. These results further demonstrate the innate, natural defense responses of the cotton plant and its ability to defend itself upon microbial attack, with the possibility to extrapolate to other seeds (Zeringue, 2002).

Harvest management

For cereals, harvest is the first stage in the production chain where moisture management becomes the dominant control measure in the prevention of mycotoxin development. Since the moisture content may vary considerably within the same field, the control of moisture in several spots of each load of the harvested grain during the harvesting operation is very important. Another equally important control measure is an effective assessment of the crop for the presence of disease such as FHB. This should be accompanied by an efficient strategy for separation of the diseased material from healthy grain. There is evidence that fungal infection can be minimized by avoiding the mechanical damage to the grain and by avoiding contact with soil at this stage.

Post-harvest management

Post-harvest strategies are important in the prevention of mycotoxin contamination and include improved drying and storage conditions, together with the use of natural and chemical agents, as well as irradiation.

Improving of drying and storage conditions

In cereals, mycotoxigenic fungal growth can arise in storage as a result of moisture variability within the grain itself or as a result of moisture migration results from the cooling of grains located near the interface with the wall of the storage container/silo (Topal et al., 1999). Thus control of adequate aeration and periodical monitoring of the moisture content of silos plays an important role in the restriction of mycotoxin contamination during the storage period (Heathcote & Hibbert, 1978). The moisture level in stored crops is one of the most critical factors in the growth of mycotoxigenic moulds and in mycotoxin production (Abramson, 1998), and is one of the main reasons for mycotoxin problems in grain produced in developing countries. Cereal grains are particularly susceptible to grow by Aspergilli in storage environments. The main toxigenic species are A. flavus and A. parasiticus for aflatoxins, and Penicillium verrucosum is the main producers in cereals for OTA (Lund & Frisvad, 2003), while A. ochraceus is tipically associated with coffee, grapes, and species, aflatoxins can be produced at aw values ranging from 0.95 to 0.99 with a minimum aw value of 0.82 for A. flavus, while the minimum aw for OTA production is 0.80 (Sweeney & Dobson, 1998). It has been reported that A. flavus will not invade grain and oilseeds when their moisture contents are in equilibrium with a relative humidity of 70% or less. The moisture content of wheat at this relative humidity is about 15%, and around 14% for maize, but it is lower for seeds containing more oil, approximately 7% and 10% for peanuts and cottonseeds, respectively (Heathcote & Hibbert, 1978), while A.

parasiticus has been reported to produce aflatoxins at 14% moisture content in wheat grains after 3 months of storage (Atalla et al., 2003). The second critical factor influencing post-harvest mould growth and mycotoxin production is temperature. Both the main aflatoxin producing Aspergillus strains A. flavus and A. parasiticus can grow in the temperature range from 10-12°C to 42-43°C, with an optimum in the 32 to 33°C range, with several studies highlighting the relatively high incidence of mycotoxins such as aflatoxins and ochratoxins in foods and feeds in tropical and subtropicals regions (Soufleros et al., 2003). The control of temperature of the stored grain at several fixed time intervals during storage may be important in determining mould growth. A temperature rise of 2-3°C may indicate mould growth or insect infestation. Until recently, little if any work has been carried out on monitoring how spoilage fungi interact with each other in the stored grain ecosystem, and the effect that this has on mycotoxin production. Magan et al. (2003), have shown that the system is in a state of dynamic flux with niche overlap altering in direct response to temperature and aw levels. It appears that the fungi present tended to occupy separate niches, based on resources utilization, and this tendency increased with drier conditions. Initially, A. flavus and other Aspergillus spp. were considered exclusively storage fungi, and aflatoxin contamination was believed to be primarily a storage problem. This is very severe in many rural areas that lack of infrastructure for drying and other appropriate storage conditions. Usually, corncobs are harvested at moisture contents that vary between 25-30% and are dried under the sunlight to reach 12-14% moisture content. Research has been conducted to determine the optimum temperature and moisture content of grains during storage to prevent Aspergillus spp. growth and aflatoxin production. In maize inoculated with A. flavus and stored at 27°C for 30 days with varying moisture contents, an association between moisture content and aflatoxin levels was established. At 16% moisture, aflatoxin levels reach 116 µg/kg while a 22% moisture 2166 µg/kg aflatoxin levels were obtained (Moreno-Martínez et al., 2000). In this same study, the authors tested the protective effects of propionic acid salts (6.5-12.5 L/t) on fungal growth and aflatoxin production. All grains treated with ammonium, calcium, or sodium propionates yielded very low Aspergillus flavus growth and aflatoxin levels (2 - 5.6 µg/kg) at all moisture contents. It is well established that rapid crop drying may be useful in controlling aflatoxin contamination in storage and that in addition that crops containing different moisture values are not stored together. It is also well established that mould invasion is facilitated as a result of increased moisture levels of stored commodities. Moisture abuse can even occur in crops with very low moisture content. Another factor to bear in mind is the fact that if fungal growth does occur in storage, moisture will be released during metabolism, which will be released during metabolism, leading to the growth of other fungal species and to the production of mycotoxins such as OTA.

DETECTION OF MYCOTOXINS IN FOOD

Aflatoxigenic fungi can contaminate food commodities, including cereals, peanuts, spices and figs. Foods and feeds are especially susceptible to colonization by aflatoxigenic Aspergillus species in warm climates where they may produce aflatoxins at several stages in the food chain, i.e. either at pre-harvest, processing, transportation or storage (Ellis et al., 1991). The level of mold infestation and identification of the governing species are important indicators of raw material quality and predictors of the potential risk of mycotoxin occurrence (Shapira et al., 1996). Traditional methods for the identification and detection of these fungi in foods include culture in different media and morphological studies. This approach, however, is tim-consuming, laborious and requires special facilities and mycological expertise (Edwards et al., 2002). Moreover, these methods have a low degree of sensitivity and do not allow the specification of mycotoxigenic species (Zhao et al., 2001). PCR-based methods that target DNA are considered a good alternative for rapid diagnosis due to their high specificity and sensitivity, and have been used for the detection of aflatoxigenic strains of A. flavus and A. parasiticus (Somashekar et al., 2004). However, as yet, none of these methods can reliably differentiate A. flavus from other species of the A. flavus group. In particular, A. flavus and A. parasiticus have different toxigenic profiles, A. flavus produces aflatoxin B1 (M1), B2, cyclopiazonic acid, aflatrem, 3-nitropropionic acid, sterigmatocystin, verdsicolorin A and aspetoxin, whereas A. parasiticus produced aflatoxin B1 (M1), B2, G1, G2 and versicolorin A. Another important fact is that A. flavus and A. fumigatus are responsible for 90% of the aspergillosis in human beings (González-Salgado et al., 2008). It is evident that one fundamental solution to the problem of mycotoxins in food would be to ensure that no contamination of edible crops occurred during harvesting and storage. It is equally clear, however, that such a solution is virtually unattainable, and hence that the presence of mycotoxins in food will have to be accommodated. Three approaches to the problem are most widely encountered; one involves physico-chemical methods of analysis, other relies on biological assays, and another one is microscopic examination. The former approach has found most widespread acceptance for routine purposes, but some authorities feel that a chemical diagnosis should be supported with some form of demonstration that the detected material is, in fact biologically toxic. The validity of this requirement is open to debate, but, for specific legal purposes, it may well become obligatory (Robinson, 1975).

REFERENCES

1. Abbas A, Vales H, Dobson ADW. 2009. Analysis of the effect of nutritional factors on OTA and OTB biosynthesis and polychetide syntase gene expression in Aspergillus ochraceus. Int. J. Food Microbiol, 135:22-27

2. Abbas HK, Weaver MA, Zablotowics RM, Horn BW, Shier WT. 2005. Relationships between aflatoxin production, sclerotia formation and source among Mississippi Delta Aspergillus isolates. Eur J Plant Pathol, 112:283-287

3. Abbas HK, Wilkinson JR, Zablotowics RM, Accinelli C, Abel CA, Bruns HA, Weaver Ma. 2009. Ecology of Aspergillus flavus, regulation of aflatoxin production, and management strategies to reduce aflatoxin contamination of corn. Toxin Reviews, 28:142-153

4. Abbas HK, Zablotowics RM, Locke MA. 2004a. Spatial variability of Aspergillus flavus soil populations under different crops and corn grain colonization and aflatoxins. Botany, 82:1768-1775.

5. Abbas HK, Zablotowics RM, Weaver MA, Horn BW, Xie W, Shier WT. 2004b. Comparison of cultural and analytical methods for determination of aflatoxin production by Mississippi Delta Aspergillus isolates. Can. J. Microbiol. 50:193-199

6. Abdel-Wahhab MA, Kholif AM. 2008. Mycotoxins in animal feeds and prevention strategies: A review. Asian Journal Of Animal Sciences, 2(1):7-25

7. Abramson D. 1998. Mycotoxin formation and environmental factors. In: Sinha, K.K., and Bhatnagar D., Eds., Mycotoxins in Agriculture and Food Safety. Marcel Dekker, Inc, New York, 255-277

8. Accinelli C, Abbas HK, Zablotowicz RM, Wilkinson JR. 2008. Aspergillus flavus aflatoxin occurrence and expression of aflatoxin biosynthesis genes in soil. Can. J. Microbiol. 54:371-379

9. Aflatoxin Mitigation Center of Excellence (AMCE). 2010. Preventing Health Hazards and Economic Losses from Aflatoxin. Texas Corn Producers. Aldred D, Magan N. 2004. Prevention strategies for trichothecenes. Toxicol. Lett., 153:165- 171.

10. Andresen I, Becker W, Schluter K, Burges J, Parthier B. 1992. The identification of leaf thionin as one of the main jasmonate-induced proteins of barley (Hordeum Vulgare). Plant Mol. Biol. 19:193–204.

11. Angle JS. 1986. Aflatoxin decomposition in various soils. J. Environ. Sci. Health B. 21:277-288 Atalla MM, Hassanein NM, El-Beith AA, Youssef YA. 2003.

12. Mycotoxin production in wheat grains by different Aspergilli in relation to different relative humidities and storage periods. Nahrung, 47:6-10

13. Atroshi F, Rizzo A, Wastermack T, Ali-Vehmas T. 2002. Antioxidant nutrients and mycotoxins. Toxicol., 180:151-167

14. Bell E, Mullet JE. 1991. Lipoxygenase gene expression is modulated in plants by water deficit, wounding, and methyl jasmonate. Mol. Gen. Genet. 230:456–462.

15. Benndorf D Müller A, Bock K, Manuwald O, Herbarth O, Van Bergen M. 2008. Identification of spore allergens from the indoor mold Aspergillus versicolor. Allergy, 63:454-460

16. Bennett JW, Chang PK, and Bhatnagar D. 1997. One gene to whole pathway: the role of norsolorinic acid in aflatoxin research. Adv. Appl. Microbiol. 45:1–15.

17. Bennett JW, Kale S, Yu J. 2007. Aflatoxins: Backround, Toxicology and Molecular Biology.

18. From Infectious Disease: Foodborne Diseases Edited by S.Simjee. Human Press Inc., Totowa, NJ. 355-374.

19. Bennett JW, Klich M. 2003. Mycotoxins. Clin. Microbiol. Rev., 16:497-516

20. Bent AF, Mackey D (2007). Elicitors, effectors, and R genes: the new paradigm and a lifetime supply of questions. Annu. Rev. Phytopathol. 45: 399-436

21. Berthiller F, Schumacher R, Adam G, Krska R. 2009. Formation, determination, and significance of masked and other conjugated mycotoxins. Anal Bioanal Chem,395:1243-1252

22. Bhat RV, Vasanthi S. 2003. Mycotoxin food safety risks in developing countries, food safety

23. in food security and food trade. Vision 2020, Agriculture and Environment, Focus

24. 10, pp: 1-2

25. Bhatnagar D, Ehrlich KC, Yu J, Cleveland TE. 2003. Molecular genetic analysis and regulation of aflatoxin biosynthesis. Appl Microbiol Biotechnol. 61:83-93.

26. Bhatnagar D, Cary JW, Ehrlich KC, Yu J, Cleveland TE (2006). Understanding the genetics of regulation of aflatoxin production and Aspergillus flavus development. Mycopathologia 162:155-166

27. Bhatnagar D, Proctor R, Payne GA, Wilkinson J, Yu J, Cleveland TE, Nierman WC. 2006.

28. Genomics of mycotoxigenic fungi. In: Barug D, Bhatnagar D, van Egmond HP, van der Kamp JW, van Osenbruggen WA, Visconti A, eds. The mycotoxigenic factbook

29. (Food & Feed Topics). Wagningen, The Netherlands: Wagningen Academic Publishers, pp.157-178

30. Bhatnagar D, Yu J, Ehrlich KC. 2002. Toxins in filamentus fungi. In: Breitenbach M, Crameri R, Lehrer SB (Eds.), Fungal Allergy and Pathogenicity. Chem. Immunol. Basel, Karger 81:167-206.

31. Bhatnagar RK, Ahmad SK, Mukerji G. 1986. Nitrogen metabolism in Aspergillus parasiticus NRRL 3240 and A. flavus NRRL 3537 in relation to aflatoxin production. J. Appl. Bacteriol., 60:203-211

32. Bhatnagar, D. 2010. Elimination of postharvest and preharvest aflatoxins contamination; 10th International working conference on stored product protection, Section: Microbiology, mycotoxins and food safety: 425.

33. Bley NC. 2009. Economic Risks of Aflatoxin Contamination in the Production and Marketing of Peanut in Benin. Thesis Submitted to the Graduate Faculty of Auburn University in Partial Fulfillment of the Requirements for the Degree of Master of Science. Blout WP. 1961. Turkey "X" disease. Turkeys, 52:55-58

34. Bohnert M, Wackler B, Hoffmeister D. 2010. Spotlights on advances in mycotoxin research. Appl Microbiol. Biotechnol. 81:1-7

35. Bräse S, Encinas A, Keck J, Nising CF. 2009. Chemistry and biology of mycotoxins and related fungal metabolites. Chem Rev, 109:3903-3990

36. Brown RL, Cotty PJ, Cleveland TE. 1991. Reduction in aflatoxin content of maize by atoxigenic strains of Aspergillus flavus. J. Food Prot., 54(8):623-626.

37. Buchi G, Foulkes DM Kurono M, Mitchell GF, Schneider RS. 1967. The total synthesis of racemic aflatoxin B1. Journal of the American Chemical Society, 89:6745-6753

38. Calvo AM, Wilson RA, Bok JW, Keller NP (2002). Relationship between secondary metabolism and fungal development. Microb. Mol. Biol. Rev. 66:447–459.

39. Campbell BC, Molyneux RJ, Schatzki TF. 2003. Current research on reducing pre- and postharvest aflatoxin contamination of U.S. almond, pistachio, and walnut. Journal Of Toxicology, Toxin Reviews, 22:225-266.

40. Carbone I, Ramirez-Prado JH, Jakobek JL, Horn BW (2007). Gene duplication, modularity and adaptation in the evolution of the aflatoxin gene cluster. BMC Evol. Biol. 7: 111.

41. Cardwell, K.F., D. Desjardins, S. H. Henry, et al. 2004. The Cost of Achieving Food Security and Food Quality. http://www.apsnet.org/online/ festure/mycotoxin/ top.html.

42. Cary JW, Calvo AM. 2008. Regulation of Aspergillus mycotoxin biosynthesis. Toxin Reviews, 27:347-370

43. Cary JW, Ehrlich K. 2006. Aflatoxigenicity in Aspergillus: molecular genetics, phylogenetic relationships and evolutionary implications. Mycopathologia,162:167-177

44. Cary JW, Szerszen L, Calvo AM. 2009. Regulation of Aspergillus flavus aflatoxin biosynthesis and development. American Chemical Society, 13:183-203

45. Catangui MA, Berg RK. 2006. Western beat cutworm, Striacosta albicosta (Smith) (Lepidoptera:Noctuidae), as a potential pest of transgenic Cry1Ab Bacillus thuringensis corn hybrids in South Dakota. Environ Entomol. 35:1439-1452

46. Center for Disease Control and Prevention (CDC). 2004. Outbreak of aflatoxin poisoningEastern and Central provinces. Kenya, January-July, 2004.

47. Chang PK. 2003. The Aspergillus parasiticus protein AFLJ interacts with the aflatoxin pathway-specific regulator AFLR. Mol. Genet. Genomics 268: 711–719.

48. Chen ZY, Brown RL, Cleveland TE, Damann KE, Russin JS. 2001. Comparison of constitutive and inducible maize kernel proteins of genotypes resistant or susceptible to aflatoxin production. J. Food Prot., 64:1785-1792

49. Cleveland T, Dowd PF, Desjardins AE, Bhatnagar D, Cotty PJ. 2003. United States Department of Agriculture-Agricultural research service research on preharvest prevention of mycotoxins and mycotoxigenic fungi in US crops. Pest Manag. Sci., 59:629-642

50. Cleveland TE, Yu J, Fedorova N, Bhatnagar D, Payne GA, Nierman WC, Bennett JW. 2009. Potential of Aspergillus flavus genomics for applications in biotechnology. Trends in Biotechnology 27:151-157.

51. Codex Alimantarius Commission. 2002. Proposed draft code of practice for the prevention (reduction) of mycotoxin contamination in cereals, including annexes on ochratoxin A, zearalenone, fumonisins, and trichothecens, CX/FAC 02/21, Joint FAO/WHO

52. Food Standards Programme, Rotterdam, the Netherlands. Cole RJ, Cox RH. 1981. Handbook of Toxic Fungal Metabolites. New York: Academic Press, 937 pp.

53. Cole RJ, Cotty PJ. 1990. Biocontrol of aflatoxin production by using biocompetitive agents. In Robens, J., Huff, W. and Richard, J. (eds.) A Perspective on Aflatoxin in Field Crops and Animal Food Products in the United States: A Symposium; ARS-83. U.S. Department of Agricul-ture, Agricultural Research Service, Washington, D.C., pp. 62-66.

54. Cotty P. 1988. Aflatoxin and sclerotial production by Aspergillus flavus: influence of pH. Phytopathology, 78:1250-1253

55. Cotty, P.J. 1989. Virulence and cultural characteristics of two Aspergillus flavus strains pathogenic on cotton. Phytopathology 79, 808-814.

56. Cotty PJ. 1994. Influence of field application of an atoxigenic strain of Aspergillus flavus on the population of A. flavus infecting cotton bolls and on the aflatoxin content of cottonseed. Phyto-pathology 84, 1270-1277.

57. Cotty PJ. 1996. Aflatoxin contamination of commercial cottonseed caused by the S strain of Asper-gillus flavus. Phytopathology 86, S71.

58. Cotty PJ, Probst C, Jaime-Garcia R. 2008. Etiology and Management of Aflatoxin from Contamination.Mycotoxins: detection methods, management, public health and agricultural trade. ISBN: 978-1-84593-082-0. DOI: 10.1079/9781845930820.0287

59. Coulibaly O, Hell K, Bandyopadhyay R, Hounkponou S, Leslie JF. 2008. "Mycotoxins: Detection Methods, Management, Public Health and Agricultural Trade", Published by CAB International, ISBN 9781845930820.

60. Creelman RA, Tierney ML, Mullet JE. 1992. Jasmonic acid/methyl jasmonate accumulate in wounded soybean hypocotyls and modulate wound gene expression. Proc. Natl. Acad. Sci. U.S.A. 89, 4938–4941.

61. Cuero RG, Duffus E, Osuji G, Pettit R. 1991. Aflatoxin control in preharvest maize: effects of chitosan and two microbial agents. J. Agr. Sci. 117:165–169.

62. D'Mello JPF, McDonald AMC, Postel D, Dijksma WTP, Dujardin A, Placinta CM. 1998. Pesticide use and mycotoxin production in Fusarium and Aspergillus phytopathogenes. Eur. J. Plant Pathol., 104:741:751

63. Dauch AL, Watson AK, Jabaji-Hare SH. 2003. Detection of the biological control agent Colletotrichum coccodes (183088) from the target weed velvetleaf and soil by strain specific PCR markers. J. Microbiol. Methods, 55:51-64

64. Degola F, Berni E, Dall'Asta C, Spotti E, Marchelli R, Ferrero I, Restivo FM. 2007. A multiplex RT-PCR approach to detect aflatoxigenic strains of Aspergillus flavus. J. Appl. Microbiol. 103:409-417

65.   Diamond H, Cooke BM. 2003. Preliminary studies on biological control of the Fusarium ear blight complex of wheat. Crop Prot., 22:99-107

66.   Diener UL, Cole RJ, Sanders TH, Payne GA, Lee LS, Klich MA (1987). Epidemiology of aflatoxin formation by Aspergillus flavus. Annu Rev Phypathol. 25:249-270

67.   Dill-Macky R, Jones RK. 2000. The effect of previous crop residues and tillage on Fusarium head blight of wheat. Plant Dis., 84:71-76

68.   Dohlman, E. 2003. "Mycotoxin Hazards and Regulations: Impacts on Food and Animal Feed Crop Trade," International Trade and Food Safety: Economic Theory and Case Studies, Jean Buzby (editor), Agricultural Economic Report 828. USDA, ERS.

69.   Dorner, J. W., Cole, R. J., Blankenship, P. D. (1998). Effect of inoculum rate of biological control agents on preharvest aflatoxin contamination of peanuts. Biol. Control 12:171–176.

70.   Dorner, J. W., Cole, R. J., Wicklow, D. T. (1999). Aflatoxin reduction in corn through field application of competitive fungi. J. Food Prot. 62:650–656.

71.   Dorner, J.W. 2004. Biological Control of Aflatoxin Contamination of Crops. Journal of Toxicology-Toxin Reviews. Vol. 23, Nos. 2 & 3, pp. 425–450, 2004.

72.   Dowd PF. 2003. Insect management to facilitate preharvest mycotoxin management. J.Toxicol. Toxin Rev. 22(2):327-350

73.   Du W, O'brian GR, Payne GA (2007). Function and regulation of aflJ in the accumulation of aflatoxin early pathway intermediate in Aspergillus flavus. Food Addit. Contam. 24: 1043–1050.

74.   Edwards SG. 2004. Influence of agricultural practices on Fusarium infection of cereals and subsequent contamination of grain by trichothecene mycotoxins. Toxicol. Lett., 153:29-35

75.   Ehrlich KC, Cotty PJ (2002). Variability in nitrogen regulation of aflatoxin production by Aspergillus flavus strains. Appl. Microbiol. Biotechnol. 60: 174– 178.

76.   Ehrlich KC, Montalbano BG, Cary JW, Cotty PJ. 2002. Promoter elements in the aflatoxin pathway polyketide synthase gene. Biochim. Biophys. Acta 1576:171-175

77.   Ellis WO, Smith JP, Simpson BK. 1991. Aflatoxin in food: Occurrence, biosynthesis, effects on organisms, detection, and methods of control.

78. Epple P, Apel K, Bohlmann H. 1995. An Arabidopsis thaliana thionin gene is inducible via a signal transduction pathway different from that for pathogenesis-related proteins. Plant Physiol. 109:813–820.

79. Fabbri AA, Fanelli C, Panfili G, Passi S, Fasella P. 1983. Lipoperoxidation and aflatoxin biosynthesis by Aspergillus parasiticus and A. flavus. Gen Microbiol, 29: 3447-3452.

80. Fanelli C, Fabbri AA. 1989. Relationship between lipids and aflatoxin biosynthesis. Mycopathologia 107:115-120.

81. Farmer EE, Ryan CA. 1992. Octadecanoid precursors of jasmonic acid activate the synthesis of wound inducible proteinase inhibitors. Plant Cell 4:129–134.

82. Fernandes M, Keller NP, Adams TH. 1998. Sequence-specific binding by Aspergillus nidulans AflR, a C6 zinc cluster protein regulating mycotoxin biosynthesis. Mol. Microbiol. 28: 1355–1365.

83. Food and Agriculture Organization (FAO). 1997. Worldwide Regulations for Mycotoxins 1995: A compendium. FAO Food and Nutrition Paper. No. 64. Rome, Italy. Food and Agriculture Organization FAO. 2004. Manual Técnico: Manejo Integrado de Enfermedades en Cultivos Hidropónicos. Oficina Regional para América Latina y el Caribe.

84. Galvano F, Piva A, Ritieni A, Galvano G. 2001. Dietary strategies to counteract the effects of mycotoxins: A review. J. Food Prot., 64:120-131.

85. Garcia-Brugger AG, Lamotte O, Vandelle E, Bourque S, Lecourieux D, Poinssot B, Wendehenne D, Pugin A (2006). Early signaling events induced by elicitors of plant defenses. MPMI, 19(7): 711-724.

86. Garman H, Jewett HH.1914. The life-history and habits of the corn-ear worm (Chloridae obsoleta). Kentucky Agricultural Experimental Station Bulletin. 187:388-392

87. Gebrehiwet Y, Ngqangweni S, Kirsten JF. 2007. Quantifying the trade effect of sanitary and phytosanitary regulations of OECD countries on South African foods exports. Agrekon, 46:23-38

88. Georgianna DR, Payne GA. 2009. Genetic regulation of aflatoxin biosynthesis: from gene to genome. Fungal Genet Biol., 46:113-125

89. Goodrich-Tanrikulu, M., Mahoney, N. E. and Rodriguez, S.B. 1995. The plant growth regulator methyl jasmonate inhibits aflatoxin production by Aspergillus flavus. Microbiology. 141: 2831-2837.

90. Groopman JD, Kensler TW. 1999. The light at the end of the tunnel for chemical-specific biomarkers: Daylight or headlight?, Carcinogenesis, 20:1-11

91. Gundlach, H., Muller, M.J., Kutchan, T.M., Zenk, M.H., 1992. Jasmonic acid is a signal transducer in elicitor-induced plant cell structures. Proc. Natl. Acad. Sci. U.S.A. 89, 2389–2393.

92. Heathcote JG, Hibbert JR. 1978. Aflatoxin chemical and biological aspects. Elsevier Scientific Publishing Company, Amsterdam.

93. Holopainen JK, Heijari J, Nerg AM, Vuorinen M, Kainulainen P (2009). Potential for the use of exogenous chemical elicitors in disease and insect pest management of conifer seedling production. Open. For. Sci. J. 2: 17-24.

94. Hormisch D, Brost I, Kohring GW, Gifthorn F, Krooppenstedt E, Farber P, Holzapfel WH. 2004. Mycobacterium fluoranthenivorans sp. nov., a fluoranthene and aflatoxin B1 degrading bacterium from contaminated soil of a former coal gas plant. Syst. Appl. Microbiol. 27:653-660

95. Horn BW, Greene RL, Dorner JW. 1995. Effect of corn and peanut cultivation on soil

96. populations of Aspergillus flavus and A. parasiticus in southwestern Georgia. Appl Environ Microbiol, 61:2472-2475

97. Horn BW. 2007. Biodiversity of Aspergillus section Flavi in the United States: a review. Food Addit. Contam. 24:1088-1101.

98. Hua SS. 2002. Biological Control of Aflatoxin in Almond and Pistachio by Preharvest Yeast Application in Orchards. In: Special Issue: Aflatoxin/ Fumonisin Elimination and Fungal Genomics Workshops. Phoenix, Arizona, October 23–26, 2001. Mycopathologia, 65.

99. Hua SS, Baker T, Flores-Espiritu M. 1999. Interactions of saprophytic yeasts with a nor mutant of Aspergillus flavus. Appl. Environ. Microbiol. 65:2738–2740. Hussein HS, Brasel JM. 2001. Toxicity, metabolism, and impact of mycotoxins on humans and animals. Toxicol., 167:101-134

100. Huyn VL, Lloyd AB. 1984. Synthesis and degradation of aflatoxins by Aspergillius parasiticus. I. Synthesis of aflatoxin B1 by young mycelium and its subsequent degradation in aging mycelium. Aust. J. Biol. Sci. 37:37-43

101. Ioos R, Belhadj A, Menez M, Faure A. 2005. The effects of fungicideson Fusarium spp. and Microdochium nivale and their associated trichothecene mycotoxins in French naturally-infected cereal grains. Crop Prot., 24:894-902

102. Jaime-Garcia R, Cotty PJ. 2004. Aspergillus flavus in soils and corncobs in South Texas: implications for management of aflatoxins in corn-cotton rotations. Plant Dis. 88:1366-1371.

103. Jolly CM, Bayard B, Awuah RT, Fialor SC, Williams JT. 2009. "Examining the Structure of Awareness and Perceptions of Groundnut Aflatoxin among Ghanaian Health and Agricultural Professionals and its influence on their Actions" The Journal of SocioEconomics, 38:280-287.

104. Jones RK, Duncan HE, Payne GA, Leonard KJ. 1980. Factors influencing infection by Aspergillus flavus in silk-inoculated corn. Plant. Dis. 64:859-863

105. Kaaya AN, Warren HL. 2005. A Review of Past and Present Research on Aflatoxin in Uganda. African Journal of Food Agriculture Nutrition and Development (AJFAND) 5(1):1-18.

106. Kabak B, Dobson ADW, Var I. 2006. Strategies to prevent mycotoxin contamination of food and animal feed: A review. Critical Reviews in Food Science and Nutrition, 46:593- 619

107. Kelkar HS, Skloss TW, Haw JF, Keller NP, Adams TH. 1997. Aspergillus nidulans stcL encodes a putative cytochrome P-450 monooxygenase required for bisfuran desaturation during aflatoxin/sterigmatocystine biosynthesis. Journal of Biological Chemistry, 272(3): 1589-1594

Chapter 6

BIOENGINEERING THERMODYNAMICS OF BIOLOGICAL CELLS

Umberto Lucia

Dipartimento Energia, Politecnico di Torino, Corso Duca degli Abruzzi 24, 10129 Torino, Italy

ABSTRACT

Background

Cells are open complex thermodynamic systems. They can be also regarded as complex engines that execute a series of chemical reactions. Energy transformations, thermo-electro-chemical processes and transports phenomena can occur across the cells membranes. Moreover, cells can also actively modify their behaviours in relation to changes in their environment.

Methods

Different thermo-electro-biochemical behaviours occur between health and disease states. But, all the living systems waste heat, which is no more than the result of their internal irreversibility. This heat is dissipated into the environment. But, this wasted heat represent also a sort of information, which outflows from the cell toward its environment, completely accessible to any observer.

Results

The analysis of irreversibility related to this wasted heat can represent a new approach to study the behaviour of the cells themselves and to control their behaviours. So, this approach allows us to consider the living systems as black boxes and analyze only the inflows and outflows and their changes in relation to the modification of the environment. Therefore, information on the systems can be obtained by analyzing the changes in the cell heat wasted in relation to external perturbations.

Conclusions

The bioengineering thermodynamics bases are summarized and used to analyse possible controls of the calls behaviours based on the control of the ions fluxes across the cells membranes.

BACKGROUND

Nature, from a physical, biological, chemical and mathematical point of view, is a complex system, while from an engineering point of view, it is the "first" engineer! In particular, cells can be modelled as an adaptive thermal and chemical engines which convert energy in one form to another by coupling metabolic and chemical reactions with transport processes [1–5], by consuming irreversibly [6–8] free energy for thermal and chemical processes, transport of matter, energy and ions.

Energy is a thermodynamic property of any system in relation to a reference state, which changes during any process, while its total amount remains constant in relation to the universe, considering it as the system together with its environment. In cells, many processes such as replication, transcription and translation need to convert molecular binding energy, chemical bond hydrolysis and electromagnetic gradients into mechanical work, related to conformational changes and displacements [9]. The biomechanical analysis of DNA has pointed out the connections among forces, thermodynamics, nano-mechanical and electromagnetic behaviour of biological structures and kinetics [10].

Engineering thermodynamics is the science which studies both energy and its best use in relation to the available energy resources with particular regards to energy conversion, including power production, refrigeration and relationships among the properties of matter, including also living matter. So, engineering thermodynamics can be introduced in the mechanobiological and system biological approach in order to improve these sciences by analysing the biosystems also from a thermal point of view: a new engineering science could be considered, the bioengineering thermodynamics. Indeed, the first law of thermodynamics expresses the conservation of energy, while the second law states that entropy continuously increases for the system and its environment and introduces the statistical and informational meaning of global quantities [11–14].

In this paper we develop the bioengineering thermodynamic of biological cells, with particular regards to possible control of the cells growth by a control of the ions transport across the cell membrane. To do so, we consider that cells spontaneously exchange heat, and this heat is related to their biochemical and biophysical behaviour. This wasted heat represents the interaction between the cell and its environment, a sort of "spontaneous communication" towards environment. This interaction is fundamental to developing a thermodynamic study of the cell. Indeed, cells are too complex to understand the contribution of each process to the global result, and the study of cells as black boxes allows us to simplify the analysis by considering only the inflow and outflow balances [15]. Moreover, it is easier to have access to the cell environment than to the living cell itself. These considerations allow us to introduce the bases of the bioengineering thermodynamic approach introduced in the study of the cells:

- An open irreversible real linear or non-linear system is considered;
- Each process has a finite lifetime ;
- What happens in each instant in the range [0,] cannot be known, but what has happened after time (the result of the process) is well-known (at least it is sufficient to wait and observe): local equilibrium is not necessarily required;
- The balance equations are balance of fluxes of energy, mass and ions.

The fundamental quantity used in this analysis is the global entropy [16, 17], related to systems changes, highlighted as the only effective criterion for spontaneity of change in any system, with particular regards to the entropy variation due to irreversibility, named entropy generation [18], which is the result of the global effect of the entropy variation

- due to the interaction with the environment
- within the system itself.

The introduction of entropy generation comes from the need to avoid inequalities: entropy results as a state function, so nothing is really produced or generated. Therefore, entropy is nothing more than a parameter characterising the thermodynamic state, and the term due to irreversibility, S_g, measures how far the system is from the state that will be attained in a reversible way [12]. It is always $S_g \geq 0$.

Recently, it has been highlighted that any effect in Nature is always the consequence of the dynamic balances of the interactions between the real systems and their environments [12] and the real systems evolution is always related to the decrease of their free energy, in the least time [19–21]. So, bioengineering thermodynamics is based just on two fundamental concepts of physics: interactions and flows. The result is the analytical formulation of flow-based analysis in thermodynamics, which can play the role of a "rallying point" of the different modelling approach to biosystems. Indeed, if we consider natural systems we can highlight that they are always open systems, which means that they can exchange heat and mass with their environment.

So, the interaction with the environment is a fundamental concept for the thermodynamic analysis.

We consider the environment as a thermostat and the system, together with its environment, is an adiabatic closed system [18]. But, for an adiabatic close system, the total entropy, defined as:

$$dS = d_iS + d_eS$$

(1)

it always increases, as a consequence of the second law [18]. In relation (1) dS is the variation of the total entropy elementary, d_eS is the entropy variation for interaction between the open system considered and its environment, and d_iS is the entropy variation due to irreversibility, such that:

$$\frac{dS}{dt} \geq 0$$

(2)

Now, we can write the relation (1) as [22]:

$$\frac{dS}{dt} = \int_V \left[-\nabla \cdot \left(\frac{Q}{T} \right) + \dot{s}_s \right] dV$$

(3)

where Q is the heat flow, T is the temperature, V is the volume, t is the time and $_g$ is the density of the entropy generation rate. Now, we consider that the stationary states of the open system correspond to the equilibrium states of the adiabatic closed system. Considering the system together with its environment, we are analyzing an adiabatic closed system, so the entropy variation for the volume considered is maximum at the equilibrium [23]:

$$dS = 0 \quad \Rightarrow \quad \left[-\nabla \cdot \left(\frac{Q}{T} \right) + \dot{s}_g \right] = 0$$

(4)

and

$$\nabla \cdot \left(\frac{Q}{T} \right) = \dot{s}_g$$

(5)

This last relation allows us to state that the flows between the open system and its environment cause the entropy generation rate density, so the interaction between system and environment is responsible of irreversibility. But, we cannot state if the cause of changes is the change of the entropy inside the cell or the fluxes across the cell membrane. We can only highlight the relation between changes and fluxes, but this approach doesn't allow us to establish if are the fluxes to cause entropy changes or if entropy changes causes fluxes.

Now, considering that the entropy generation rate density can be written as [22]:

$$\dot{s}_g = \sum_k J_k \cdot X_k$$

$$(6)$$

where J_k is the flow of the k-th quantity involved in the process considered and X_k is the related thermodynamic force. Now, considering that:

$$\nabla \cdot \left(\frac{Q}{T}\right) = Q \cdot \nabla \left(\frac{1}{T}\right) + \frac{1}{T} \nabla \cdot Q = \sum_k J_k \cdot X_k$$

$$(7)$$

the relation (5) becomes:

$$\frac{1}{T} \nabla \cdot Q = \sum_k J_k \cdot X_k - Q \cdot \nabla \left(\frac{1}{T}\right)$$

$$(8)$$

in agreement with Le Chatelier's principle [24], for which any change in concentration, temperature, volume, or pressure generates a readjustment of the system in opposition to the effects of the applied changes in order to establish a new equilibrium, or stationary state. It follows that the fundamental imperative of Nature is to consume free energy in least time. Any readjustment of the state of the system can be obtained only by generating fluxes of free energy which entail any process where the system evolves from one state to another.

RESULTS AND DISCUSSION

The existence of bioelectric signalling among most cell types suggests a wide field of applicability of these electro-magnetical signals. Here, we provide bioengineering thermodynamic theory that suggest how to explain the effects of energy, mass and ionic flows across cell membranes and, consequently, to control the cell behaviour by a control of ion fluxes.

Living cells are separated from their environment by the lipid bilayer membrane, which presents a different concentration of specific ion species on both sides. As a consequence, a charge separation across the membrane is generated by the electro-diffusion of ions down their electrochemical gradient. These ions move into a negative (inside the cell) membrane potential of around -70 to -100 mV. The hydrophobic component of the lipid bilayers behaves as a capacitor dielectric, which maintains the ionic gradients across the membrane; in some instances, the action of ATP-driven ionic pumps supports this effect by separating the charges. The cell function is regulated by the membrane proteins, sensitive to electric field; indeed, changes in the electric field are transduced into a conformational change that accomplishes the function of the membrane protein with consequences for the regulation of cell functions. The charged species, their arrangements, the local field strength, charges and dipoles disposition and movements can vary with the result of changing the electric field which is tranduced into a conformational change related to the protein functions themselves [32].

These considerations suggest that control and regulation of the membrane's electric field could represent a new approach to therapies against diseases such as cancer. To understand how to control the fluxes across the membrane we consider the concentration of the ions on the opposite sides of the membrane [33]:

$$c_{outside} = c_{inside} \ \exp\left(\frac{\Phi_{outside} - \Phi_{inside}}{RT}\right)$$

(9)

where c is the molar concentration of the chemical species, R is the universal constant of gas, T is the temperature and Φ is the electric potential energy. As a consequence of this concentration difference the cell can move the ions, and change the pH inside and outside its membrane. The ion drift velocity v_{drift} across the cell membrane can be obtained by using the classical kinetic theory [34] as:

$$v_{drift} = \frac{Ze}{m} \frac{\phi}{d} \tau_{drift}$$

(10)

where Ze is the electric charge of the ion, m is the ion mass, ϕ is the electric potential across the membrane, d is the length of the membrane and τ_{drift} is the mean time between two collisions [33]:

$$v_{drift} = \frac{Ze}{m} \frac{\phi}{d} \tau_{drift}$$

(11)

where σ is the electric conductivity. Consequently, an electric current I occurs for each ion i=H^+, Na^+, K^+, Ca^{2+}, Cl^-, Mg^{2+}, etc.:

$$I_i = n_i A Z_i e v_{drift}$$

(12)

where A is the mean surface area of the membrane. Now, considering the equivalent RC electric circuit for a membrane it is possible to state that the resonant frequency for such a circuit results in $(2\pi RC)^{-1}$, where R is the electric resistivity for the ion considered and C is the membrane capacity.

It follows, that if we want to control the cross-membrane flux we must impact the current. The easier physical way to interact with a current is to use an electromagnetic wave of the resonant frequency for the membrane, in relation to the ion considered, with its amplitude being related to the entropy generation as just obtained in Ref. [25–30].

In Figs. 1 and 2, it is represented an example of this kind of control. Figure 1 represents the natural behaviour of cell requirement of energy to grow. Figure 2 represents the cell requirement of energy by cell to grow when they are inside an electromagnetic field. It represents the ratio between the variation in percentage of the energy used by a cancer in a magnetic field (50 µT, 40 Hz) respect the energy used by a cancer outside of the field, related to the energy used by the cancer outside the field vs the growth of the cancer in terms of volume growth (ratio between the cancer volume during the cancer growth and the initial volume). It has been obtained by using the entropy generation approach described in the following section on methods. It is possible to highlight how the different ions have different effect. The positive ions determines a decreasing of the energy used while the negative ion increase it. So the positive ions determine an opposition to the growth. The more effective ion is Ca^{2+}. It means that a control of calcium ion can determine a control of the volume growth of a cancer. Here, the control is suggested by the use of an electromagnetic field. The field induces in the cell a greater use of energy to obtain the same growth.

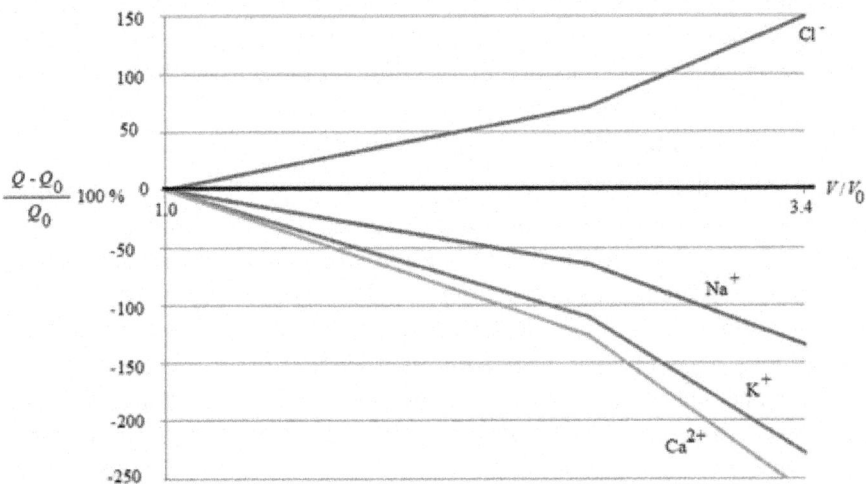

Figure. 1: Effect of ions flux control on the energy required by cells to growth in natural conditions.

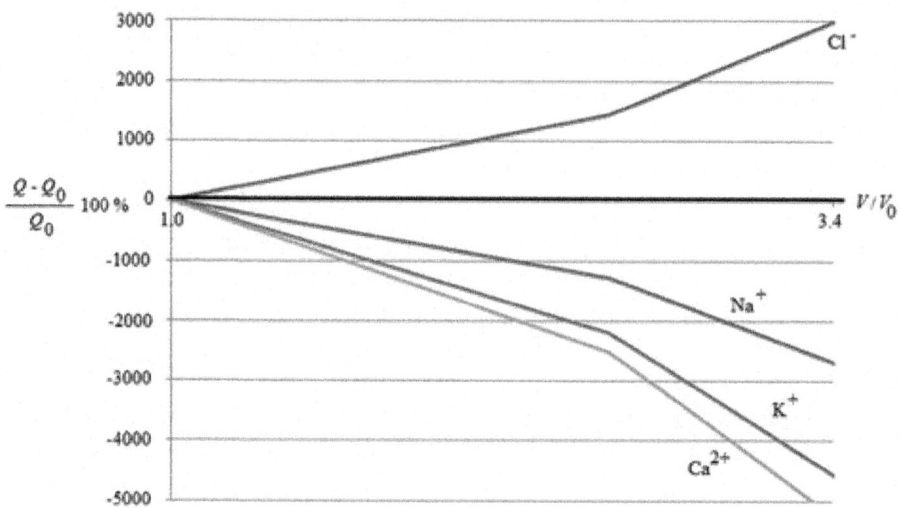

Figure. 2: Effect of ions flux control on the energy required by cells to grow in conditions controlled by electromagnetic field.

CONCLUSIONS

Life is an organisational and thermodynamic process that tends towards the maximum conversion of available energy. The biochemical reactions produce or consume external metabolites, and they connect internal metabolites, in constant concentrations in the cells at their steady states. To do so, the cell must exchange energy and matter through its membrane. The fundamental phenomena used by cells to reach their optimality consist of a redistributing of the flow patterns through their metabolic network.

By using the bioengineering thermodynamics, it has been highlighted how the different ions have different effect on the use of energy by the cell to grow. To do so, a control of the cells behaviours is introduced. Here, an electromagnetic field is used as a control system, but other field could be used. Cells inside and outside an electromagnetic field have been considered. The positive ions determines a decreasing of the energy used by the cancer, such that the cancer cannot grow as outside the field. On the other hand, the negative ion increase the use of energy. It means that a control of ions can determine a control of the volume growth of a cancer. This result can be extended to all the molecular fluxed across the cell membrane, obtaining a possible bioengineering thermodynamic approach to control the cancer growth.

METHODS

The approach previously used is based on the following considerations:

- The energy lost by a system is gained by the environment, consequently, the information lost by the system is gained by the environment: here the problem is to codify this information;

- The environment is completely accessible by any observer, so it is easy to collect data on the lost energy of any system;

- The flows cause entropy generation variations, consequently we can evaluate the entropy generation to obtain information to the flows, even when we are unable to evaluate the flows themselves;

- The entropy generation is a global quantity, so we can obtain global information on the cells, but from a biomedical point of view just the global cells behaviour is the useful information.

Biological systems are very interesting because they are able to adapt to the variation of environmental conditions; indeed, cells attain their "optimal" performance by a selection process driven by their environmental interactions. The resultant effect is a redistribution of energy, ions and mass flows in their metabolic network, by using regulatory proteins.

The bioengineering thermodynamic approach to biological systems consists in the analysis of the biological optimization process realized by Nature. It is no more than the classical and engineering thermodynamic analysis of the steady-state flux distribution, which, for a cell, are no more than the metabolic flows. So, starting from Equation (1) and considering the second law for the open systems [18]:

$$S_g = \int_0^\tau \dot{S}_g dt = \Delta S - \sum_i \frac{Q_i}{T_i} - \int_0^\tau \left(\sum_{in} G_{in} s_{in} - \sum_{out} G_{out} s_{out} \right) dt$$

(13)

where Q is the heat exchanged, T is the temperature of the thermal source, s is the specific entropy, G is the mass flow and τ is the lifetime of the process. But, for any open system, the entropy balance in a local form results [22]:

$$\frac{\partial s}{\partial t} + v\nabla \cdot \left[\frac{\mathbf{Q}}{T} + \sum_i \rho_i s_i \left(\dot{\mathbf{x}}_i - \dot{\mathbf{x}}_B \right) \right] = v\sigma$$

(14)

where s = S/m, is the specific entropy, S is the entropy, σ is the entropy production density, v is the specific volume, $\mathbf{Q}$ is the heat flow, $_i$ is the relative velocity in relation to the centre of mass reference, and $_B$ is the centre of mass velocity. Now, considering that [22]:

$$T\frac{ds}{dt} = \frac{du}{dt} + p\frac{dv}{dt} - \sum_i \mu_i \frac{dc_i}{dt}$$

$$\frac{du}{dt} = \frac{dq}{dt} - p\frac{dv}{dt} - v\Pi : \nabla \dot{\mathbf{x}}_B + v\sum_k \mathbf{J}_k \cdot \mathbf{F}_k$$

$$\frac{dq}{dt} = -v\nabla \cdot \mathbf{J}_q$$

$$\frac{dc_i}{dt} = -\frac{1}{v}\nabla \cdot \mathbf{J}_i + \frac{1}{v}\sum_j v_{ij}J_j$$

(15)

where s is the specific entropy, u is the internal specific energy, v is the specific volume, p is the pressure, μ_i are the chemical potentials, c_i is the concentrations, T is the temperature, $d/dt = \partial/\partial t +_B \cdot \nabla$, q is the heat per unit mass, $\Pi = \mathbf{P} - p\,\mathbf{I}$ with Π total pressure tensor, p hydrostatic pressure and $\mathbf{I}$ identity matrix of which the elements are $I_{jk} = \delta_{jk} = 1$ if $j = k$ and 0 in the other cases, $\mathbf{a}:\mathbf{b} = \sum_{ij} a_{ij} b_{ji}$ is the product between two tensors $\mathbf{a}$ and $\mathbf{b}$, $\mathbf{J}_k = \rho_k(_i - _B)$ is the diffusion flows and $\mathbf{F}_k$ are the forces, J_j is the chemical reaction rate of the j-th chemical reaction and v_{ij} are quantities such that if they are divided by the molecular mass of the i-th component they are proportional to the stoichiometric coefficients. Now, introducing the electro-chemical affinity $\tilde{A} = A + Z\Delta$ related also to pH variation and the electric field variation, with $A_j = \sum_k v_{kj}\mu_j$, Z the electric charge per unit mass, the electrostatic potential, the relation (AA) holds [25–28]:

$$S_g = \int_V dV \left(-\int_o^{r_1} \frac{v}{T^2}\mathbf{J}_q \cdot \nabla T dt - \int_o^{r_2} v\sum_k \mathbf{J}_k \cdot \nabla\left(\frac{\mu_k}{T}\right) dt - \int_0^{r_3} \frac{v}{T}\Pi : \nabla\dot{\mathbf{x}}_B dt - \int_0^{r_4} \frac{v}{T}\sum_j J_j A_j + \int_0^{r_5} \frac{v}{T}\sum_k \mathbf{J}_k \cdot \mathbf{F}_k \right) =$$

$$= S_{g,tf} + S_{g,dc} + S_{g,vg} + S_{g,cr} + S_{g,de}$$

(16)

where [25–27]:

- $S_{g,tf}$ is the entropy generation due to the thermal flux driven by temperature difference;

- $S_{g,dc}$ is the entropy generation due to the diffusion current driven by chemical potential gradients, with $\mu{\sim}\mu{\sim} = \mu + Z$ electrochemical potential, μ chemical potential;

- $S_{g,vg}$ is the entropy generation due to the velocity gradient coupled with viscous stress;

- $S_{g,cr}$ is the entropy generation due to the chemical reaction rate driven by affinity, always positive;
- $S_{g,de}$ is the entropy generation due to the dissipation due to work by interaction with the environment;

and τ_i, $i \in [1, 5]$, are the lifetimes of any process and the volume of the cell is evaluated by a characteristic length, in transport phenomena usually considered the diameter of the cell approximated as the diameter of a sphere $L=(6V/\varpi)^{1/3}=2r$, with r being the cell radius;

1. the mean environmental temperature can be assumed as $T_0=310\,K$ and the mean cell temperature has been estimated to be $T_0+\Delta T$. The quantity ΔT would be experimentally evaluated for different cells lines in relation to their metabolism;

2. the internal energy density results in $u=3.95 \times 10^7\,Jm^{-3}$, being calculated as the ratio between the ATP energy, $U=3 \times 10^{-7}$ J and the mean value of the cell inside the human body, $V=7600\,\mu m^3$. It must be emphasized that this is an approximation because the cell volume inside the human body is in the range of 200–$15000\,\mu m^3$;

3. the thermal molecular mean velocity inside the cytoplasm is considered to be $=5 \times 10^{-5}$ m s^{-1};

4. the membrane volume is calculated with $V_m=\frac{4}{3}\pi r^3-\frac{4}{3}\pi(r-d_e)^3=\frac{4}{3}\pi r^3-\frac{4}{3}\pi(r-0.2r)^3=0.992V$ $V_m=\frac{4}{3}\pi r^3-\frac{4}{3}\pi(r-d_e)^3=\frac{4}{3}\pi r^3-\frac{4}{3}\pi(r-0.2r)^3=0.992V$ being $d_e=0.2\,r$;

5. the chemical potential gradient can be approximated through the ratio between the mean value of the chemical potential $\mu=1.20\times10^{-9}$ J kg^{-1} and the membrane length $d_m=0.01\,\mu m$, with the mean density being $\rho=1000$ kg m^{-3};

6. the viscosity is taken to be 6.91×10^{-3} N s m^{-2};

7. $\eta \sim 2.07 \times 10^{-3}$ N s m^{-2} at 30 °C;

8. $_B$ is set as 3.0×10^{-6} m s^{-1};

9. τ_1 is the time related to the thermal flow driven by temperature difference. It can be assessed considering that the time constant of the thermal transient for heat conduction is $\tau_{cv} \approx pcV/(hA)$ with $\rho \approx 1000$ kg m^{-3} density, V the cell volume, A the external cell surface, $c \approx 4186$ J kg^{-1}K^{-1} specific heath, and h the convection heat transfer coefficient evaluated as: $h \approx 0.023Re^{0.8}$ Pr$^{0.35}$ λ/L, where $\lambda \approx 0.6$ W m^{-1}K^{-1} of heat conductibility, L the characteristic dimension of the cell (here we have considered the diameter), Re≈ 0.2 the Reynolds number and Pr≈ 7 the Prandtl number. The process would have occurred in a time $\tau_1 \approx 5\,\tau_{cv}$. For human cells this value can be considered in the range 15–269 ms;

10. τ_2 is the time related to the diffusion current driven by chemical potential gradients. It can be evaluated as $\tau_2 \approx d/D$, with $d = 0.01$ μm, i.e., the length of the membrane, and D being the diffusion coefficient. Considering that the diffusion coefficient of glucose is approximately 10^{-9} m²s⁻¹ it follows that $\tau_2 \approx 10$ s;

11. τ_3 is the time related to the velocity gradient coupled with viscous stress. This time can be evaluated as the propagating time of a mechanical wave on the surface of the cell $\tau3 \approx 2\pi rc\tau3 \approx 2\pi rc$ with $c \sim 1540$ m s⁻¹ the sound velocity, considered to be the same in biological tissue;

12. τ_4 is the time related to the chemical reaction rate driven by affinity and it can be evaluated considering the magnitude order of a chemical reaction in a cell ($\sim 10^{-7}$ mol s⁻¹l⁻¹). Moreover, we consider that the moles number is proportional to the density of the chemical species (for glucose 1540 kg m⁻³) and the volume of the cell itself. It follows that this time is in the range 17–1283 ns;

13. τ_5 is the time related to the dissipation due to work by interaction with the external forces. It depends on the interaction considered;

L is a characteristic length, introduced as usually done in transport phenomena.

An experiment has been developed to obtain also a direct proof [31]. Therefore the spontaneous heat exchanged by the cell represents the interaction or the spontaneous communication between the cell and its environment. The proposed thermodynamic theory predicts that the temperature difference between cells with distinct metabolic characteristics can be amplified by an altered interaction with the external environment, due to the entropy generation term related to the interaction of the system with the external fields. The experiments carried out on cells exposed to low frequency electromagnetic waves consolidate the thermodynamic approach. Indeed, through infrared thermography an adimensional number, maned thermal dispersion index, was evaluated. This adimensional number represents the inability of the cells to fit their thermal power to environmental changes. Primary fibroblasts display a high dispersion index, with a maximal value of 800 % vs NIH3T3 immortalized line, which means that the primary fibroblasts adjust more efficiently their thermal production or dissipation than the NIH3T3 cells. This significant difference implies that, when exposed to selected environmental conditions, transformed cells dissipate heat more slowly than their normal counterpart. The results of this experimental approach demonstrate that selecting environmental conditions it is possible to appreciate distinct cellular phenotypes; these differences can be evaluated by thermal dispersion patterns measured by infrared thermography. The experiment confirmed the bioengineering thermodynamics theoretical results.

The results obtained can be improved by considering other approach to bioengineering thermodynamics devoted to the study of organization in living systems and by linking each others. Indeed, evolution over the long term requires a constant generation of new alternative forms, a biological behavior named mutation [35]. The cooperative effect of mutation and selection consists in different processes on different time scales:

- Microevolution: changes within a natural populations, in the composition of populations due to mutation and natural selection. It occurs on a time scale of generations, it represents an adaptive change;

- Macroevolution: changes between populations, in the composition of lineages due to speciation and extinction. It occurs in a geological time scale.

- Macroevolution has its origin in microevolution as the result of natural selection acting on genotypic and phenotypic variation [35–38]. These natural processes can be described by introducing mathematical models, based on two thermodynamic actions [35, 39]:

- The acquisition of resources from the external environment and its conversion into energy storage;

The transformation of the metabolic energy into useful work.

The bases of these processes is the interaction between bio-system and environment [40]. This brings to non equilibrium states, and the mathematical formalisms developed to the biosystems analysis was the dynamical systems, based on the studies of Bowen [41], Ruelle [42] and Sinai [43], who provided new perspectives in the analysis of far from equilibrium systems by the discovery of certain connections between non-equilibrium statistical mechanics and the ergodic theory of dynamical systems. In this context the fundamental concept is the entropy and just this concept represents the link between the dynamical systems approach and the thermodynamic approach here developed. Indeed, following Ruelle [44], considering a classical system with isokinetic time evolution described by the equation:

$$\frac{d}{dt}\begin{pmatrix} \mathbf{p} \\ \mathbf{q} \end{pmatrix} = \begin{pmatrix} \xi - \alpha\mathbf{p} \\ \mathbf{p}/m \end{pmatrix} \Longleftrightarrow \frac{d\mathbf{x}}{dt} = \mathbf{F}_\xi(\mathbf{x})$$

$$(17)$$

with

$$\mathbf{x} = \begin{pmatrix} \mathbf{p} \\ \mathbf{q} \end{pmatrix} \quad \text{and} \quad \mathbf{F}_\xi(\mathbf{x}) = \begin{pmatrix} \xi - \alpha\mathbf{p} \\ \mathbf{p}/m \end{pmatrix}$$

$$(18)$$

$p \in R^N$ and $q \in R^N$ momentum and position respectively, ξ a nongradient time independent force, m mass and $(-\alpha \, \mathbf{p})$ the isokinetic thermostat mathematical expression with α defined as:

$$\alpha(\mathbf{x}) = \frac{\mathbf{p} \cdot \xi(\mathbf{q})}{\mathbf{p} \cdot \mathbf{p}}$$

(19)

so that [44]:

$$\frac{d}{dt}\left(\frac{\mathbf{p} \cdot \mathbf{p}}{2m}\right) = 0$$

(20)

Under these conditions Ruelle defined the entropy increment as [44]

$$S(\xi + \Delta\xi) = \int_0^\infty dt \int_{-\infty}^t d\tau \rho_\xi \left(\nabla_{\mathbf{x}}\left(d\mathbf{x} \circ f_{\xi_0}^{t-\tau}\right) \cdot \delta_\tau \mathbf{F}(\mathbf{x})\right) \Phi(\mathbf{x})$$

(21)

with $\delta_\tau \mathbf{F}$ is a time-dependent small perturbation of

$$F, \rho_\xi \left(\nabla_{\mathbf{x}}\left(d\mathbf{x} \circ f_{\xi_0}^{t-\tau}\right) \cdot \delta_\tau \mathbf{F}(\mathbf{x})\right)$$

probability distribution, $f_\xi^{t-\tau}$ the solution of the equation (18) at the time $t-\tau$ corresponding to the initial conditions ξ_0, $\Phi(\mathbf{x}) = (N - 1)\alpha$. Then, Denbigh [18, 45] expressed the fundamental processes of living systems, introducing an entropy approach:

$$dS = dS_{\text{int}} + dS_{\text{ext}}$$

(22)

where dS is the total entropy elementary variation, dS_{int} is the entropy elementary production within the system due to its metabolism of ingested exergy and dS_{ext} is the entropy exchange with the environment. Entropy is a path independent state function, and the overall reaction entropy ΔS_R can be evaluated by the macroscopic reaction stoichiometry between external metabolites:

$$\Delta S_R = \sum_{i=1}^n p_i \Delta S_i = \sum_{i=1}^n p_i \sum_{l=1}^k \nu_l s_{li} = c \sum_{i=1}^n p_i \ln p_i$$

(23)

where $\Delta S_i = -c \ln p_i$ is the entropy of reaction, $s_{li} = (h_{li} - g_{li})/T$, with h_{li} molar enthalpy and g_{li} Gibbs molar energy, are the molar entropies of the k reactants and products, ν_l are the stoichiometry coefficients, p_i is the probability of the i-th mode and c is a constant related to the numbers of elementary modes and on their reaction entropies. It represents the state of the fully evolved metabolic network [46]. When the living systems increase in organization, they increase their entropy and, far from equilibrium, they have a high exergy content [47]; indeed, considering two systems with the same mass and the same chemical composition, the one, that has a large amount of organization, has also higher exergy content. During their evolution, the living systems, and also ecosystems, increase their structure in organization, which is a working information useful for resilience and integrity, and also their efficiency in converting exergy to entropy, in order to reduce the applied exergy gradient, while their internal entropic state continue to decrease [48, 49]. Then, while dS_{int} is always positive defined ($dS_{int} \geq 0$), dS_{ext} can have any sign.

The inner entropy generation rate σ is defined as the local first time derivative of the [50] internal component of the entropy:

$$\sigma = \frac{dS_{int}}{dt}$$

(24)

If the irreversible processes are sufficiently slow, the Gibbs equation can be applied to any subsystem [50]:

$$TdS = dU + pdv - \sum_i \mu_i dn_i$$

(25)

and the entropy can be expressed in terms of fluxes J_i and conjugated generalized forces X_i [50]:

$$T\sigma = \sum_i J_i X_i$$

(26)

The non-equilibrium stationary states, which are the states whose variables are independent of time, play a fundamental role in the irreversible processes. After a characteristic time, the system achieves the equilibrium if no restraints are imposed on it, while if a number of constant restraints are imposed, a steady state is attained [50]. In any steady state the total entropy is independent of time, consequently:

$$\frac{dS}{dt} = \frac{dS_{int}}{dt} + \frac{dS_{ext}}{dt} = 0 \Rightarrow \frac{dS_{ext}}{dt} = -\frac{dS_{int}}{dt}$$

(27)

but

$$\frac{dS_{int}}{dt} \geq 0 \Rightarrow \frac{dS_{ext}}{dt} \leq 0$$

(28)

and it is possible to argue that the entropy generation rate in a stationary system must be compensated by the liberation of entropy to the surroundings. This means also that non-equilibrium steady states cannot occur in isolated systems because these last systems do not allow exchange of entropy between the systems and the surroundings [8]. Prigogine proved that [51–53]:

$$do \leq 0 \Rightarrow \frac{d^2 S_{\text{int}}}{dt^2} \leq 0$$

$$(29)$$

On the use of the Prigogine's results there is little doubt that a mature organism may reached a stationary state; indeed, the homeostasis of all self regulating systems is interpreted as tendency to return from a perturbed state to that of highest stability compatible with biological constraints [50].

Moreover, considering an irreversible and open system, it is composed by N elementary volumes. Every i-th element of this system is located by a position vector $\mathbf{x}_i$, it has a velocity $_i$, a mass m_i and a momentum $\mathbf{p}_i = m_i{}_i$. The total mass of the system is $m = \sum_i m_i$ and its density is $\rho = m/V$ with $V = \sum_i V_i$ total volume. The position of the centre of mass is $\mathbf{x}_B$ and its velocity results $_B = \sum_i m_i{}_i/m$, while the diffusion velocity is $\mathbf{u}_i = {}_i - {}_B$. The total mass of the system is conserved, so the following relation $\rho' + \rho\nabla\cdot\mathbf{x}'B=0\rho' + \rho\nabla\cdot\mathbf{x}'B=0$ is satisfied together with its local expression $\rho'i+\rho\nabla\cdot\mathbf{x}'i=\rho\Xi i\rho'i+\rho\nabla\cdot\mathbf{x}'i=\rho\Xi i$, related to the density of the i-th elementary volume of density ρ_i and a source Ξ generated by matter transfer, chemical reactions or thermodynamic transformations. For an open system, as just described in macroscopic way, the equation of the entropy balance is [22]:

$$\frac{\partial s}{\partial t} + v\nabla \cdot \mathbf{J}_S = \dot{s}$$

$$\dot{s} = v\sigma$$

$$(30)$$

where $s = S/m$, is the specific entropy, S entropy, σ the density of the entropy generation rate, v the specific volume and $\mathbf{J}_S$ is the entropic flux defined as:

$$\mathbf{J}_S = \frac{\mathbf{Q}}{T} + \sum_i \rho_i s_i (\dot{\mathbf{x}}_i - \dot{\mathbf{x}}_B)$$

$$(31)$$

with $\mathbf{Q}$ heat flux.

Any dynamical state of this system can be described by the 3N canonical coordinates $\{\mathbf{x}_i \in R^3, i \in [1,N]\}$ and their conjugate momenta $\{\mathbf{p}_i \in R^3, i \in [1,N]\}$. The 6N−dimensional space spanned by $\{(\mathbf{p}_i,\mathbf{x}_i), i \in [1,N]\}$ is the phase space Ω of the open system considered. Any point $\mathbf{q}_i = (\mathbf{p}_i,\mathbf{x}_i), \mathbf{q}_i \in R^{6N}$ in the phase space Ω, represents a state of the entire N−elements system [54]. Any family $\{\xi(t), t \in R\}$ is called stochastic process in the phase space Ω and it can be represented by a family of equivalent classes of random variables $\xi(t)$ on Ω, $\{\gamma(\sigma(t)) : t \in R\}$. The point function $\gamma(\mathbf{q}(t))$ is called trajectory of the stochastic process $\xi(t)$: a description of a physical system in terms of a trajectory of a stochastic process corresponds to a point dynamics, while its description in terms of equivalent classes of trajectories and their associated probability measure corresponds to an ensemble dynamics [55]. So it is considered a non-equilibrium system moving in the Ω-space between two states, which are in two elementary cells of a given partition of the phase space. We use the concept of path of classical mechanics: if the motion of the system is regular, or if the phase manifold has positive or zero Riemannian curvature, there will be only a fine bundle of paths which track each other between the initial and the final cells [13]. For a system in chaotic motion, or when the Riemannian curvature of the phase manifold is negative, two points indistinguishable in the initial cell can separate from each other exponentially [54]. Then, between two given phase cells, there may be many possible paths γ_k, $k \in [1,\omega]$ with ω number of all the paths, with different travelling time $t_{\gamma k}$ of the system and different probability $p_{\gamma k}$ for the system to take the path k, called path probability distribution [56–59]. It is considered an ensemble of a large number Lof identical systems, all moving in the phase space from two cells with ω possible paths, and L_k systems travelling on the path γ_k. The probability $p_{\gamma k}$ that the system take the path γ_k is defined as usual by $p_{\gamma k} = L_k/L$. If $\omega_k = 1$ then $p_{\gamma k} = 1$. By definition, $p_{\gamma k}$ is the transition probability from the two states considered. These trajectories must be the paths minimizing action according to the principle of least action [54]. Since 1962, Jaynes argued that Gibbs' formalism of equilibrium statistical mechanics could be generalised in a statistical inference theory for non-equilibrium systems [60]. Jaynes developed the non-equilibrium statistical mechanics for the stationary state constraint on the basis of maximum entropy; his approach consists of maximising the path Shannon information entropy written for the path, $S_I = -\Sigma_\gamma p_\gamma \ln p_\gamma$, with respect to p_γ of the path γ, with the probability subject to the actual constraints. According to Shannon, 'the information entropy is the logarithm of the number of the outcomes i with non-negligible probability p_i', while in 'non-equilibrium statistical mechanics it is the logarithm of the number of microscopic phase-space paths γ having non-negligible probability p_γ' [60]. Jaynes' approach consists of finding the 'most probable macroscopic path realised by the greater number of microscopic paths compatible with the imposed constrained' [60], in analogy with the Boltzmann microstate counting: 'paths rather then states are the central objects of interest in non-equilibrium systems, because of the presence of non-zero macroscopic fluxes whose statistical description requires considering the underlying microscopic behaviour over time' [60] which implies that 'the macroscopic behaviour is reproducible under given constraints' and it is 'characteristic of each of the great number of microscopic paths compatible with those constraints' [60]. Following this approach and these considerations, the statistical expression of the entropy generation has been written as [56–59]:

$$S_g = -k_B \sum_k p_{\gamma k} \ln p_{\gamma k}$$

$$(32)$$

It can be also interpreted as the missing information necessary for predicting which path a system of the ensemble takes during the transition from a state to another.

In the theory of probability the stochastic order is introduced. Two random variables X and Y are in stochastic order if there exists a random variable Z and functions ψ_1 and ψ_2 such that $X = \psi_1(Z)$ and $Y = \psi_2(Z)$, with $\psi_1(Z) \leq \psi_2(Z)$ [61]. Now, the set of paths $\{\gamma_k, k \in [1,\omega]\}$ is considered, with ω number of all the paths between two thermodynamic states, represented by two points in the phase space. It is possible to define a stochastic order among the paths, saying that a path γ_i is stochastically smaller than a path γ_j if its probability $p_{\gamma i}$ is smaller that the probability of the other path, $p_{\gamma j}$ [13]:

$$\gamma_i <_{ST} \gamma_j \quad \text{if} \quad p_{\gamma i} < p_{\gamma j}$$

$$(33)$$

So, the probability of a path can be expressed in term of the first order differential of the entropy generation respect to the probability itself, as follows [13]:

$$p_{\gamma i} = \exp\left(-\frac{1}{k_B} \frac{\partial S_g}{\partial p_{\gamma i}} - 1 \right)$$

$$(34)$$

But, in the analysis of the complex systems, it was highlighted how chaotic and fractal behaviour are very widespread in nature: any numerical evaluation based on accessible states in the phase space is incomplete because of the rejected, singular or inaccessible points [56]. The basis of the incomplete information is that a part of information on complex system may not be completely accessible. The consequence is that irreversibility is the physical model by which thermodynamic phenomena can be completely described [54]. The related information is incomplete because, for complex systems, it occurs that $\sum_{j=1}^{v} {}^v p_j = \theta \leq 1$, with v number of accessible or accountable states, smaller of the total number of states, and θ incompleteness of the treatment and linked to the nature of the system, consequence of the partial knowledge of the dynamics or of the inaccessible states of the system itself [54]. Non-statistical mechanics replaces the complete probability normalisation by:

$$\sum_{j=1}^{\nu} p_j^{\vartheta} = 1$$

(35)

with ϑ incompleteness parameter such that $= 1$ if the probability distribution is complete. It can be related to the incompleteness θ by the following relation [13]:

$$\sum_{j=1}^{\nu-1} p_j^{\vartheta} + \left(\theta - \sum_{j=1}^{\nu-1} p_j\right)^{\vartheta} = 1$$

(36)

The phase space cells, which represent the stationary states, was proven to form a subset of all the cells on which the evolution acts as a one-cycle permutation: this kind of ergodicity has been defined ergodicity for irreversibility [54]. Moreover, in non-equilibrium transformation, the volume of the phase space was proven to contract indefinitely [54]. Recently, considering the expression for the probability $p_{\gamma i}$ of a path γ_i and the statistical results on the entropy generation [54], it was proven that [13]:

$$\frac{\partial S_g}{\partial p_{\gamma i}} \leq \frac{\partial S_g}{\partial p_{\gamma j}} \quad \text{if} \quad p_{\gamma i} \leq p_{\gamma j}$$

(37)

which means that the paths are statistically ordered. The stochastic order of the path proves that the evolution of the bio-systems is related to their irreversibility and the quantity useful to evaluate the allowed paths and their probability is the entropy generation. Consequently, a link between the bioengineering thermodynamic approach proposed and the dynamical system approach is obtained.Declarations

ACKNOWLEDGMENTS

The author must thank prof. Antonio Ponzetto (Università di Torino) for his scientific support in biomedical interpretation of the thermodynamic results and Prof. Bartolomeo Montrucchio (Politecnico di Torino) for his support in the design of the experimental devices.

COMPETING INTERESTS

The author declare that he has no competing interests.

AUTHORS' CONTRIBUTIONS

UL has developed the thermodynamics approach and written the paper.

AUTHORS' INFORMATION

Umberto Lucia, MSc in Physics at Turin University and PhD in Energetics at Florence University, teaches Engineering thermodynamics at Biomedical Engineering School at Politecnico di Torino. He is member of the Energy Department and develops researches in thermodynamics bio-systems, and classical, statistical and quantum thermodynamics of complex systems. After twenty years spent to study the relations between irreversibility and steady states in open systems, in the last three years he developed the bioengineering thermodynamics applied to biosystems, biological cells, cancer and non equilibrium states. This approach to biosystems has been experimentally confirmed. Consequently, he continue to improve the theory developed in order to obtain a full thermodynamic approach useful to predict the behaviour of the biological cells with particular interest to cancer and its thermodynamic control.

REFERENCES

1.   Demirel Y, Sandler SI. Thermodynamics and bioenergetics. Biophys Chem. 2002;97:87–111.

2.   Toussaint O, Schneider ED. The thermodynamic and evolution of complexity in biological systems. Comp Biochem Physiol A. 1998;120:3–9.

3.   Caplan SR, Essig A. Bioenergetics and Linear Nonequilibrium Thermodynamics, The Steady State. Cambridge: Harvard University Press; 1983.

4.   Lucia U. Entropy generation approach to cell systems. Physica A. 2014;406:1–11.

5.   Lucia U. Molecular machine as chemical-thermodynamic devices. Chem Phys Lett. 2013;556:242–4.

6.   Salerian AJ, Saleri NG. Cooling core body temperature may slow down neurodegeneration. CNS Spectr. 2008;13:227–9.

7.   Katchalsky A, Curran PF. Nonequilibrium thermodynamics in biophysics. Cambridge: Harvard University Press; 1967.

8.   Lucia U. Irreversibility in biophysical and biochemical engineering. Physica A. 2012;391:5997–6007.

9.   Bustamante C, Chemla YR, Forde NR, Izhaky D. Mechanical processes in biochemistry. Annu Rev Biochem. 2004;73:705–48.

10.   Hudspeth A, Choe Y, Mehta A, Martin P. Putting ion channels to work: mechanoelectrical transduction, adaptation and amplication by hair cells.

Proc Natl Acad Sci. 2000;97(22):11765.

11. Lucia U. Irreversibility, entropy and incomplete information. Physica A. 2009;388(19):4025–33.

12. Lucia U. (2013b) Stationary open systems: A brief review on contemporary theories on irreversibility. Physica A. 2013;392(5):1051–62.

13. Lucia U. Thermodynamic paths and stochastic order in open systems. Physica A. 2013;392(18):3912–9.

14. Lucia U. Thermodynamics and cancer stationary states. Physica A. 2013;392(17):3648–53.

15. Lucia U. Transport processes in biological systems: tumoral cells and human brain. Physica A. 2014;393:327–36.

16. Denbigh KG. Note on entropy, disorder and disorganization. Brit J Phil Sci. 1989;40:323–32.

17. Denbigh KG. The many faces of irreversibility. Brit J Phil Sci. 1989;40:501–18.

18. Bejan A. Advance engineering thermodynamics. New York: John Wiley; 2006.

19. Lucia U. Some considerations on molecular machines and Loschmidt paradox. Chem Phys Lett. 2015;623:98–100.

20. Lucia U. A link between nano- and classical thermodynamics: dissipation analysis (the entropy generation approach in nano-thermodynamics). Entropy. 2015;17(2):1309–28.

21. Lucia U. Entropy production and generation: clarity from nanosystems considerations. Chem Phys Lett. 2015;629:87–90.

22. de Groot SG, Mazur P. Non-equilibrium thermodynamics. Amsterdam: North-Holland Publishing; 1984.

23. Zemansky MW. Heat and thermodynamics. New York: McGraw-Hill; 1966.

24. Atkins PW. The elements of physical chemistry. 3rd ed. Oxford: Oxford University Press; 1993.

25. Lucia U. Entropy generation and cell growth with comments for a thermodynamic anticancer approach. Physica A. 2014;406:107–18.

26. Lucia U. Thermodynamic approach to nano-properties of cell membrane. Physica A. 2014;407:185–91.

27. Lucia U. Transport processes and irreversible thermodynamics analysis in tumoral systems. Physica A. 2014;410:380–90.

28. Lucia U. The gouy-stodola theorem in bioenergetic analysis of living

systems (Irreversibility in bioenergetics of living systems). Energies. 2014;7:5717–39.

29. Lucia U, Ponzetto A, Deisboeck TS. A thermo-physical analysis of the proton pump vacuolar-ATPase: the constructal approach. Sci Rep. 2014;4:6763.

30. Lucia U, Ponzetto A, Deisboeck TS. A thermodynamic approach to the 'mitosis/apoptosis' ratio in cancer. Physica A. 2015;436:246–55.

31. Lucia U, Grazzini G, Montrucchio B, Grisolia G, Borchiellini R, Gervino G, et al. Constructal thermodynamics combined with infrared experiments to evaluate temperature differences in cells. Sci Rep. 2015;5:11587.

32. Tuszynski JA, Kurzynski M. Introduction to molecular biophysics. Raton: Taylor & Francis; 2003.

33. Newman J. Physics of the life sciences. Berlin: Springer; 2008.

34. Kittel C, Kroemer H. Thermal physics. London: W.H. Freeman and Company; 1980.

35. Demetrius LA, Gundlach VM. Directionality theory and the entropic principle in natural selection. Entropy. 2014;16:5428–522.

36. Mayr E. What evolution is. New York: Weidenfeld and Nicholson; 2002.

37. Mayr E. Evolution and the diversity of life. Harvard: Harvard University Press; 1976.

38. Demetrius L. Directionality principles in thermodynamics and evolution. Proc Natl Acad Sci U S A. 1997;94:3491–8.

39. Lehninger A. Bioenergetics. New York: WA Benjamin Inc.; 1965.

40. Lewontin RC. Gene, organism and environment. In: Bendall DS, editor. Evolution from molecules to men. Cambridge: Cambridge University Press; 1983. p. 273–85.

41. Bowen R. Equilibrium states and the ergodic theory of anosov diffeomorphisms. Vol. 470 lecture notes in math. New York: Springer; 1975.

42. Ruelle D. Thermodynamic formalism. Vol. 5 encyclopedia of mathematics and its applications. Reading: Addison-Wesley; 1978.

43. Sinai YG. Gibbs measures in ergodic theory. Russ Math Surv. 1972;4:21–69.

44. Ruelle D. Extending the definition of entropy to nonequilibrium steady states. Proc Natl Acad Sci U S A. 2003;100:3054–8.

45. Denbigh KG. Note on entropy, disorder and disorganization. Brit J Phil Sci. 1989;40:323–32.

46. Sandler SI, Orbey H. On the thermodynamics of microbial-growth processes. Biotech Bioeng. 1991;38:697–718.

47. Swenson R. Emergent attractors and the law of maximum entropy prpduction: foundations to a theory of general evolution. Systems Research. 1989;6:187–97.

48. Brooks DR, Collier J, Maurer BA, Smith JDH, Wiley EO. Entropy and information in evolving biological systems. Biology & Philosophy. 1989;4:407–32.

49. Günther F, Folke C. Characteristics of nested living systems. J Biological Systems. 1993;1:257–74.

50. Katchalsky A, Kedem O. Thermodynamics of flow processes in biological systems. Biophys J. 1962;2:53–78.

51. Glansdorff P, Prigogine I. Thermodynamic theory of structure, stability and fluctuations. New York: John Wiley & Sons; 1971.

52. Prigogine I. Etude Thermodynamique des Phénomènes Irréversibles. Liège: Desoer; 1947.

53. Prigogine I. Introduction to thermodynamics of irreversible processes. New York: Interscience; 1961.

54. Lucia U. Probability, ergodicity, irreversibility and dynamical systems. Proc Royal Soc A. 2008;464:1089–184.

55. Primas H. Basic elements and problems of probability theory. J Sci Explor. 1999;13(4):579–613.

56. Lucia U. Statistical approach of the irreversible entropy variation. Physica A. 2008;387(14):3454–60.

57. Lucia U. Irreversibility, entropy and incomplete information. Physica A. 2009;388:4025–33.

58. Lucia U. Maximum entropy generation and ϕ-exponential model. Physica A. 2010;389:4558–63.

59. Lucia U. Irreversibility entropy variation and the problem of the trend to equilibrium. Physica A. 2007;376:289–92.

60. Dewar R. Information theory explanation of the fluctuation theorem, maximum entropy production and self-organized criticality in non-equilibrium stationary states. J Phys A: Math Gen. 2003;36:631–41.

61. Shaked M. Stochastic orders. New York: Springer; 2006.

Chapter 7

ENZYMES IN FOOD PROCESSING: A CONDENSED OVERVIEW ON STRATEGIES FOR BETTER BIOCATALYSTS

Pedro Fernandes

Institute for Biotechnology and Bioengineering (IBB), Centre for Biological and Chemical Engineering, Instituto Superior Técnico, Avenue Rovisco Pais, 1049-001 Lisboa, Portugal

ABSTRACT

Food and feed is possibly the area where processing anchored in biological agents has the deepest roots. Despite this, process improvement or design and implementation of novel approaches has been consistently performed, and more so in recent years, where significant advances in enzyme engineering and biocatalyst design have fastened the pace of such developments.

This paper aims to provide an updated and succinct overview on the applications of enzymes in the food sector, and of progresses made, namely, within the scope of tapping for more efficient biocatalysts, through screening, structural modification, and immobilization of enzymes. Targeted improvements aim at enzymes with enhanced thermal and operational stability, improved specific activity, modification of pH-activity profiles, and increased product specificity, among others. This has been mostly achieved through protein engineering and enzyme immobilization, along with improvements in screening. The latter has been considerably improved due to the implementation of high-throughput techniques, and due to developments in protein expression and microbial cell culture. Expanding screening to relatively unexplored environments (marine, temperature extreme environments) has also contributed to the identification and development of more efficient biocatalysts. Technological aspects are considered, but economic aspects are also briefly addressed.

INTRODUCTION

Food processing through the use of biological agents is historically a well-established approach. The earliest applications go back to 6,000 BC or earlier, with the brewing of beer, bread baking, and cheese and wine making, whereas the first purposeful microbial oxidation dates from 2,000 BC, with vinegar production [1–3]. Coming to modern days, in the late XIX, century Christian Hansen reported the use of rennet (a mixture of chymosin and pepsin) for cheese making, and production of bacterial amylases was started at Takamine (latter to become part of Genencor). Pectinases were used for juice clarification in the 1930s, and for a short period during World War II, invertase was also used for the production of invert sugar syrup in a process that pioneered the use of immobilized enzymes in the sugar industry [1]. Still, the large-scale application of enzymes only became really established in the 1960s, when the traditional acid hydrolysis of starch was replaced by an approach based in the use of amylases and amyloglucosidases (glucoamylases), a cocktail that some years latter would include glucose (xylose) isomerase [1, 2, 4, 5]. From then on, the trend for the design and implementation of processes and production of goods anchored in the use of enzymes has steadily increased. Enzymes are currently among the well established products in biotechnology [6], from US \$1.3 billion in 2002 to US \$4 billion in 2007; it is expected to have reached US \$5.1 billion in a rough 2009 year, and is anticipated to reach \$7 billion by 2013 [3, 5, 7–9].

In the overall, this pattern corresponds to a rise in global demand slightly exceeding 6% yearly [7, 9]. Part of this market is ascribed to enzymes used in large-scale applications, among them are those used in food and feed applications [10]. These include enzymes used in baking, beverages and brewing, dairy, dietary supplements, as well as fats and oils, and they have typically been dominating one, only bested by the segment assigned to technical enzymes [11, 12]. The latter includes enzymes in the detergent, personal care, leather, textile and pulp, and paper industries [10, 13]. A recent survey on world sales of enzymes ascribes 31% for food enzymes, 6% for feed enzymes and the remaining for technical enzymes [11]. A relatively large number of companies are involved in enzyme manufacture, but major players are located in Europe, USA and Japan. Denmark is dominating, with Novozymes (45%) and Danisco (17%), moreover after the latter taking over Genencor (USA), with DSM (The Netherlands) and BASF (Germany) lagging behind, with 5% and 4% [10, 11, 14]. The pace of development in emerging markets is suggestive that companies from India and China can join this restricted party in a very near future [15–17].

RELEVANT ENZYMES: TAPPING FOR IMPROVED BIOCATALYSTS

General Aspects and the Screening Approach

Roughly all classes of enzymes have an application within the food and feed area, but hydrolases are possibly the prevalent one. Representative examples of the enzymes and their role in food and feed processing are given in Table 1. The widespread use of enzymes for food and feed processing is easily understandable, given their unsurpassed specificity, ability to operate under mild conditions of pH, temperature and pressure while displaying high activity and turnover numbers, and high biodegradability. Enzymes are furthermore generally considered a natural product [18, 19].

Table 1: An overview of enzymes used in food and feed processing (adapted from [10,12, 13, 68])

Class	Enzyme	Role
Oxidoreductases	Glucose oxidase	Dough strengthening
	Laccases	Clarification of juices, flavor enhancer (beer)
	Lipoxygenase	Dough strengthening, bread whitening
Transferases	Cyclodextrin Glycosyltransferase	Cyclodextrin production
	Fructosyltransferase	Synthesis of fructose oligomers
	Transglutaminase	Modification of viscoelastic properties, dough processing, meat processing
Hydrolases	Amylases	Starch liquefaction and sacharification
		Increasing shelf life and improving quality by retaining moist, elastic and soft nature
		Bread softness and volume, flour adjustment, ensuring uniform yeast fermentation
		Juice treatment, low calorie beer
	Galactosidase	Viscosity reduction in lupins and grain legumes used in animal feed, enhanced digestibility
	Glucanase	Viscosity reduction in barley and oats used in animal feed, enhanced digestibility
	Glucoamylase	Saccharification
	Invertase	Sucrose hydrolysis, production of invert sugar syrup
	Lactase	Lactose hydrolysis, whey hydrolysis
	Lipase	Cheese flavor, in-situ emulsification for dough conditioning, support for lipid digestion in young animals, synthesis of aromatic molecules
	Proteases (namely, chymosin, papain)	Protein hydrolysis, milk clotting, low-allergenic infant-food formulation, enhanced digestibility and utilization, flavor improvement in milk and cheese, meat tenderizer, prevention of chill haze formation in brewing
	Pectinase	Mash treatment, juice clarification
	Peptidase	Hydrolysis of proteins (namely, soy, gluten) for savoury flavors, cheese ripening
	Phospholipase	In-situ emulsification for dough conditioning
	Phytases	Release of phosphate from phytate, enhanced digestibility
	Pullulanase	Saccharification
	Xylanases	Viscosity reduction, enhanced digestibility, dough conditioning
Lyases	Acetolactate decarboxylase	Beer maturation
Isomerases	Xylose (Glucose) isomerase	Glucose isomerization to fructose

The whole contributes for developing sustainable and environmentally friendly processes, since there is a low amount of by-products, hence reducing the need for complex downstream process operations, and the energy requirements are relatively low. Life-cycle assessment (LCA) has confirmed, that within the range of given practical case studies, including food and feed processing, the implementation of enzyme-based technology has a positive impact on the environment [3]. LCA is a methodology used to compare the environmental impact of alternative production technologies while providing the same user benefits [20].

Some of the broad generalizations on the limitations of enzymes for application as biocatalysts in commercial scale, namely, their high cost, low productivity and stability, and narrow range of substrates, have been rebutted [21, 22]. Aiming at improving the performance of biocatalysts for food and feed applications, particular care has been given to increasing thermal stability, enhancing the range of pH with catalytic activity and decreasing metal ions requirements, as well as to overcoming the susceptibility to typical inhibitory molecules. Some examples of strategies taken to improve the performance of relevant enzymes for food and feed are given in Table 2. Along with these different strategies focused on the enzyme molecule (namely, protein engineering, enzyme immobilization), the developments in recombinant DNA technology that occurred in the 1980s also had a huge impact on the application of enzymes in food and feed. By allowing gene cloning in microorganisms compatible with industrial requirements, this methodology enabled cost-feasible production of enzymes that were naturally produced in conditions that prevented large-scale application (namely, enzymes from plant or animal cells, such as transglutaminase or even slow-growing microorganisms). When successfully implemented, the undertaken approaches allow:

- continuous operations at relatively high temperatures;
- eased implementation of enzyme cascade, given the reduced need for processing the reaction media (pH adjustments; metal ion removal/addition) throughout the intermediate steps of a multistep biotransformation (namely, starch to high fructose syrup); and
- the use of raw substrates, preferably as high-concentrated solutions, hence cutting back in costs related to upstream processing and increasing productivity [4, 23, 24].

Methodologies with a high level of parallelization, anchored in computer-monitored microtiter plates equipped with optic fibers and temperature control have also been developed. These provide high-throughput capability for a speedy and detailed characterization of the performance of enzymes [25]. Particular focus was given to the prediction of the long-term stability of enzymes under moderate conditions using short-term runs (up to 3 hours).

Table 2: Some examples of strategies undertaken to improve the performance of enzymes with applications in food and feed

Enzyme	Role	Targeted improvement	Strategy/comments	Reference
α-amylase	Starch liquefaction	Thermostability	Protein engineering through site-directed mutagenesis. Mutant displayed increased half-life from 15 min to about 70 min (100°C).	[70]
	Starch liquefaction	Activity	Directed evolution. After 3 rounds the mutant enzyme from *S. cerevisiae* displayed a 20-fold increase in the specific activity when compared to the wild-type enzyme.	[71]
	Baking	pH-activity profile	Protein engineering through site-directed mutagenesis	[72]
L-arabinose isomerase	Tagatose production	pH-activity profile	Protein engineering through directed evolution	[73]
Glucoamylase	Starch saccharification	Substrate specificity, thermostability and pH optimum	Protein engineering through site-directed mutagenesis	[74]
Lactase	Lactose hydrolysis	Thermostability	Immobilization	[75]
Pullulanase	Starch debranching	Activity	Protein engineering through directed evolution	[76]
Phytase	Animal feed	pH-activity profile	Protein engineering through site-directed mutagenesis	[77]
Xylose (glucose) isomerase	Isomerization/epimerization of hexoses, pentoses and tetroses	pH-activity profile	Protein engineering through directed evolution. The turnover number on D-glucose in some mutants was increased by 30%–40% when compared to the wild type at pH 7.3. Enhanced activities are maintained between pH 6.0 and 7.5.	[78]
		Substrate specificity	Protein engineering through site-directed mutagenesis. The resulting mutant displayed a 3-fold increase in catalytic efficiency with L-arabinose as substrate.	[79]

One of the methodologies to obtain improved biocatalyst relies on in-vitro modifications, which will be addressed latter in this paper; another approach relies on screening efforts, which has been consistently undertaken, as summarized recently [26–31]. Some focus is given to extremophiles, particularly thermophiles, since operation at high temperatures (roughly above 45–50 °C) minimizes the risk of microbial contamination, a particularly delicate matter under continuous operation. Furthermore, the extension of some reactions in relevant food applications is favored at relatively high temperatures (namely, isomerization of glucose to fructose), although care should be taken to avoid an operational environment that may lead by-product formation (namely, Maillard reactions).

Examples of screened enzymes include the isolation of amylases, with some of them being calcium independent [32–38]; amylopullulanases [39]; fructosyltransferases [40]; glucoamylases [41]; glucose (xylose) isomerases [42, 43]; glucosidases [44, 45]; inulinases [46–49]; levansucrases [50]; pullulanases [51, 52]; and xylanases [53, 54]. Other examples of these enzymes, with some of which able to retain stability under temperatures of 90 °C or higher, were reviewed by Gomes and Steiner [55]. The majority of enzymes used in food and feed processing is of terrestrial microbial origin, and screening-efforts for isolation of promising enzyme-producing strains have accordingly been performed in such background [3, 5, 56].

From some years now, marine environment has also been tapped as a source for useful enzymes from either microbial or higher organisms origin [57–60]. This latter environment has allowed the isolation of some promising biocatalysts, such as the heat-stable invertase/inulinase from Thermotoga neapolitana DSM 4359 or inulinase from Cryptococcus aureus [61–63], amylolytic enzymes, glucosidases and proteases from severalgenera[32, 44, 45, 64, 65], esterase from Vibrio fischeri [66],and glycosyl hydrolases [67, 68]. Other examples of useful enzymes for food and feed, but isolated from higher organisms [59, 69], are given in Table 3. Some of these enzymes are actually psychrophiles, hence performing best at low temperatures [30].

Table 3: Examples of enzymes isolated from various marine higher organisms with potential of application in food and feed (adapted from [68, 69])

Class	Enzyme	Source
Transferases	Transglutaminase	Muscles of atka mackerel (*Pleurogrammus azonus*), botan shrimp (*Pandalus nipponensis*), carp (*Cyprinus carpio*), rainbow trout (*Oncorhynchus mykiss*), scallop (*Patinopecten yessoensis*).
Hydrolases	Amylase	Gilt-head (sea) bream (*Sparus aurata*), found in Mediterranean sea and coastal North Atlantic Ocean.
		Turbot (*Scophthalmus maximus*), found mostly in Northeast Atlantic Ocean, Baltic, Black and Mediterranean seas, and Southeast the Pacific Ocean
	Chymotrypsin	Deepwater redfish (Sebastes mentella, found in North Atlantic).
		Atlantic cod (*Gadus morhua*), crayfish, white shrimp.
	Pepsin	Arctic capelin (*Mallotus villosus*), Atlantic cod (*Gadus morhua*).
	Protease	Marine sponges *Spheciospongia vesperia*, found in Caribbean sea and South Atlantic, close to Brazil, and *Geodia cydonium*, found in Northeast Atlantic Ocean and Mediterranean sea.
		Mangrove crab (*Scylla serrata*), found in estuaries and mangroves of Africa, Asia and Australia.
		Sardine Orange roughy (*Hoplostethus atlanticus*)

Operation at low temperatures is also welcome since it also reduces the risk of microbial contamination, enables some processes to be carried out with minimal deterioration of the raw material. These include protein processing, such as cheese maturing and milk coagulation with proteases [59, 80]; milk processing with lactase for lactose-free milk [81–83]; clarification of fruit juices with pectinases to produce clear juice [84]; or production of oligosaccharides [85].

Since extremophiles are often difficult to grow under typical laboratory conditions if not nonculturable at all, different approaches have been developed in order to assess the potential of enzymes from such microorganisms. One approach relies on the generation and screening of target genes from DNA libraries, which can be obtained from mixed microbial population from environmental samples. Recombinant microorganisms can then be obtained using mesophiles as hosts where the genes of interest from extremophiles have been expressed [86]. In order to screen the huge number of DNA-libraries typically generated for the intended property, high-throughput methods have been implemented [87]. These methods are also widely used when protein engineering is carried out.

This will be addressed in the following section.

Several enzymes (namely,α- amylases; pullulanases) currently used in food processing, namely, in starch hydrolysis, are actually produced by recombinant microorganisms. Despite some complexity in the implementation of their use in large-scale applications, partly resulting from lack of uniformity in the US and EU legislation, quite a few enzyme preparations have been accepted for industrial use [88, 89].

IMPROVING BIOCATALYSTS: BEYOND SCREENING

Taking advantage of the knowledge gathered on molecular biology, high-throughput processing, and computer-assisted design of proteins, in-vitro improvement of biocatalysts have been consistently implemented [90–93]. Some of the research efforts in this area has focused on the biochemical and molecular mechanisms underlying the stability of enzymes from extremophiles [31, 94–96]. Such knowledge is also particularly useful for protein engineering of known enzymes, aiming at enhancing stability without compromising catalytic activity [97]. Enhancing the stability of enzymes is of paramount importance when implementation of industrial processes is foreseen, since it allows for reducing the amount of enzyme used in the process. Given that thermostability is determined by a series of short- and long-range interactions, it can be improved by several substitutions of amino acids in a single mutant, where the combination of each individual effect is usually roughly additive [98]. The targeted improvements have not been restricted to thermostability, but they have also addressed other features, such as broadening the range of pH where the enzyme is active, or lessening the temperature of operation while retaining high activity [91, 99].

Two methodologies can be used for protein engineering [97].(i)The first is directed evolution of enzymes, through random mutagenesis and recombination, where the environmental adaptation is reproduced in-vitro in a much hastened timescale, towards the optimization of the intended property. In order to control the pathway of the process, either a screening test for the assessed feature is performed after each round of modification, or selective pressure is applied [100–102]. This methodology, which allows for a high throughput, has been extensively applied, aiming for more efficient biocatalysts [103–106]. Some relevant examples in the area of food and feed processing include the following.

(1)The first is the enhancement of the activity of the hyperthermostable glucose (xylose) isomerase fromThermotoga neapolitana at relatively low temperature and pH, without decay in thermostability [107]. The enzyme from the parent strain is highly active at 97 $^{\circ}$C but it retains only 10% of its activity at 60°C and requires neutral pH for optimal activity. This pattern is often reported when glucose isomerases from hyperthermophilic strains operate in mesophilic environments. Large-scale glucose isomerization is carried out at 55–60°C and slightly alkaline pH [1, 31]. This set of conditions results from the optimal range of pH (typically 7.0 to 9.0) and temperature (60 to 80 °C for glucose isomerization displayed by most of the glucose isomerases used, combined with process boundary conditions. The latter result from by-product and color formation occurring when the reaction is carried out at alkaline pH and high temperatures [31, 108]. There is therefore interest in selecting an enzyme able to operate efficiently at temperatures close to those currently used but at a lower pH. The mutant glucose isomerase 1F1 obtained by Sriprapundh and coworkers displayed a roughly 5-fold higher activity at 60°C and pH 5.5, when compared with the parent T. neapolitana isomerase, and was more thermostable than the wild type isomerase [104, 107]. The activation energy required by the triple 1F1 mutant (V185T/L282P/F186S) was roughly half of the wild-type, hence allowing for high activity at relatively low temperatures [107]. The encouraging results obtained suggest the soundness of the approach to obtain a mutant glucose isomerase competitive with those currently used, while being able to operate in a slightly acidic environment and 60°C The second is the enhancement of the thermostability of the maltogenic amylase from Thermus sp. IM6501 [109], of the amylosucrase from Neisseria polysaccharea [110], of the glucoamylase from Aspergillus niger [111], of a phytase from Escherichia coli [112, 113], and of a xylanase from Bacillus subtilis [114]. Amylases and glucoamylases are enzymes used in starch processing, which involves temperatures typically in excess of 60°C hence, improving thermal stability without decreasing enzyme activity is of relevance. Starch liquefaction is performed at 105°C in the presence of α amylase, upon which the effluent reaction stream has to be cooled to 60°C so that glucoamylases can be used. In order to avoid, or at least minimize, the cooling step, thermostable glucoamylases are aimed at. Wang and coworkers obtained a multiply-mutated enzyme (N20C, A27C, S30P, T62A, S119P, G137A, T290A, H391Y), which displayed a 5.12 kJ mol^{-1} increase in the free energy of thermal inactivation, when compared to the wild type, thus resulting in the enhanced thermal stability of the mutant. Furthermore specific activities and catalytic efficiencies remained unaltered, when mutant and wild type were compared [111]. Kim and coworkers obtained also a multiply-mutated amylase (R26Q, S169N, I333V, M375T, A398V, Q411L, P453L) which displayed an optimal reaction temperature 15°C higher than that of the wild-type and a half-life of roughly 170 min at 80 °C a temperature at which the wild-type ThMA was fully inactivated in less than 1 minute. However, one of the mutations most accountable for enhanced thermal stability, M375T, close to the active site, also led to a 23% decrease in specific activity, as compared to the wild type [109].

The amylosucrase engineered by Emond and coworkers was a double mutant (R20C/A451T), displaying a 10-fold increase in the half-life at 50°C compared to the wild-type enzyme. Actually, the mutant was claimed to be the only amylosucrase usable at 50°C At the latter temperature, the mutant enabled the synthesis of amylose chains twice as long as those obtained by the wild-type enzyme at 30 °C for sucrose concentrations of 600 mM. The mutant thus allowed for a process with increased yield in amylose chains (31 g L^{-1}), lower risk of contamination, enhanced substrate and product solubility and overall productivity [110]. Phytases are added to animal feeds to improve phosphorus nutrition and to reduce phosphorus excretion, by promoting the hydrolysis of phytate into myoinositol and inorganic phosphate. Thermal stable enzymes are needed, since feed pelleting is carried out at high temperature (60 to 80 °C)

Phytases produced by thermophiles do not provide a suitable approach, since they have low activity at the physiological temperature of animals [115]. E. coli phytases, which are appealing to industrial application, due to the acidic pH optimum, specificity phytate, and resistance to pepsin digestion, were thus engineered in order to improve their thermal stability, without compromising the kinetic parameters. As a result, mutants were obtained, with roughly 20% increased thermostability at 80 °C improved overall catalytic efficiency (k $_{cat}$ Turnover number/km

Michaelis constant) within 50 to 150%, as compared to the wild type. No significant changes in the pH activity profile were observed, but for some mutants, containing a K46E substitution, that displayed a decrease in activity at pH 5.0 [112, 113]. Xylanases catalyze the cleavage of β1,4 bonds in xylan polymers. Accordingly, these enzymes can be used in dough making, in baking, in brewing and in animal feed compositions. When the latter contain cereals (namely, barley, maize, rye or wheat), or cereal by-products, xylanases improve the break-down of plant cell walls, which favors the ingestion of plant nutrients by the animals and consequently enhances feed consumption and growth rate. Furthermore, the use of xylanases decreases the viscosity of xylan-containing feeds [116, 117]. As referred for phytases, the formulation of commercial feed often involves steps at high temperatures. Xylanases added to the the formulations hence have to withstand these conditions, while they are to display high activity at about , 40 °C which is the temperature in the intestine of animals. However, most xylanases are inactive a t temperatures exceeding 60°C, hence the need for enhancing thermal stability [114, 117]. Miyazaki and coworkers obtained a triple-mutant xylanase (Q7H, N8F, and S179C) which retained full activity for 2 hours at 60°C, whereas the wild-type enzyme was inactivated within 5 minutes under the same conditions. The mutation also led to a 10°C increase in the optimal temperature for reaction and enhanced activity at higher temperatures, albeit at the cost of decreased activity at lower temperatures, as compared to the wild-type enzyme [114].

(3)Third is the enhancement of the activity of the amylosucrase from Neisseria polysaccharea [118]. Amylosucrases can be used for the modification or synthesis of amylose-type polymers from sucrose, but their industrial application is somehow thwarted by the low catalytic efficiency on sucrose and by side reactions leading to the formation of sucrose isomers. Van der Veen and co-works engineered mutant enzymes through error-prone PCR that displayed increases in activity up to 5-fold and in overall catalytic efficiency up to 2-fold, when compared to the wild-type enzyme. Furthermore, the mutants were able to produce amylose polymers from 10 mM sucrose on, unlike the wild-type enzyme [118]. Their work provides an illustrative example on the use of random mutagenesis and recombination for the enhancement of the catalytic properties of enzymes with application on food and feed. Another example was provided by Tian and coworkers who engineered a phytase fromAspergillus niger 113 through gene shuffling, to obtain mutants with enhanced catalytic properties [119]. Hence, K41E and E121F substitutions allowed for increases in the specific activity of 2.5- and 3.1-fold, and of affinity for sodium phytate, as expressed by decreases in KM of roughly 35% and 25%, as compared to the wild-type enzyme. Furthermore, the overall catalytic efficiency of the mutants increased 1.4- and 1.6-fold as compared to the wild type.

Other examples can be found elsewhere [120, 121].(ii)The second methodology underlines that rational pinpoint modifications in one or more amino acids are made, where these changes are predicted to bring along the envisaged improvement in the targeted enzyme function. The alterations promoted are performed based on the growing knowledge on the structure and functions of enzyme. Information on this matter mostly comes from bioinformatics, which provides data on amino-acid propensities and on protein sequences. Adequate processing of the data enable the output of generalized rules predicting the effect of mutations on enzyme properties. Also used are molecular potential functions, which, once implemented, enable the prediction of the effect of mutations in enzyme structure [97].

Computational tools used for enzyme engineering have been recently reviewed [122]. Enzyme engineering through molecular simulations requires structural data from the native enzyme, which can be preferably obtained from crystallography or NMR. Otherwise a model is built based on known enzyme structures with homologous sequences [90]. Computational methods are also welcome in directed evolution, as a tool to better lead the random mutagenesis [97]. Ultimately this approach is put into practice by producing a site-directed mutant, where selected amino acids are replaced with those suggested from the outcome of modeling.Some relevant examples of this strategy in the area of food and feed processing are given. These mostly aim to improve thermal stability and/or catalytic efficiency and/or to modify the range of pH/temperature where the enzyme is active—goals that were already referred to when examples of enzyme modifications using random mutagenesis were addressed.

(1)   The first example underlines the enhancement of the thermostability of the recombinant glucose (xylose) isomerase from Actinoplanes missouriensis [123, 124] and of glucose (xylose) isomerase fromStreptomyces diastaticus [125]; of amylases from Bacillus spp. [126, 127]; and of glucoamylase fromAspergillus awamori [128]. The mutant isomerase from A. missouriensis displayed an enhanced thermal stability, alongside with improved stability at different pH, as compared with the original enzyme, with no changes in catalytic properties [123, 124]. The double mutant isomerase (G138P, G247D) displayed a 2.5-fold increase in half-life, and additionally a 45% increase in the specific activity, when compared to the wild type. Such features were ascribed to increased molecular rigidity due to the introduction of a proline in the turn of a random coil [125]. Multiply-mutated amylases obtained by Declerck and coworkers displayed considered enhanced thermal stability. Based on the temperature at which amylase initial activity is reduced by 50% for a 10-minute incubation, this parameter went as high as 106°C, as compared to 83.°C for the wild-type strain. Furthermore, the thermal stabilization was not accompanied by a decrease in the catalytic activity [126]. The work by Lin and coworkers on amylase mutants from Bacillus sp. strain TS-23 highlighted the relevance of E219 for the thermal stability of the enzyme [127].

The mutated glucoamylases engineered by Liu and Wang allowed to establish the role of several intermolecular interactions in thermal stability of these enzymes. Thermostable enzymes were obtained through the introduction of disulfide bonds in highly flexible region in the polypeptide chain of the enzyme, as well as by the introduction of more hydrophobic residues-stabilized α-helices. Data gathered also showed that care had to be taken not to disrupt the hydrogen bond and salt linkage network in the catalytic center as a result of mutagenesis, for this could lead to a decrease in the specific activity and overall catalytic efficiency [128].

(2)   The second example underlines the enhancement of the pH-activity profile and of the thermostability of phytase from A. niger. This was achieved by combining several individual mutations that allowed for mutants that were quite active at pH 3.5. Efficient operation in the stomach of simple-stomached animals where phytate hydrolysis mostly occurs at a pH around 3.5, and the wild type was ineffective, was thus enabled. Furthermore, the hydrolytic activity of the mutants at pH 3.5 exceeded in roughly 1.5-fold that of the parent one at pH 5.5, which was the optimum of the latter. Mutants also retained higher residual activity after incubation within 70 to , as compared to the wild type. The work demonstrates that cumulative improvements in pH activity and thermostability through mutation are compatible in this phytase; see [129].

(3)   The third example underlines the modification of the temperature- and pH activity profile of the L-arabinose isomerase from Bacillus stearothermophilus US100 [130].

L-Arabinose isomerases catalyze the conversion of L-arabinose to L-ribulose in-vivo, but in-vitro they also isomerize D-galactose intoD-tagatose [130]. The latter keto-hexose is being used as a low-calorie bulk sweetener, since its taste and sweetness are roughly equivalent to sucrose, but the caloric value is only 30% of that of sucrose [131, 132]. Although several thermostable L-arabinose isomerases have been isolated and characterized, most of these display an alkaline pH optimum. For industrial application this presents the same drawbacks of by-product and color formation referred to when the random mutation of glucose isomerases was addressed. Hence, again arises the need for enzymes able to isomerize L-arabinose in an acidic environment and at relatively low temperature, 60 to 70°C . Operation within the latter temperature range also rules away the use of divalent ions, which stabilize isomerases at high temperatures [133, 134]. Rhimi and coworkers engineered two individual mutants, harboring each N175H and Q268K mutations. These led to broader optimal temperature range within 50 to 65°C and to enhanced stability in acidic media, respectively, when compared to the wild type. An engineered double mutant, harboring both modifications, displayed optimal activity within a pH range of 6.0 to 7.0 and a temperature range within 50 to 65°C Such set of operational conditions matches the targeted goals and again shows that the basis for pH-activity profile and thermostability in L-arabinose isomerase are quite independent and compatible. Cumulative enhancements in both properties in the same enzyme were thus possible [134]. A similar pattern was also observed in the previous example dedicated to a mutant phytase.

(4)   The fourth example underlines the modification of the product profile of inulosucrase fromLactobacillus reuteri [135] and from B. subtilis [136]. Inulosucrases are used to synthesize fructooligosaccharides or fructan polymer from sucrose. The transglycosylation catalyzed by the inulosucrase from L. reuteri leads to a wide range of fructooligosaccharides alongside with minor amounts of an inulin polymer. In order to minimize the dispersion in the products obtained, mutants R423K and W271N were obtained, which allowed the synthesis of a significant amount of polymer and a lower amount of oligosaccharide, without significantly affecting the catalytic activity, when compared with the wild type. The data gathered showed that the -1 subsite in the inulosucrase fromL. reuteri has a key role in the determination of the size of the products obtained [135].

Ortiz-Soto and coworkers also showed that the product profile of transfructosylation reactions could be adequately tuned through modification of target residues of an inulosucrase from B. subtilis. These authors established the effect of mutations on the reaction specificity (hydrolysis/transfructosylation), molecular weight and acceptor specificity. For example, engineered mutants R360S, Y429N and R433A only synthesized oligosaccharides, whereas the wild type synthesized levan, since the former are more hydrolytic. On the other hand these mutations reduced the affinity for sucrose, and thermal stability, when compared to the wild type [136].

(5)   The fifth example underlines the enhancement of the product profile of cyclodextrin glycosyltransferases (CGTase) from differentgenera[137, 138]. These enzymes promote the production of cyclodextrins,α $(1\rightarrow4)$ linked oligosaccharides form starch, through an intramolecular transglycosylation reaction. In the process, a starch oligosaccharide is cleaved and cleaved and the resulting reducing-end sugar is transferred to the non-reducing-end sugar of the same chain [137]. The resulting cyclodextrin may consist of six, seven or eight, which are accordingly termed α, β, or γ-cyclodextrin, respectively. Given their ability to form inclusion complexes with small hydrophobic molecules, they are of interest for both industrial and research applications. Wild-type CGTases typically produce a mixture of the three cyclodextrins when incubated with starch.

The purification of a given cyclodextrin from the reaction mixture requires several additional steps, including selective complexation with organic solvents, which may prove restrictive for cyclodextrin applications involving human consumption [139, 140]. There is therefore a clear interest in obtaining a mutant CGTase capable of producing a particular type of cyclodextrin in a high rate. Van der Veen and coworkers engineered a double-mutant (Y89D/S146P) of CGTase from Bacillus circulans which displayed a 2-fold increase in the production of α-cyclodextrin and a marked decrease in β-cyclodextrin when compared to the wild type. From the data gathered, the authors suggested that hydrogen bonds (S146) and hydrophobic interactions (Y89), are likely to play a key role in to the size of cyclodextrin products formed, and that changes in sugar-binding subsites −3 and −7 may result in mutant CGTases with altered product specificity [137]. Li and coworkers were also able to obtain CGTase mutants from Paenibacillus macerans strain JFB05-01 with increased specificity for α-cyclodextrin, through mutations at subsite −3. In particular, double mutant D372K/Y89R displayed a 1.5-fold increase in the production of α-cyclodextrin, and a significant (roughly 45%) decrease in the production of β-cyclodextrin when compared to the wild-type enzyme [138].

The two methods are not mutually exclusive and methodologies for engineering of enzymes can assemble both strategies [141].

Upon identification of the most adequate enzyme, this can be formulated adequately for better process integration. One of the most widely considered approaches for such formulation is enzyme immobilization.

IMMOBILIZATION

There are several issues that can be lined up to sustain enzyme immobilization. It allows for high-enzyme load with high activity within the bioreactor, hence leading to high-volumetric productivities; it enables the control of the extension of the reaction; downstream process is simplified, since biocatalyst is easily recovered and reused; the product stream is clear from biocatalyst; continuous operation (or batch operation on a drain-and-fill basis) and process automation is possible; and substrate inhibition can be minimized. Along with this, immobilization prevents denaturation by autolysis or organic solvents, and can bring along thermal, operational and storage stabilization, provided that immobilization is adequately designed [142,143].

Immobilization has some intrinsic drawbacks, namely, mass transfer limitations, loss of activity during immobilization procedures, particularly due to chemical interaction or steric blocking of the active site; the possibility of enzyme leakage during operation; risk of support deterioration under operational conditions, due to mechanical or chemical stress; and a (still) relative empirical methodology, which may hamper scale up. Economical issues are furthermore to be taken into consideration when commercial processes are envisaged, although immobilization can prove critical for economic viability if costly enzymes are used. Still, the cost of the support, immobilization procedure and processing the biocatalyst once exhausted, up- and downstream processing of the bioconversion systems, and sanitation requirements have to be taken into consideration. In the overall, the enhanced stability allowing for consecutive reuse leads to high specific productivity ($mass_{product}^{-1}$ $mass_{biocatalyst}^{-1}$), which influences biocatalyst-related production costs [1, 142]. A typical example is the output of immobilized glucose isomerase, allowing for 12,000–15,000 kg of dry-product high-fructose corn syrup (containing 42% fructose) per kilogram of biocatalyst, throughout the operational lifetime of the biocatalyst [144]. Increased thermal stability, allowing for routine reactor operation above 60°C minimizes the risks of microbial growth, hence leading to lower risks of microbial growth and to less demanding sanitation requirements, since cleaning needs of the reactor are less frequent [1, 144]. A rule of thumb suggesting that the enzyme costs should be a few percent of the total production costs has been established [142].

The half-life of the bioreactor is also a critical issue when evaluating the economical feasibility of a bioconversion process, longer half-lives favoring process economics. Examples of commercial bioreactors depict half-lives of several months to years, and the same packing can work throughout some months to years. Among this group, are immobilized enzyme reactors packed with glucose isomerase for the production of high-fructose corn syrup; lactase for lactose hydrolysis, for the production of whey hydrolysates and for the production of tagatose; aminoacylase for the production of amino acids; isomaltulose synthase for the production of isomaltulose; invertase for the production of inverted sugar syrup; lipases for the interesterification of edible oils, ultimately targeted at the production of trans-free fat, of cocoa butter equivalents, and of modified triacylglycerols; and β-fructofuranosidase for the production of fructooligosaccharides [144–146]. On the other hand, despite the technical advantages of immobilization, the large-scale liquefaction of starch to dextrins by α-amylases is performed by free enzymes, given the low cost of the enzyme [18].

Immobilization can be performed by several methods, namely, entrapment/ microencapsulation, binding to a solid carrier, and cross-linking of enzyme aggregates, resulting in carrier-free macromolecules [142]. The latter presents an alternative to carrier-bound enzymes, since these introduce a large portion of noncatalytic material. This can account to about 90% to more than 99% of the total mass of the biocatalysts, resulting in low space-time yields and productivities, and often leads to the loss of more than 50% native activity, which is particularly noticeable at high enzyme loadings [142]. A broad, generalized overview of the advantages and drawbacks of the different immobilization approaches is given in Table 4. A typical example of the patterns suggested by data in Table 4 was observed by Abdel-Naby when evaluating the immobilization of α-amylase through different methods [147]. Details on the different methods, as well as some illustrative examples of their applications, are given hereafter.

Table 4: A generalized characterization of immobilization methods

| Parameter | Immobilization method | | | | |
	Covalent	Carrier binding Ionic	Adsorption	CLEAs, CLECs	Entrapment
Activity	High	High	Low	Intermediate/High	High
Range of application	Low	Intermediate	Intermediate	Low	Intermediate/High
Immobilization efficiency	Low	Intermediate	High	Intermediate	Intermediate
Cost	Low	Low	High	Intermediate	Low
Preparation	Easy	Easy	Difficult	Intermediate	Intermediate/Difficult
Substrate specificity	Cannot be changed	Cannot be changed	Can be changed	Cannot be changed	Can be changed
Regeneration	Possible	Possible	Impossible	Impossible	Impossible

Entrapment/(micro)encapsulation, where the enzyme is contained within a given structure. This can be: a polymer network of an organic polymer or a sol-gel; a membrane device such as a hollow fiber or a microcapsule; or a (reverse) micelle. Apart from the hollow fiber, the whole process of immobilization is performed in-situ. The polymeric network is formed in the presence of the enzyme, leading to supports that are often referred to as beads or capsules. Still, the latter term could preferably be used when the core and the boundary layer(s) are made of different materials, namely, alginate and poly-L-lysine. Although direct contact with an adverse environment is prevented, mass transfer limitations may be relevant, enzyme loading is relatively low, and leakage, particularly of smaller enzymes from hydrogels (namely, alginate, gelatin), may occur.

This may be minimized by previously cross-linking the enzyme with multifunctional agent (namely, glutaraldehyde) [148, 149] or by promoting cross-linkage of the matrix after the entrapment [150]. The use of LentiKats, a polyvinyl-alcohol-based support in lens-shaped form, has been used for several applications in carbohydrate processing. Among these are the synthesis of oligosaccharides with dextransucrase [149], maltodextrin hydrolysis with glucoamylase [151], lactose hydrolysis with lactase [152], and production of invert sugar syrup with invertase [153]. In these processes the biocatalyst could be effectively reused or operated in a continuous manner. Methodologies for large scale production of these supports have been implemented [154, 155]. Flavourzyme, (a fungal protease/peptidase complex) entrapped in calcium alginate [156], k-carragenan, gellan, and higher melting-fat fraction of milk fat [157], was effectively used in cheese ripening, in order to speed up the process, while avoiding the problems associated with the use of free enzyme. These include deficient enzyme distribution, reduced yield and poor-quality cheese, partly ascribed to excessive proteolysis and whey contamination. The enzyme complex is released in a controlled manner due to pressure applied during cheese curd [156].

Calcium alginate beads were also used to immobilize glucose isomerase [158] and α-amylase for starch hydrolysis to whey [159]. In the latter work, the authors observed that increasing the concentration of $CaCl_2$ and of sodium alginate to 4% and 3%, respectively, enzyme leakage was minimized (a common drawback of hydrogels) while allowing for high activity and stability. This effect was also observed in a previous work where alginate-entrapped inulinase was used for sucrose hydrolysis [160]. The stability of an amylase immobilized biocatalyst was further enhanced with the addition of 1% silica gel to the alginate prior to gelation, as reflected by the use of the biocatalyst in 20 cycles of operation, while retaining more than 90% of the initial efficiency [159]. Several enzymes, namely, chymosin, cyprosin, lactase, Neutrase, trypsin, have also been immobilized in liposomes, [161]. In a particularly favored technique immobilization of enzymes in liposomes, known as dehydration-rehydration vesicles (DRVs), small (diameters usually below 50 nm) unilamellar vesicles (SUVs) is prepared in distilled water and mixed with an aqueous solution of the enzyme to be encapsulated.

The resulting vesicle suspension is then dehydrated under freeze drying or equivalent method. Upon rehydration, the resulting DRVs are multilamellar and larger (from 200 nm to a little above 1000 nm) than the original SUVs, and can capture solute molecules [161, 162]. Recent work in this particular application has used lactase as enzyme model and has focused on the optimization and characterization of the liposome-based immobilized system [163, 164]. If liposome-based biocatalysts are used in a process under continuous operation, biocatalyst separation has to be integrated (namely, using an ultra-filtration membrane). In a different concept, based in batch mode, liposome-encapsulated lactase was incorporated in milk. After ingestion, the vesicles are disrupted in the stomach by the presence of bile salts, allowing in-situ degradation of lactose [165]. Cocktails of enzymes, namely, Flavourzyme, bacterial proteases and Palatase M (a commercial lipase preparation), were immobilized in liposomes and successfully used to speed up cheddar cheese ripening [166]. Encapsulation in lipid vesicles has been proved a mild method, providing high protection against proteolysis. There is however some lack of consensus on the feasibility of its application on large scale, as well as on the effectiveness of the methodology for controlled release of enzymes [156, 157,161, 163, 167]. Containment within an ultra-filtration (UF) membrane allows the enzyme to perform in a fully fluid environment; hence, with little loss (if any) of catalytic activity. However, the membrane still presents a boundary for overall mass transfer of substrate/products and enzyme molecules are prone to interact with the membrane material.

This feature is enhanced along with the hydrophobicity of the membrane, hence immobilization in membrane devices may have some adsorptive nature, a feature that will be addressed in (ii). Besides, regular replacement of the membrane may be required. Enzyme containment by a membrane has been used for the continuous production of galactooligosaccharides from lactose. The reaction, with up to 80% lactose conversion out of a substrate concentration of 250 gL^{-1}, was carried out in a perfectly mixed reactor and enzyme was recovered in a 10 kDa nominal molecular weight cutoff. The resulting product presented some similarities to the commercially available Vivinal prebiotic [168]. Within the same methodology, a hollow-fiber module was used to contain lactase, in order to carry out lactose hydrolysis in continuous operation. A conversion rate close to 95% in skim milk was observed for an initial substrate concentration close to 40 gL^{-1} [169].

Binding to a solid carrier, where enzyme-support interaction can be of covalent, ionic, or physical nature. The latter comprehends hydrophobic and van der Waals interactions. These are of weak nature and easily allow for enzyme leakage from the support, namely, after environmental shifts in pH, ionic strength, temperature or even as a result of flow rate or abrasion. On the other hand, desorption can be turned into an advantage if performed under a controlled manner, since it enables the expedite removal of spent enzyme and its replacement with fresh enzyme [170]. A recent paper by Gopinath andSugunanillustrates the increased trend for leakage when adsorption is compared with covalent binding, using α-amylase as model enzyme [171]. Curiously, the first reported application of enzyme immobilization was of invertase onto activated charcoal [172]. Recently invertase was immobilized in different types of sawdust, aiming at its application for sucrose hydrolysis. When wood shavings were used as support, the immobilized invertase retained 90% of the original activity after 20 cycles of 15 minutes, each under consecutive batch operation; and it retained 65% of the original activity after 10 hours of continuous operational regime in a column reactor [173]. Anther example is the immobilization of pectinase in egg shell for the preparation of low-methoxyl pectin. The immobilized biocatalyst could be reused for 32 times at , and it was used in a fluidized-bed reactor, operated at an optimum flow rate of 5 mL h^{-1} and 30^{0}C [174].

Other examples are the surface immobilizations of α-amylase on alumina [175] and in zirconia [176]. Covalent binding is the strongest form of enzyme linking to a solid support. It involves chemically reactive sites of the protein such as amino groups, carboxyl groups, and phenol residues of tyrosine; sulfhydryl groups; or the imidazole group of histidine. The binding can be carried out by several methods; among them are amide bond formation, alkylation and arylation, or UGI reaction. However, this often brings along loss of activity during the process of immobilization, due to support binding to critical residues for enzyme activity, and steric hindrance, among others. Examples include the immobilization of α-amylase [177] and of levansucrase [178] on glutaraldehyde-treated chitosan beads, through the glutaraldehyde reaction between the free amino groups of chitosan and the enzyme molecule; the immobilization of pectinase onto Amberlite IRA900 Cl through glutaraldehyde cross-linking [179]; glucoamylase onto dried oxidized bagasse [180], onto polyglutaraldehyde-activated gelatin [181], or onto macroporous copolymer of ethylene glycol dimethacrylate and glycidyl methacrylate through the carbohydrate moiety of the enzyme [182]; glucoamylase or invertase immobilized onto montmorillonite K-10 activated with aminopropyltriethoxysilane and glutaraldehyde [183, 184]; and invertase immobilized on nylon-6 microbeads, previously activated with glutaraldehyde and using PEI as spacer [185, 186]; on polyurethane treated with hydrochloric acid, polyethylenimine and glutaraldehyde [187]; on poly(styrene-2-hydroxyethyl methacrylate) microbeads activated with epichlorohydrin [188]; or on poly(hydroxyethyl methacrylate)/glycidyl methacrylate films [189]. Within this methodology for immobilization, highlight should be given to the introduction of commercial supports (namely, Eupergit, Sepabeads) with a high density of epoxide functional groups aimed at multipoint attachment, typically with the -amino group of lysine, to confer high rigidity to the enzyme molecule, hence enhancing stabilization [190, 191]. This methodology has been used for lactase immobilization in magnetic poly(GMA-MMA), formed from monomers of glycidylmethacrylate and ethylmethacrylate, and cross-linked with ethyleneglycol dimethacrylate [192]; for the immobilization of cyclodextrin glycosyltransferases to glyoxylagarose supports for the production of cyclodextrins [193]; or for the immobilization of dextransucrase on Eupergit C [194]. Ionic binding to a carrier involves interaction of negatively or positively charged groups of the carrier with charged amino-acid residues on the enzyme molecules [195].

Ionic interaction may be favored if enzyme leakage is not an issue, since it allows for support regeneration, unlike immobilization by covalent binding. Ion-exchanger resins are typical supports for ionic binding; among them are derivatives of cross-linked polysaccharides, namely, carboxymethyl-

(CM-) cellulose, CM-Sepharose, diethylaminoethyl- (DEAE-) cellulose, DEAE-Sephadex, quaternary aminoethyl anion exchange- (QAE-) cellulose, QAE-dextran, QAE-Sephadex; derivatives of synthetic polymers, namely, Amberlite, Diaion, Dowex, Duolite; and resins coated with ionic polymers, namely, polyethylenimine (PEI) [196]. Recent examples include the immobilization of invertase in Dowex [197], in Duolite [198], in poly(glycidyl methacrylate-co-methyl methacrylate beads grafted with PEI [199], and in epoxy(amino) Sepabeads [200]; lactase immobilization in PEI-grafted Sepabeads [201]; fructosyltransferase in DEAE-cellulose for the production of fructosyl disaccharides [202]; glucose isomerase in DEAE-cellulose [203] or in Indion 48-R [204]; glucoamylase onto SBA-15 silica [205] and in epoxy(amino) Sepabeads [200]. Ionic binding to Sepabeads-like supports has acknowledged multipoint attachment nature. Enzyme molecules can be modified chemically or genetically modified to enhance immobilization efficiency, an approach followed by Kweon and coworkers, who obtained a cyclodextrin glycosyltransferase fused with 10 lysine residues to improve ionic binding to SP-Sepharose [206].

Carrier-free macroparticles, where a bifunctional reagent (namely, glutaraldehyde), is used to cross-link enzyme aggregates (CLEAs) or crystals (CLECs), leading to a biocatalyst displaying highly concentrated enzyme activity, high stability and low production costs [142, 207]. The use of CLEAs is favored given the lower complexity of the process. This approach is recent, as compared with entrapment and binding to a solid carrier, and there are still relatively few examples of its application to enzymes used in the area of food processing. Among those are following.

- First is the immobilization of Pectinex Ultra SP-L, a commercial enzyme preparation containing pectinase, xylanase, and cellulose activities [208]. The CLEA biocatalyst displayed a slight (30%) in the Vmax, maximal reaction rate/KM ratio, but a significant enhancement in thermal stability (a roughly 10-fold increase in half-life), when the pectinase activity of the immobilized biocatalyst was compared with the free form.

- Second is the immobilization of lactase for the hydrolysis of lactose, where, under similar operational conditions as for the free enzyme, the CLEA yielded 78% monosaccharides in 12 h as compared to 3.9% of the free form [209].

- Dimethylmalonimidate, dimethylsuccinimidate, and dimethyl glutarimidate, led to biocatalysts with improved thermal stability as compared to the free form (over 2-fold increase in half-lives) [210].

- Fourth, CLEAs of wild type and two mutant levansucrases were assayed for oligosaccharides/levan and for fructosyl-xyloside synthesis. Although the specific activity of the three free enzymes was 1.25- to 3-fold higher than the corresponding CLEAs, these displayed a 40- to 200-fold higher specific activity than the equivalent Eupergit-C-immobilized enzyme preparations. Furthermore, all CLEA preparations displayed enhanced thermal stability when compared with the corresponding free enzymes [211].

- Fifth are CLECs of glucose isomerase, aimed at the conversion of glucose into fructose for the production of high fructose corn syrup. When placed in a packed-bed, the resulting enzyme preparation allowed for flow rates that matched or even exceeded those processed by commercially available enzyme preparations (either free, carrier free, or carrier-bound), while achieving the same 45% yield in fructose, under similar operational condition [212].

- Sixth, CLECs of glucose isomerase packed in a column were also used for the concentration/purification of xylitol from dilute or impure solutions. The approach was based on the high specificity of the enzyme crystals towards xylitol, allowing its separation from other sugars, including the natural substrates, xylose and glucose. Recovery of the adsorbed xylitol was achieved by elution with $CaCl_2$ solutions, with Ca^{2+} being acknowledged to inactivate glucose isomerase [213].

Each method for enzyme immobilization has a unique nature. Therefore, despite the potential of immobilization to improve enzyme performance by enhancing activity, stability, or specificity, no specific approach tackles simultaneously these different features. A careful evaluation and characterization of the methodology addressed is thus required, which can be significantly fastened by high-throughput approaches [214]. Again, the feasibility of its application to reactor configuration and mode of operation has also to be considered in the selection process of the most adequate immobilized biocatalyst for a given bioconversion.

Typical Bioreactors

The most common form of enzymatic reactors for continuous operation is the packed-bed setup, basically a cylindrical column holding a fixed bed of catalyst particles (Figure 1). These should not have sizes below 0.05 mm, in order to keep the pressure drop within reasonable limits. Commercially available carriers such as Eupergit C have particle sizes of roughly 0.1 mm [215]. Commonly operated in down-flow mode, the range of flow rates used must be such as to provide a compromise between reasonable pressure drop, minimal diffusion layer and high conversion yield. Minimization of external mass-transfer resistances with enhanced flow rates can be considered, leading to the fluidized-bed reactor. This is basically a variation of the packed-bed reactor, but operated in up-flow mode, where the biocatalyst particles are not in close contact which each other; hence, pressure drop is low, and accordingly are pumping costs. The residence time allowed by the flow rates required for fluidization may however result in low conversion yields. This can be overcome by operating a battery of reactor or by operation in recycle mode [216]. Bioconversions with free enzymes are carried out in stirred tanks. When on their own, they are restricted to batch mode, but when coupled to a membrane setup with suitable cutoff, they can be integrated in a continuous process, since the enzymes are rejected by the membrane, which acts as an immobilization device, whereas the product (and unconverted substrate) freely permeates. Shear stress induced by stirring creates a hazardous environment for immobilized biocatalysts, particularly when hydrogels are considered, since they are prone to abrasion. In order to overcome this, a basket reactor was developed, but is seldom used, possibly due to mass transfer resistances associated [18].

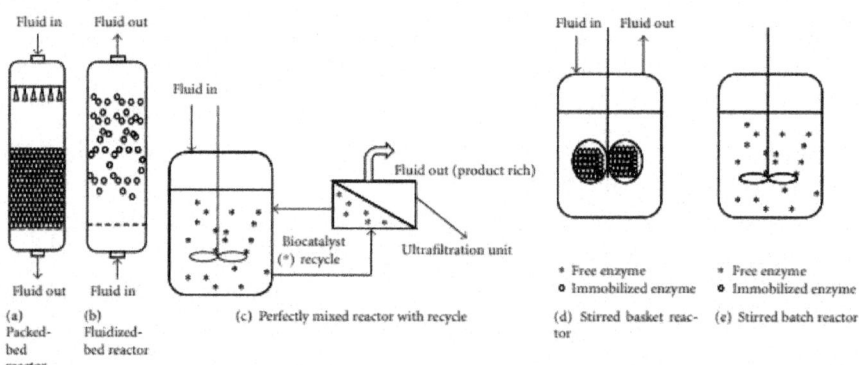

Figure 1: Examples of bioreactor configurations commonly used in bioconversion processed involving free or immobilized enzymes. Reactors (a) to (d) are depicted under continuous mode of operation, whereas reactor (e) is depicted.

CONCLUSIONS AND FUTURE PERSPECTIVES

The integration of enzymes in food and feed processes is a well-established approach, but evidence clearly shows that dedicated research efforts are consistently being made as to make this application of biological agents more effective and/or diversified. These endeavors have been anchoring in innovative approaches for the design of new/improved biocatalysts, more stable (to temperature and pH), less dependent on metal ions and less susceptible to inhibitory agents and to aggressive environmental conditions, while maintaining the targeted activity or evolving novel activities. This is of particular relevance for application in the food and feed sector, for it allows enhanced performance under operational conditions that minimize the risk of microbial contamination. It also favors process integration, by allowing the concerted use of enzymes that naturally have diverse requirements for effective application. Such progresses have been made through the ever-continuing developments in molecular biology, the accumulated evolutionary enzyme engineering expertise, the (bio)computational tools, and the implementation of high-throughput methodologies, with high level of parallelization, enabling the efficient and timely screening/characterization of the biocatalysts. Alongside with these strategies, the immobilization of enzymes has also been a key supporting tool for rendering these proteins fit for industrial application, while simultaneously enabling the improvement of their catalytic features. Again, and despite the developments made in this particular field, there is still the lack of a set of unanimously applicable rules for the selection of carrier and method of enzyme immobilization, which furthermore encompass both technical and economic requirements. The latter can be particularly restrictive in the food and feed sector, since most products are of relatively low added value. Therefore, there is no universal support and method for enzyme immobilization aimed at application in food and feed (let alone the overall range of possible fields of use), and the immobilized biocatalyst fit for a given process and product may be totally unsuitable for another. Given the diversity of enzyme nature and applications this pattern is unlikely to be reversed. Hence, it can be foreseen that efforts will be towards the development of immobilized biocatalyst with suitable chemical, physical, and geometric characteristics, which can be produced under mild condition, that can be used in different reactor configurations and that comply with the economic requirements for large-scale application. All these strategies either isolated or preferably suitably integrated have been put into practice in food and feed, to improve existing processes or to implement new ones, with the latter often combined with the output of new goods, resulting from novel enzymatic activities. Given the recent developments in this field, this trend is foreseen to be further implemented.

ACKNOWLEDGMENT

Pedro Fernandes acknowledges Fundação para a Ciência e a Tecnologia (Portugal) for financial support under program Ciência 2007.

REFERENCES

1. D. Vasic-Racki, "History of industrial biotransformations—dreams and realities," in Industrial Biotransformations, A. Liese, K. Seelbach, and C. Wandrey, Eds., pp. 1–35, Wiley-VCH, Weinheim, Germany, 2nd edition, 2006.

2. P. B. Poulsen and H. Klaus Buchholz, "History of enzymology with emphasis on food production," inHandbook of Food Enzymology, J. R. Whitaker, A. G. J. Voragen, and D. W. S. Wong, Eds., pp. 11–20, Marcel Dekker, New York, NY, USA, 2003.

3. T. Schäfer, O. Kirk, T. V. Borchert, et al., "Enzymes for technical applications," in Biopolymers, S. R. Fahnestock and S. R. Steinbüchel, Eds., pp. 377–437, Wiley-VCH, Weinheim, Germany, 2002.

4. P. Fernandes, "Enzymes in sugar industries," in Enzymes in Food Processing: Fundamentals and Potential Applications, P. Panesar, S. S. Marwaha, and H. K. Chopra, Eds., pp. 165–197, I.K. International Publishing House, New Delhi, India, 2010.

5. M. Leisola, J. Jokela, O. Pastinen, O. Turunen, and H. Schoemaker, "Industrial use of enzymes," inEncyclopedia of Life Support Systems (EOLSS), O. O. P. Hänninen and M. Atalay, Eds., pp. 1–25, EOLSS, Oxford, UK, 2002.

6. J. Norus, "Building sustainable competitive advantage from knowledge in the region: the industrial enzymes industry," European Planning Studies, vol. 14, no. 5, pp. 681–696, 2006.·

7. E. P. S. Bon and M. A. Ferrara, Bioethanol production via enzymatic hydrolysis of cellulosic biomass, Document prepared for "The Role of Agricultural Biotechnologies for Production of Bioenergy in Developing Countries" an FAO seminar held in Rome on 12 October 2007,http://www.fao.org/biotech/seminaroct2007.htm.

8. H. El Enshasy, A. Abuoul-Enein, S. Helmy, and Y. El Azaly, "Optimization of the industrial production of alkaline protease by Bacillus licheniformis in different production scales," Australian Journal of Applied Science, vol. 2, pp. 583–593, 2008.

9. Freedonia Group Inc. World Enzymes—Industry Study with Forecasts for 2013 & 2018: Study #2506, August 2009, http://www.freedoniagroup.com/brochure/25xx/2506smwe.pdf.

10. P. Binod, R. R. Singhania, C. R. Soccol, and A. Pandey, "Industrial enzymes," in Advances in Fermentation Technology, A. Pandey, C. Larroche, C. R. Soccol, and C.-G. Dussap, Eds., pp. 291–320, Asiatech Publishers, New Delhi, India, 2008.

11. R. M. Berka and J. R. Cherry, "Enzyme biotechnology," in Basic Biotechnology, C. Ratledge and B. Kristiansen, Eds., pp. 477–498, Cambridge University Press, Cambridge, UK, 3rd edition, 2006.

12. O. Kirk, T. V. Borchert, and C. C. Fuglsang, "Industrial enzyme applications," Current Opinion in Biotechnology, vol. 13, no. 4, pp. 345–351, 2002.

13. T. Schäfer, T. W. Borchert, V. S. Nielsen et al., "Industrial enzymes," Advances in Biochemical Engineering/Biotechnology, vol. 105, pp. 59–131, 2006.

14. J. Ogawa and S. Shimizu, "Industrial microbial enzymes: their discovery by screening and use in large-scale production of useful chemicals in Japan," Current Opinion in Biotechnology, vol. 13, no. 4, pp. 367–375, 2002.

15. K. Chandel, R. Rudravaram, L. V. Rao, P. Ravindra, and M. L. Narasu, "Industrial enzymes in bioindustrial sector development: an Indian perspective," Journal of Commercial Biotechnology, vol. 13, no. 4, pp. 283–291, 2007.

16. D. Carrez and W. Soetaert, "Looking ahead in Europe: white biotech by 2025," Industrial Biotechnology, vol. 1, pp. 95–101, 2005.

17. Research and markets (2010). Future of Enzymes in China to 2020,http://www.researchandmarkets.com/reportinfo.asp?cat_id=0&report_id=1202421&q=future%20of%20enzymes&p=1.

18. Illanes, Enzyme Biocatalysis—Principles and Applications, Springer, New York, NY, USA, 2008.

19. S. Bommarius and B. R. Riebel, Biocatalysis: Fundamentals and Applications, Wiley-VCH, Weinheim, Germany, 2004.

20. K. Oxenbøll and S. Ernst, "Environment as a new perspective on the use of enzymes in the food industry," Food Science and Technology, vol. 22, no. 1, pp. 35–37, 2008.

21. J. D. Rozzell, "Commercial scale biocatalysis: myths and realities," Bioorganic and Medicinal Chemistry, vol. 7, no. 10, pp. 2253–2261, 1999.

22. H. E. Schoemaker, D. Mink, and M. G. WubboLts, "Dispelling the myths—biocatalysis in industrial synthesis," Science, vol. 299, no. 5613, pp. 1694–1697, 2003.

23. R. H. Sajedi, H. Naderi-Manesh, K. Khajeh et al., "A Ca-independent α-amylase that is active and stable at low pH from the Bacillus sp. KR-8104," Enzyme and Microbial Technology, vol. 36, no. 5-6, pp. 666–671, 2005.

24. X. D. Liu and Y. Xu, "A novel raw starch digesting α-amylase from a newly isolated Bacillus sp. YX-1: purification and characterization," Bioresource Technology, vol. 99, no. 10, pp. 4315–4320, 2008.

25. K. Rachinskiy, H. Schultze, M. Boy, U. Bornscheuer, and J. Büchs, ""Enzyme Test Bench," a high-throughput enzyme characterization technique including the long-term stability," Biotechnology and Bioengineering, vol. 103, no. 2, pp. 305–322, 2009.

26. M. M. C. Andrade, N. Pereira Jr., and G. Antranikian, "Extremely thermophilic microorganisms and their polymerhydrolytic enzymes," Revista de Microbiologia, vol. 30, no. 4, pp. 287–298, 1999. ·

27. H. Sun, P. Zhao, X. Ge et al., "Recent advances in microbial raw starch degrading enzymes," Applied Biochemistry and Biotechnology, vol. 160, no. 4, pp. 988–1003, 2009.

28. Bertoldo and G. Antranikian, "Starch-hydrolyzing enzymes from thermophilic archaea and bacteria," Current Opinion in Chemical Biology, vol. 6, no. 2, pp. 151–160, 2002.·

29. J. Synowiecki, B. Grzybowska, and A. Zdziebło, "Sources, properties and suitability of new thermostable enzymes in food processing," Critical Reviews in Food Science and Nutrition, vol. 46, no. 3, pp. 197–205, 2006.

30. M .M. Kristjánsson and A. Ásgeirsson, "Properties of extremophilic enzymes and their importance in food science and technology," in Handbook of Food Enzymology, J. R. Whitaker, A. G. J. Voragen, and D. W. S. Wong, Eds., pp. 77–100, Marcel Dekker, New York, NY, USA, 2003.

31. Vieille and G. J. Zeikus, "Hyperthermophilic enzymes: sources, uses, and molecular mechanisms for thermostability," Microbiology and Molecular Biology Reviews, vol. 65, no. 1, pp. 1–43, 2001.

32. S. H. Brown, H. R. Costantino, and R. M. Kelly, "Characterization of amylolytic enzyme activities associated with the hyperthermophilic archaebacterium Pyrococcus furiosus," Applied and Environmental Microbiology, vol. 56, no. 7, pp. 1985–1991, 1990.

33. N. Goyal, J. K. Gupta, and S. K. Soni, "A novel raw starch digesting thermostable α-amylase fromBacillus sp. I-3 and its use in the direct hydrolysis of raw potato starch," Enzyme and Microbial Technology, vol. 37, no. 7, pp. 723–734, 2005.

34. B. Arikan, "Highly thermostable, thermophilic, alkaline, SDS and chelator resistant amylase from a thermophilic Bacillus sp. isolate A3-15," Bioresource Technology, vol. 99, no. 8, pp. 3071–3076, 2008.

35. G. D. Haki, A. J. Anceno, and S. K. Rakshit, "Atypical Ca2+-independent, raw-starch hydrolysing α-amylase from Bacillus sp. GRE1: characterization and gene isolation," World Journal of Microbiology and Biotechnology, vol. 24, no. 11, pp. 2517–2524, 2008.

36. M. Ballschmiter, O. Fütterer, and W. Liebl, "Identification and characterization of a novel intracellular alkaline α-amylase from the hyperthermophilic bacterium Thermotoga maritima MSB8," Applied and Environmental Microbiology, vol. 72, no. 3, pp. 2206–2211, 2006.

37. P. Dheeran, S. Kumar, Y. K. Jaiswal, and D. K. Adhikari, "Characterization of hyperthermostable α-amylase from Geobacillus sp. IIPTN," Applied Microbiology and Biotechnology, vol. 86, pp. 1857–1866, 2010.

38. J. L. Uma Maheswar Rao and T. Satyanarayana, "Biophysical and biochemical characterization of a hyperthermostable and Ca2+-independent α-amylase of an extreme thermophile Geobacillusthermoleovorans," Applied Biochemistry and Biotechnology, vol. 150, no. 2, pp. 205–219, 2008.

39. S. M. Noorwez, M. Ezhilvannan, and T. Satyanarayana, "Production of a high maltose-forming, hyperthermostable and Ca2+-independent amylopullulanase by an extreme thermophile Geobacillusthermoleovorans in submerged fermentation," Indian Journal of Biotechnology, vol. 5, no. 3, pp. 337–345, 2006.

40. S. Hernalsteens and F. Maugeri, "Properties of thermostable extracellular FOS-producing fructofuranosidase from cryptococcus sp," European Food Research and Technology, vol. 228, no. 2, pp. 213–221, 2008.

41. M.-S. Kim, J.-T. Park, Y.-W. Kim et al., "Properties of a novel thermostable glucoamylase from the hyperthermophilic archaeon sulfolobus solfataricus in relation to starch processing," Applied and Environmental Microbiology, vol. 70, no. 7, pp. 3933–3940, 2004.

42. K. Srih-Belghith and S. Bejar, "A thermostable glucose isomerase having a relatively low optimum pH: study of activity and molecular cloning of the corresponding gene," Biotechnology Letters, vol. 20, no. 6, pp. 553–556, 1998.

43. R. K. Bandlish, J. M. Hess, K. L. Epting, C. Vieille, and R. M. Kelly, "Glucose-to-fructose conversion at high temperatures with xylose (glucose) isomerases from Streptomyces murinus and two hyperthermophilic Thermotoga species," Biotechnology and Bioengineering, vol. 80, no. 2, pp. 185–194, 2002.

44. H. R. Costantino, S. H. Brown, and R. M. Kelly, "Purification and characterization of an α-glucosidase from a hyperthermophilic

archaebacterium, Pyrococcus furiosus, exhibiting a temperature optimum of 105 to 115°C," Journal of Bacteriology, vol. 172, no. 7, pp. 3654–3660, 1990. ·

45. S. W. M. Kengen, E. J. Luesink, A. J. M. Stams, and A. J. B. Zehnder, "Purification and characterization of an extremely thermostable β-glucosidase from the hyperthermophilic archaeon Pyrococcus furiosus," European Journal of Biochemistry, vol. 213, no. 1, pp. 305–312, 1993. ·

46. K. Kato, T. Araki, T. Kitamura, N. Morita, M. Moori, and Y. Suzuki, "Purification and properties of a thermostable inulinase (β-D-fructan fructohydrolase) from bacillus stearothermophilus KP1289,"Starch/Stärke, vol. 51, no. 7, pp. 253–258, 1999.

47. P. K. Gill, R. K. Manhas, and P. Singh, "Comparative analysis of thermostability of extracellular inulinase activity from Aspergillus fumigatus with commercially available (Novozyme) inulinase,"Bioresource Technology, vol. 97, no. 2, pp. 355–358, 2006.

48. D. Sharma and P. K. Gill, "Purification and characterization of heat-stable exo-inulinase from Streptomyces sp," Journal of Food Engineering, vol. 79, no. 4, pp. 1172–1178, 2007. ·

49. M. Lima, R. Q. Oliveira, A. P.T. Uetanabaro, A. Góes-Neto, C. A. Rosa, and S. A. Assis, "Thermostable inulinases secreted by yeast and yeast-like strains from the Brazilian semi-arid region,"International Journal of Food Sciences and Nutrition, vol. 60, supplement 7, pp. 63–71, 2009.

50. Y. B. Ammar, T. Matsubara, K. Ito et al., "Characterization of a thermostable levansucrase fromBacillus sp. TH4-2 capable of producing high molecular weight levan at high temperature," Journal of Biotechnology, vol. 99, no. 2, pp. 111–119, 2002.

51. S. H. Brown and R. M. Kelly, "Characterization of amylolytic enzymes, having both α-1,4 and α-1,6 hydrolytic activity, from the thermophilic archaea Pyrococcus furiosus and Thermococcus litoralis,"Applied and Environmental Microbiology, vol. 59, no. 8, pp. 2614–2621, 1993. ·

52. Kunamneni and S. Singh, "Improved high thermal stability of pullulanase from a newly isolated thermophilic Bacillus sp. AN-7," Enzyme and Microbial Technology, vol. 39, no. 7, pp. 1399–1404, 2006.

53. Winterhalter and W. Liebl, "Two extremely thermostable xylanases of the hyperthermophilic bacterium Thermotoga maritima MSB8," Applied and Environmental Microbiology, vol. 61, no. 5, pp. 1810–1815, 1995.

54. R. Khandeparkar and N. B. Bhosle, "Purification and characterization of thermoalkalophilic xylanase isolated from the Enterobacter sp. MTCC

5112," Research in Microbiology, vol. 157, no. 4, pp. 315–325, 2006.

55. J. Gomes and W. Steiner, "The biocatalytic potential of extremophiles and extremozymes," Food Technology and Biotechnology, vol. 42, no. 4, pp. 223–235, 2004.

56. S. Linko, "Novel approaches in microbial enzyme production," Food Biotechnology, vol. 3, pp. 31–43, 1989.

57. N. F. Haard, "A review of proteotlytic enzymes from marine organisms and their application in the food industry," Journal of Aquatic Food Product Technology, vol. 1, pp. 17–35, 1992.

58. M. Chandrasekaran, "Industrial enzymes from marine microorganisms: the Indian scenario," Journal of Marine Biotechnology, vol. 5, no. 2-3, pp. 86–89, 1997.

59. Shahidi and Y. V. A. Janak Kamil, "Enzymes from fish and aquatic invertebrates and their application in the food industry," Trends in Food Science and Technology, vol. 12, no. 12, pp. 435–464, 2001.

60. R. S. Rasmussen and M. T. Morrissey, "Marine biotechnology for production of food ingredients,"Advances in Food and Nutrition Research, vol. 52, pp. 237–292, 2007.

61. L. Dipasquale, A. Gambacorta, R. A. Siciliano, M. F. Mazzeo, and L. Lama, "Purification and biochemical characterization of a native invertase from the hydrogen-producing Thermotoga neapolitana (DSM 4359)," Extremophiles, vol. 13, no. 2, pp. 345–354, 2009.

62. W. Liebl, D. Brem, and A. Gotschlich, "Analysis of the gene for β-fructosidase (invertase, inulinase) of the hyperthermophilic bacterium Thermotoga maritima, and characterisation of the enzyme expressed in Escherichia coli," Applied Microbiology and Biotechnology, vol. 50, no. 1, pp. 55–64, 1998.

63. J. Sheng, Z. Chi, J. Li, L. Gao, and F. Gong, "Inulinase production by the marine yeast Cryptococcus aureus G7a and inulin hydrolysis by the crude inulinase," Process Biochemistry, vol. 42, no. 5, pp. 805–811, 2007.

64. B. R. Mohapatra, M. Bapuji, and A. Sree, "Production of industrial enzymes (Amylase, Carboxymethylcellulase and Protease) by bacteria isolated from marine sedentary organisms," Acta Biotechnologica, vol. 23, no. 1, pp. 75–84, 2003.

65. E. Legin, C. Ladrat, A. Godfroy, G. Barbier, and F. Duchiron, "Thermostable amylolytic enzymes of thermophilic microorganisms from deep sea hydrothermal vents," Comptes Rendus de l'Academie des Sciences - Serie III, vol. 320, no. 11, pp. 893–898, 1997.

66. P. Ranjitha, E. S. Karthy, and A. Mohankumar, "Purification and partial characterization of esterase from marine vibrio fischeri," Modern Applied Science, vol. 3, pp. 73–82, 2009.

67. Giordano, G. Andreotti, A. Tramice, and A. Trincone, "Marine glycosyl hydrolases in the hydrolysis and synthesis of oligosaccharides," Biotechnology journal, vol. 1, no. 5, pp. 511–530, 2006.

68. Tramice, E. Pagnotta, I. Romano, A. Gambacorta, and A. Trincone, "Transglycosylation reactions using glycosyl hydrolases from Thermotoga neapolitana, a marine hydrogen-producing bacterium,"Journal of Molecular Catalysis B, vol. 47, no. 1-2, pp. 21–27, 2007.

69. Gudmundsdóttir and J. B. Bjarnason, "Applications of cold adapted proteases in the food industry," in Novel Enzyme Technology for Food Application, R. Rastall, Ed., pp. 205–221, Woodhead Publishing Limited, Cambridge, UK, 2008.

70. M. B. Ali, B. Khemakhem, X. Robert, R. Haser, and S. Bejar, "Thermostability enhancement and change in starch hydrolysis profile of the maltohexaose-forming amylase of Bacillusstearothermophilus US100 strain," Biochemical Journal, vol. 394, no. 1, pp. 51–56, 2006.

71. D. W. S. Wong, S. B. Batt, C. C. Lee, and G. H. Robertson, "High-activity barley α-amylase by directed evolution," Protein Journal, vol. 23, no. 7, pp. 453–460, 2004.

72. S. Danielsen and H. Lundqvist, Bacterial alpha-amylase variants. WO Patent 2008/000825, 2008.

73. D. K. Oh, H. J. Oh, H. J. Kim, J. Cheon, and P. Kim, "Modification of optimal pH in l-arabinose isomerase from Geobacillus stearothermophilus for d-galactose isomerization," Journal of Molecular Catalysis B, vol. 43, no. 1–4, pp. 108–112, 2006.

74. M. J. Allen, T,-Y. Fang, Y. Li, et al., "Protein engineering of glucoamylase to increase pH optimum, substrate specificity and thermostability," United States Patents No. 6,537,792, 2003.

75. Dwevedi and A. M. Kayastha, "Stabilization of β-galactosidase (from peas) by immobilization onto Amberlite MB-150 beads and its application in lactose hydrolysis," Journal of Agricultural and Food Chemistry, vol. 57, no. 2, pp. 682–688, 2009.

76. England, M. Kolkman, B. S. Miller, and C. Vroeman, Pullulanase variants with increased productivity. Patent WO 2008024372A2, 2008.

77. Tomschy, R. Brugger, M. Lehmann et al., "Engineering of phytase for

improved activity at low pH,"Applied and Environmental Microbiology, vol. 68, no. 4, pp. 1907–1913, 2002.

78. J. Cha and C. A. Batt, "Lowering the pH optimum of D-xylose isomerase: the effect of mutations of the negatively charged residues," Molecules and Cells, vol. 8, no. 4, pp. 374–382, 1998.

79. J. Karimäki, T. Parkkinen, H. Santa et al., "Engineering the substrate specificity of xylose isomerase,"Protein Engineering, Design and Selection, vol. 17, no. 12, pp. 861–869, 2004.

80. Q.-F. Wang, Y.-H. Hou, Z. Xu, J.-L. Miao, and G.-Y. Li, "Purification and properties of an extracellular cold-active protease from the psychrophilic bacterium Pseudoalteromonas sp. NJ276,"Biochemical Engineering Journal, vol. 38, no. 3, pp. 362–368, 2008.

81. T.Nakagawa,R.Ikehata,M.Uchino,T.Miyaji,K.Takano,andN.Tomizuka, "Cold-active acid β-galactosidase activity of isolated psychrophilic-basidiomycetous yeast Guehomyces pullulans,"Microbiological Research, vol. 161, no. 1, pp. 75–79, 2006.

82. P. Hildebrandt, M. Wanarska, and J. Kur, "A new cold-adapted β-D-galactosidase from the Antarctic Arthrobacter sp. 32c—gene cloning, overexpression, purification and properties," BMC Microbiology, vol. 9, article 151, 2009.

83. Hoyoux, J.-M. François, and P. Dubois, Cold-active beta-galactosidase, the process for its preparation and the use thereof. Patent US 6727084, 2004.

84. T. Nakagawa, T. Nagaoka, S. Taniguchi, T. Miyaji, and N. Tomizuka, "Isolation and characterization of psychrophilic yeasts producing cold-adapted pectinolytic enzymes," Letters in Applied Microbiology, vol. 38, no. 5, pp. 383–387, 2004.

85. R. D. Mahmoud and A. W. Helmy, "A novel cold-active and alkali-stable β-glucosidase gene isolated from the marine bacterium martelella mediterrânea," Australian Journal of Basic and Applied Sciences, vol. 3, pp. 3808–3817, 2009.

86. S. Fujiwara, "Extremophiles: developments of their special functions and potential resources," Journal of Bioscience and Bioengineering, vol. 94, no. 6, pp. 518–525, 2002.

87. T. H. Richardson, X. Tan, G. Frey et al., "A novel, high performance enzyme for starch liquefaction. Discovery and optimization of a low pH, thermostable α-amylase," The Journal of Biological Chemistry, vol. 277, no. 29, pp. 26501–26507, 2002.

88. Z. S. Olempska-Beer, R. I. Merker, M. D. Ditto, and M. J. DiNovi, "Food-processing enzymes from recombinant microorganisms-a review," Regulatory Toxicology and Pharmacology, vol. 45, no. 2, pp. 144–158, 2006.

89. Spök, "Safety regulations of food enzymes," Food Technology and Biotechnology, vol. 44, no. 2, pp. 197–209, 2006.

90. J. Vieceli, J. Müllegger, and A. Tehrani, "Computer-assisted design of industrial enzymes: the resurgence of rational design and in silico mutagenesis," Industrial Biotechnology, vol. 2, no. 4, pp. 303–308, 2006.

91. S. Bommarius and B. R. Riebel, Biocatalysis. Fundamentals and Applications, Wiley-VCH, Weinheim, Germany, 2004.

92. P. Fernandes and J. M. S. Cabral, "Applied biocatalysis: an overview," in Industrial Biotechnology, W. Soetaert and E. J. Vandamme, Eds., pp. 227–250, Wiley-VCH, Weinheim, Germany, 2010.

93. S. Bershtein and D. S. Tawfik, "Advances in laboratory evolution of enzymes," Current Opinion in Chemical Biology, vol. 12, no. 2, pp. 151–158, 2008.

94. W. F. Li, X. X. Zhou, and P. Lu, "Structural features of thermozymes," Biotechnology Advances, vol. 23, no. 4, pp. 271–281, 2005.

95. S. Trivedi, H. S. Gehlot, and S. R. Rao, "Protein thermostability in Archaea and Eubacteria," Genetics and Molecular Research, vol. 5, no. 4, pp. 816–827, 2006.

96. S. D'Amico, J.-C. Marx, C. Gerday, and G. Feller, "Activity-stability relationships in extremophilic enzymes," The Journal of Biological Chemistry, vol. 278, no. 10, pp. 7891–7896, 2003.

97. P. A. Dalby, "Engineering enzymes for biocatalysis," Recent Patents on Biotechnology, vol. 1, no. 1, pp. 1–9, 2007.

98. M. Lehmann and M. Wyss, "Engineering proteins for thermostability: the use of sequence alignments versus rational design and directed evolution," Current Opinion in Biotechnology, vol. 12, no. 4, pp. 371–375, 2001.

99. D. W. S. Wong, "Recent advances in enzyme development," in Handbook of Food Enzymology, J. R. Whitaker, A. G. J. Voragen, and D. W. S. Wong, Eds., pp. 379–387, Marcel Dekker, New York, NY, USA, 2003.

100. V. G. H. Eijsink, S. Gåseidnesa, T. V. Borchert, and B. van den Burg, "Directed evolution of enzyme stability," Biomolecular Engineering, vol. 22, no. 1–3, pp. 21–30, 2005.

101. A. Tracewell and F. H. Arnold, "Directed enzyme evolution: climbing fitness peaks one amino acid at a time," Current Opinion in Chemical Biology, vol. 13, no. 1, pp. 3–9, 2009.

102. S. Sen, V. V. Dasu, and B. Mandal, "Developments in directed evolution for improving enzyme functions," Applied Biochemistry and Biotechnology, vol. 143, no. 3, pp. 212–223, 2007.

103. F. Valetti and G. Gilardi, "Directed evolution of enzymes for product chemistry," Natural Product Reports, vol. 21, no. 4, pp. 490–511, 2004.

104. S. B. Rubin-Pitel and H. Zhao, "Recent advances in biocatalysis by directed enzyme evolution,"Combinatorial Chemistry and High Throughput Screening, vol. 9, no. 4, pp. 247–257, 2006.

105. M. Adamczak and S. H. Krishna, "Strategies for improving enzymes for efficient biocatalysis," Food Technology and Biotechnology, vol. 42, no. 4, pp. 251–264, 2004.

106. N. J. Turner, "Directed evolution drives the next generation of biocatalysts," Nature Chemical Biology, vol. 5, no. 8, pp. 567–573, 2009.

107. Sriprapundh, C. Vieille, and J. G. Zeikus, "Directed evolution of Thermotoga neapolitana xylose isomerase: high activity on glucose at low temperature and low pH," Protein Engineering, vol. 16, no. 9, pp. 683–690, 2003.

108. S. H. Bhosale, M. B. Rao, and V. V. Deshpande, "Molecular and industrial aspects of glucose isomerase," Microbiological Reviews, vol. 60, no. 2, pp. 280–300, 1996.

109. Y.-W. Kim, J.-H. Choi, J.-W. Kim et al., "Directed evolution of Thermus maltogenic amylase toward enhanced thermal resistance," Applied and Environmental Microbiology, vol. 69, no. 8, pp. 4866–4874, 2003.

110. S. Emond, I. André, K. Jaziri et al., "Combinatorial engineering to enhance thermostability of amylosucrase," Protein Science, vol. 17, no. 6, pp. 967–976, 2008.

111. Y. Wang, E. Fuchs, R. da Silva, A. McDaniel, J. Seibel, and C. Ford, "Improvement of Aspergillus nigerglucoamylase thermostability by directed evolution," Starch/Stärke, vol. 58, no. 10, pp. 501–508, 2006. ·

112. M.-S. Kim and X. G. Lei, "Enhancing thermostability of Escherichia coli phytase AppA2 by error-prone PCR," Applied Microbiology and Biotechnology, vol. 79, no. 1, pp. 69–75, 2008.

113. M.-S. Kim, J. D. Weaver, and X. G. Lei, "Assembly of mutations for improving thermostability ofEscherichia coli AppA2 phytase," Applied Microbiology and Biotechnology, vol. 79, no. 5, pp. 751–758, 2008.

114. K. Miyazaki, M. Takenouchi, H. Kondo, N. Noro, M. Suzuki, and S. Tsuda, "Thermal stabilization ofBacillus subtilis family-11 xylanase by directed evolution," The Journal of Biological Chemistry, vol. 281, no. 15, pp. 10236–10242, 2006.

115. Vieille and J. G. Zeikus, "Thermozymes: identifying molecular determinants of protein structural and functional stability," Trends in Biotechnology, vol. 14, no. 6, pp. 183–190, 1996.·

116. N. Kulkarni, A. Shendye, and M. Rao, "Molecular and biotechnological aspects of xylanases," FEMS Microbiology Reviews, vol. 23, no. 4, pp. 411–456, 1999.

117. M. Bauer, M. R. Bedford, and D. A. Pulliam, Microbially expresses xylanases and their use as feed additives and other uses. Patent US US2008187627A1, 2008.

118. A. van der Veen, G. Potocki-Véronèse, C. Albenne, G. Joucla, P. Monsan, and M. Remaud-Simeon, "Combinatorial engineering to enhance amylosucrase performance: construction, selection, and screening of variant libraries for increased activity," FEBS Letters, vol. 560, no. 1–3, pp. 91–97, 2004.

119. Y.-S. Tian, R.-H. Peng, J. Xu et al., "Mutations in two amino acids in phyIIs from Aspergillus niger 113 improve its phytase activity," World Journal of Microbiology and Biotechnology, vol. 26, no. 5, pp. 903–907, 2009.

120. J. M. Short, Directed evolution of thermophilic enzymes. Patent US5,830,696, 1998.

121. R. M. Kelly, L. Dijkhuizen, and H. Leemhuis, "Starch and α-glucan acting enzymes, modulating their properties by directed evolution," Journal of Biotechnology, vol. 140, no. 3-4, pp. 184–193, 2009.

122. J. Damborsky and J. Brezovsky, "Computational tools for designing and engineering biocatalysts,"Current Opinion in Chemical Biology, vol. 13, no. 1, pp. 26–34, 2009.

123. W. J. Quax, N. T. Mrabet, R. G. M. Luiten, P. W. Schuurhuizen, P. Stanssens, and I. Lasters, "Enhancing the thermostability of glucose isomerase by protein engineering," Nature Biotechnology, vol. 9, no. 8, pp. 738–742, 1991.

124. R. G. M. Luiten, W. J. Quax, P. W. Schuurhuizen, and N. Mrabet, Novel glucose isomerase enzymes and their use. Patent EP0351029 (A1), 1990.

125. P. Zhu, C. Xu, M. K. Teng et al., "Increasing the thermostability of D-xylose isomerase by introduction of a proline into the turn of a random

coil," Protein Engineering, vol. 12, no. 8, pp. 635–638, 1999.

126. N. Declerck, M. Machius, P. Joyet, G. Wiegand, R. Huber, and C. Gaillardin, "Hyperthermostabilization of Bacillus licheniformis α-amylase and modulation of its stability over a 50°C temperature range," Protein Engineering, vol. 16, no. 4, pp. 287–293, 2003.

127. L.-L. Lin, J.-S. Liu, W.-C. Wang, S.-H. Chen, C.-C. Huang, and H.-F. Lo, "Glutamic acid 219 is critical for the thermostability of a truncated α-amylase from alkaliphilic and thermophilic Bacillus sp. strain TS-23," World Journal of Microbiology and Biotechnology, vol. 24, no. 5, pp. 619–626, 2008.

128. H.-L. Liu and W.-C. Wang, "Protein engineering to improve the thermostability of glucoamylase from Aspergillus awamori based on molecular dynamics simulations," Protein Engineering, vol. 16, no. 1, pp. 19–25, 2003.

129. W. Zhang and X. G. Lei, "Cumulative improvements of thermostability and pH-activity profile of Aspergillus niger PhyA phytase by site-directed mutagenesis," Applied Microbiology and Biotechnology, vol. 77, no. 5, pp. 1033–1040, 2008.

130. M. Rhimi, N. Aghajari, M. Juy et al., "Rational design of Bacillus stearothermophilus US100 l-arabinose isomerase: potential applications for d-tagatose production," Biochimie, vol. 91, no. 5, pp. 650–653, 2009.

131. D.-K. Oh, "Tagatose: properties, applications, and biotechnological processes," Applied Microbiology and Biotechnology, vol. 76, no. 1, pp. 1–8, 2007.

132. Jørgensen, O. C. Hansen, and P. Stougaard, "Enzymatic conversion of D-galactose to D-tagatose: heterologous expression and characterisation of a thermostable L-arabinose isomerase from Thermoanaerobacter mathranii," Applied Microbiology and Biotechnology, vol. 64, no. 6, pp. 816–822, 2004.

133. P. Kim, "Current studies on biological tagatose production using L-arabinose isomerase: a review and future perspective," Applied Microbiology and Biotechnology, vol. 65, no. 3, pp. 243–249, 2004.

134. M. Rhimi, E. B. Messaoud, M. A. Borgi, K. B. khadra, and S. Bejar, "Co-expression of l-arabinose isomerase and d-glucose isomerase in E. coli and development of an efficient process producing simultaneously d-tagatose and d-fructose," Enzyme and Microbial Technology, vol. 40, no. 6, pp. 1531–1537, 2007.

135. L. K. Ozimek, S. Kralj, T. Kaper, M. J. E. C. Van Der Maarel, and

L. Dijkhuizen, "Single amino acid residue changes in subsite - 1 of inulosucrase from Lactobacillus reuteri 121 strongly influence the size of products synthesized," FEBS Journal, vol. 273, no. 17, pp. 4104–4113, 2006.·

136. M. E. Ortiz-Soto, M. Rivera, E. Rudiño-Piñera, C. Olvera, and A. López-Munguía, "Selected mutations in Bacillus subtilis levansucrase semi-conserved regions affecting its biochemical properties," Protein Engineering, Design and Selection, vol. 21, no. 10, pp. 589–595, 2008.

137. A. van der Veen, J. C.M. Uitdehaag, D. Penninga et al., "Rational design of cyclodextrin glycosyltransferase from Bacillus circulans strain 251 to increase α-cyclodextrin production," Journal of Molecular Biology, vol. 296, no. 4, pp. 1027–1038, 2000.

138. Z. Li, J. Zhang, M. Wang et al., "Mutations at subsite -3 in cyclodextrin glycosyltransferase fromPaenibacillus macerans enhancing α-cyclodextrin specificity," Applied Microbiology and Biotechnology, vol. 83, no. 3, pp. 483–490, 2009.

139. A. Van der Veen, J. C. M. Uitdehaag, B. W. Dijkstra, and L. Dijkhuizen, "Engineering of cyclodextrin glycosyltransferase reaction and product specificity," Biochimica et Biophysica Acta, vol. 1543, no. 2, pp. 336–360, 2000.

140. Z. Li, M. Wang, F. Wang et al., "γ-Cyclodextrin: a review on enzymatic production and applications,"Applied Microbiology and Biotechnology, vol. 77, no. 2, pp. 245–255, 2007.

141. K. Fujii, H. Minagawa, Y. Terada et al., "Use of random and saturation mutageneses to improve the properties of Thermus aquaticus amylomaltase for efficient production of cycloamyloses," Applied and Environmental Microbiology, vol. 71, no. 10, pp. 5823–5827, 2005.

142. R. A. Sheldon, "Enzyme immobilization: the quest for optimum performance," Advanced Synthesis and Catalysis, vol. 349, no. 8-9, pp. 1289–1307, 2007.

143. Mateo, J. M. Palomo, G. Fernandez-Lorente, J. M. Guisan, and R. Fernandez-Lafuente, "Improvement of enzyme activity, stability and selectivity via immobilization techniques," Enzyme and Microbial Technology, vol. 40, no. 6, pp. 1451–1463, 2007.

144. E. Swaisgood, "Use of immobilized enzymes in the food industry," in Handbook of Food Enzymology, J. R. Whitaker, A. G. J. Voragen, and D. M. S. Wong, Eds., pp. 359–366, Marcel Dekker, New York, NY, USA, 2003.

145. M. K. Walsh, "Immobilized enzyme technology for food applications,"

in Novel Enzyme Technology for Food Application, R. Rastall, Ed., pp. 60–84, Woodhead Publishing Limited, Cambridge, UK, 2007.

146. T. Nakakuki, "Development of functional oligosaccharides in Japan," Trends in Glycoscience and Glycotechnology, vol. 15, no. 82, pp. 57–64, 2003.

147. M. A. Abdel-Naby, A. M. Hashem, M. A. Esawy, and A. F. Abdel-Fattah, "Immobilization of Bacillussubtilis α-amylase and characterization of its enzymatic properties," Microbiological Research, vol. 153, no. 4, pp. 319–325, 1999.

148. Brady and J. Jordaan, "Advances in enzyme immobilisation," Biotechnology Letters, vol. 31, no. 11, pp. 1639–1650, 2009.

149. G. de Segura, M. Alcalde, F. J. Plou, M. Remaud-Simeon, P. Monsan, and A. Ballesteros, "Encapsulation in LentiKats of dextransucrase from Leuconostoc mesenteroides NRRL B-1299, and its effect on product selectivity," Biocatalysis and Biotransformation, vol. 21, no. 6, pp. 325–331, 2003.

150. S. A. de Assis, B. S. Ferreira, P. Fernandes, D. G. Guaglianoni, J. M. S. Cabral, and O. M. M. F. Oliveira, "Gelatin-immobilized pectinmethylesterase for production of low methoxyl pectin," Food Chemistry, vol. 86, no. 3, pp. 333–337, 2004.

151. M. Rebroš, M. Rosenberg, Z. Mlichová, L. Krištofíková, and M. Paluch, "A simple entrapment of glucoamylase into LentiKats® as an efficient catalyst for maltodextrin hydrolysis," Enzyme and Microbial Technology, vol. 39, no. 4, pp. 800–804, 2006.

152. Z. Grosová, M. Rosenberg, M. Rebroš, M. Šipocz, and B. Sedláčková, "Entrapment of β-galactosidase in polyvinylalcohol hydrogel," Biotechnology Letters, vol. 30, no. 4, pp. 763–767, 2008.

153. M. Rebroš, M. Rosenberg, Z. Mlichová, and L. Krištofíková, "Hydrolysis of sucrose by invertase entrapped in polyvinyl alcohol hydrogel capsules," Food Chemistry, vol. 102, no. 3, pp. 784–787, 2007.

154. M. Schlieker and K.-D. Vorlop, "A novel immobilization method for entrapment LentiKats®," inImmobilization of Enzymes and Cells, J. M. Guisan, Ed., pp. 333–343, Humana Press, Totowa, NJ, USA, 2nd edition, 2006.

155. R. Stloukal, M. Rosenberg, and M. Rebros, Industrial production of biocatalysts in the form of enzymes or microorganisms immobilized in polyvinyl alcohol gel. Patent US 2009/0061499 A1, 2009.

156. Anjani, K. Kailasapathy, and M. Phillips, "Microencapsulation of enzymes for potential application in acceleration of cheese ripening," International Dairy Journal, vol. 17, no. 1, pp. 79–86, 2007.

157. Kailasapathy and S. H. Lam, "Application of encapsulated enzymes to accelerate cheese ripening,"International Dairy Journal, vol. 15, no. 6-9, pp. 929–939, 2005.

158. H. Tumturk, G. Demirel, H. Altinok, S. Aksoy, and N. Hasirci, "Immobilization of glucose isomerase in surface-modified alginate gel beads," Journal of Food Biochemistry, vol. 32, no. 2, pp. 234–246, 2008.

159. Rajagopalan and C. Krishnan, "Immobilization of malto-oligosaccharide forming α-amylase fromBacillus subtilis KCC103: properties and application in starch hydrolysis," Journal of Chemical Technology and Biotechnology, vol. 83, no. 11, pp. 1511–1517, 2008.

160. R. Catana, B. S. Ferreira, J. M. S. Cabral, and P. Fernandes, "Immobilization of inulinase for sucrose hydrolysis," Food Chemistry, vol. 91, no. 3, pp. 517–520, 2005.

161. P. Walde and S. Ichikawa, "Enzymes inside lipid vesicles: preparation, reactivity and applications,"Biomolecular Engineering, vol. 18, no. 4, pp. 143–177, 2001.·

162. Z. Grosová, M. Rosenberg, and M. Rebroš, "Perspectives and applications of immobilised β-galactosidase in food industry—a review," Czech Journal of Food Sciences, vol. 26, no. 1, pp. 1–14, 2008.

163. J. M. Rodriguez-Nogales and A. Delgadillo, "Stability and catalytic kinetics of microencapsulated β-galactosidase in liposomes prepared by the dehydration-rehydration method," Journal of Molecular Catalysis B, vol. 33, no. 1-2, pp. 15–21, 2005.

164. J. M. Rodríguez-Nogales and A. D. López, "A novel approach to develop β-galactosidase entrapped in liposomes in order to prevent an immediate hydrolysis of lactose in milk," International Dairy Journal, vol. 16, no. 4, pp. 354–360, 2006.

165. C.-K. Kim, H.-S. Chung, M.-K. Lee, L.-N. Choi, and M.-H. Kim, "Development of dried liposomes containing β-galactosidase for the digestion of lactose in milk," International Journal of Pharmaceutics, vol. 183, no. 2, pp. 185–193, 1999.

166. E. Kheadr, J. C. Vuillemard, and S. A. El-Deeb, "Impact of liposome-encapsulated enzyme cocktails on cheddar cheese ripening," Food Research International, vol. 36, no. 3, pp. 241–252, 2003.

167. C.-K. Kim, H.-S. Chung, M.-K. Lee, L.-N. Choi, and M.-H. Kim,

"Development of dried liposomes containing β-galactosidase for the digestion of lactose in milk," International Journal of Pharmaceutics, vol. 183, no. 2, pp. 185–193, 1999.

168. S. Chockchaisawasdee, V. I. Athanasopoulos, K. Niranjan, and R. A. Rastall, "Synthesis of galacto-oligosaccharide from lactose using β-galactosidase from kluyveromyces lactis: studies on batch and continuous UF membrane-fitted bioreactors," Biotechnology and Bioengineering, vol. 89, no. 4, pp. 434–443, 2005.

169. S. Novalin, W. Neuhaus, and K. D. Kulbe, "A new innovative process to produce lactose-reduced skim milk," Journal of Biotechnology, vol. 119, no. 2, pp. 212–218, 2005.

170. F. Bickerstaff, "Immobilization of enzymes and cells: some practical considerations," inImmobilization of Enzymes and Cells, G. F. Bickerstaff, Ed., pp. 1–11, Humana Press, Totowa, NJ, USA, 1997.

171. S.GopinathandS.Sugunan,"Leachingstudiesoverimmobilizedα-amylase. Importance of the nature of enzyme attachment," Reaction Kinetics and Catalysis Letters, vol. 83, no. 1, pp. 79–83, 2004.

172. J. M. Nelson and E. G. Griffin, "Adsorption of invertase," Journal of the American Chemical Society, vol. 38, no. 5, pp. 1109–1115, 1916.

173. A. R. Mahmoud, "Immobilization of Invertase by a New Economical Method Using Wood Sawdust Waste," Australian Journal of Applied Science, vol. 1, pp. 364–372, 2007.

174. Nighojkar, S. Srivastava, and A. Kumar, "Production of low methoxyl pectin using immobilized pectinesterase bioreactors," Journal of Fermentation and Bioengineering, vol. 80, no. 4, pp. 346–349, 1995.

175. R. Reshmi, G. Sanjay, and S. Sugunan, "Enhanced activity and stability of α-amylase immobilized on alumina," Catalysis Communications, vol. 7, no. 7, pp. 460–465, 2006.

176. R. Reshmi, G. Sanjay, and S. Sugunan, "Immobilization of α-amylase on zirconia: a heterogeneous biocatalyst for starch hydrolysis," Catalysis Communications, vol. 8, no. 3, pp. 393–399, 2007.

177. P. Tripathi, A. Kumari, P. Rath, and A. M. Kayastha, "Immobilization of α-amylase from mung beans (Vigna radiata) on Amberlite MB 150 and chitosan beads: a comparative study," Journal of Molecular Catalysis B, vol. 49, no. 1–4, pp. 69–74, 2007.

178. A. Esawy, D. A. R. Mahmoud, and A. F. A. Fattah, "Immobilisation of Bacillus subtilis NRC33a levansucrase and some studies on its properties," Brazilian Journal of Chemical Engineering, vol. 25, no. 2,

pp. 237–246, 2008.

179. Z. S. Csanádi and C. S. Sisak, "Immobilization of Pectinex Ultra SP-L pectinase and its application to production of fructooligosaccharides,"Acta Alimentaria, vol. 35, no. 2, pp. 205–212, 2006.

180. S. Varavinit, N. Chaokasem, and S. Shobsngob, "Covalent immobilization of a glucoamylase to bagasse dialdehyde cellulose," World Journal of Microbiology and Biotechnology, vol. 17, no. 7, pp. 721–725, 2001.

181. Tanriseven and Z. Ölçer, "A novel method for the immobilization of glucoamylase onto polyglutaraldehyde-activated gelatin," Biochemical Engineering Journal, vol. 39, no. 3, pp. 430–434, 2008.

182. Milosavić, R. Prodanović, S. Jovanović, and Z. Vujčić, "Immobilization of glucoamylase via its carbohydrate moiety on macroporous poly(GMA-co-EGDMA)," Enzyme and Microbial Technology, vol. 40, no. 5, pp. 1422–1426, 2007.

183. Sanjay and S. Sugunan, "Fixed bed reactor performance of invertase immobilized on montmorillonite," Catalysis Communications, vol. 7, no. 12, pp. 1005–1011, 2006.·

184. Sanjay and S. Sugunan, "Glucoamylase immobilized on montmorillonite: influence of nature of binding on surface properties of clay-support and activity of enzyme," Journal of Porous Materials, vol. 14, no. 2, pp. 127–136, 2007.

185. L. Amaya-Delgado, M. E. Hidalgo-Lara, and M. C. Montes-Horcasitas, "Hydrolysis of sucrose by invertase immobilized on nylon-6 microbeads," Food Chemistry, vol. 99, no. 2, pp. 299–304, 2006.

186. V. Vallejo-Becerra, J. M. Vásquez-Bahena, J. A. Santiago-Hernández, and M. E. Hidalgo-Lara, "Immobilization of the recombinant invertase INVB from Zymomonas mobilis on Nylon-6," Journal of Industrial Microbiology and Biotechnology, vol. 35, no. 11, pp. 1289–1295, 2008.·

187. P. G. Cadena, R. A. S. Jeronimo, J. M. Melo, R. A. Silva, J. L. Lima Filho, and M. C. B. Pimentel, "Covalent immobilization of invertase on polyurethane, plast-film and ferromagnetic Dacron,"Bioresource Technology, vol. 101, pp. 1595–1602, 2009.

188. Altinok, S. Aksoy, H. Tümtürk, and N. Hasirci, "Covalent immobilization of invertase on chemically activated poly (styrene-2-hydroxyethyl methacrylate) microbeads," Journal of Food Biochemistry, vol. 32, no. 3, pp. 299–315, 2008.

189. Bayramoğlu, S. Akgöl, A. Bulut, A. Denizli, and M. Y. Arica, "Covalent immobilisation of invertase onto a reactive film composed

of 2-hydroxyethyl methacrylate and glycidyl methacrylate: properties and application in a continuous flow system," Biochemical Engineering Journal, vol. 14, no. 2, pp. 117–126, 2003.

190. R. Fernandez-Lafuente, "Hyperstabilization of a thermophilic esterase by multipoint covalent attachment," Enzyme and Microbial Technology, vol. 17, no. 4, pp. 366–372, 1995.·

191. C. Mateo, V. Grazú, B. C. C. Pessela et al., "Advances in the design of new epoxy supports for enzyme immobilization-stabilization," Biochemical Society Transactions, vol. 35, no. 6, pp. 1593–1601, 2007.

192. Bayramoglu, Y. Tunali, and M. Y. Arica, "Immobilization of β-galactosidase onto magnetic poly(GMA-MMA) beads for hydrolysis of lactose in bed reactor," Catalysis Communications, vol. 8, no. 7, pp. 1094–1101, 2007.

193. S. A. Ferrarotti, J. M. Bolivar, C. Mateo, L. Wilson, J. M. Guisan, and R. Fernandez-Lafuente, "Immobilization and stabilization of a cyclodextrin glycosyltransferase by covalent attachment on highly activated glyoxyl-agarose supports," Biotechnology Progress, vol. 22, no. 4, pp. 1140–1145, 2006.

194. G. de Segura, M. Alcalde, M. Yates et al., "Immobilization of dextransucrase from Leuconostoc mesenteroides NRRL B-512F on eupergit C supports," Biotechnology Progress, vol. 20, no. 5, pp. 1414–1420, 2004.

195. L. Cao, Carrier-Bound Immobilized Enzymes—Principles, Applications and Design, Wiley-VCH, Weinheim, Germany, 2005.

196. C. Mateo, B. C. C. Pessela, M. Fuentes, et al., "Very Strong But Reversible Immobilization of Enzymes on Supports Coated With Ionic Polymers," in Immobilization of Enzymes and Cells, J. M. Guisan, Ed., pp. 205–216, Humana Press, Totowa, NJ, USA, 2nd edition, 2006.

197. J. Tomotani and M. Vitolo, "Method for immobilizing invertase by adsorption on Dowex® anionic exchange resin," Brazilian Journal of Pharmaceutical Sciences, vol. 42, no. 2, pp. 245–249, 2006.

198. L. D. S. Marquez, B. V. Cabral, F. F. Freitas, V. L. Cardoso, and E. J. Ribeiro, "Optimization of invertase immobilization by adsorption in ionic exchange resin for sucrose hydrolysis," Journal of Molecular Catalysis B, vol. 51, no. 3-4, pp. 86–92, 2008.

199. Y. Arica and G. Bayramoğlu, "Invertase reversibly immobilized onto polyethylenimine-grafted poly(GMA-MMA) beads for sucrose hydrolysis," Journal of Molecular Catalysis B, vol. 38, no. 3-6, pp. 131–138, 2006.

200. C. Mateo, R. Torres, G. Fernández-Lorente et al., "Epoxy-amino groups: a new tool for improved immobilization of proteins by the epoxy method," Biomacromolecules, vol. 4, no. 3, pp. 772–777, 2003.

201. B. C. C. Pessela, R. Fernández-Lafuente, M. Fuentes et al., "Reversible immobilization of a thermophilic β-galactosidase via ionic adsorption on PEI-coated Sepabeads," Enzyme and Microbial Technology, vol. 32, no. 3-4, pp. 369–374, 2003.

202. B. Rathbone, A. J. Hacking, and P. S. J. Cheetham, Process for the preparation of fructosyl disaccharides. Patent US 4617269, 1986.

203. R. L. Antrim and A. L. Auterinen, "A new regenerable immobilized glucose isomerase," Starch/Stärke, vol. 38, pp. 132–137, 1986.

204. S. M. Gaikwad and V. V. Deshpande, "Immobilization of glucose isomerase on Indion 48-R," Enzyme and Microbial Technology, vol. 14, no. 10, pp. 855–858, 1992.

205. M. Gómez, M. D. Romero, T. M. Fernández, and S. García, "Immobilization and enzymatic activity of β-glucosidase on mesoporous SBA-15 silica," Journal of Porous Materials. In press.

206. D.-H. Kweon, S.-G. Kim, N. S. Han, J. H. Lee, K. M. Chung, and J.-H. Seo, "Immobilization of Bacillusmacerans cyclodextrin glycosyltransferase fused with poly-lysine using cation exchanger," Enzyme and Microbial Technology, vol. 36, no. 4, pp. 571–578, 2005.

207. J. Roy and T. E. Abraham, "Strategies in making cross-linked enzyme crystals," Chemical Reviews, vol. 104, no. 9, pp. 3705–3721, 2004.

208. S. Dalal, A. Sharma, and M. N. Gupta, "A multipurpose immobilized biocatalyst with pectinase, xylanase and cellulase activities," Chemistry Central Journal, vol. 1, no. 1, article 16, 2007.

209. R. Gaur, H. Pant, R. Jain, and S. K. Khare, "Galacto-oligosaccharide synthesis by immobilized Aspergillus oryzae β-galactosidase," Food Chemistry, vol. 97, no. 3, pp. 426–430, 2006.

210. Tatsumoto, K. K. Oh, J. O. Baker, and M. E. Himmel, "Enhanced stability of glucoamylase through chemical crosslinking," Applied Biochemistry and Biotechnology, vol. 20-21, no. 1, pp. 293–308, 1989.

211. E. Ortiz-Soto, E. Rudiño-Piñera, M. E. Rodriguez-Alegria, and A. L. Munguia, "Evaluation of cross-linked aggregates from purified Bacillus subtilis levansucrase mutants for transfructosylation reactions," BMC Biotechnology, vol. 9, article 68, 2009.

212. Visuri, Preparation of cross-linked glucose isomerase crystals. Patent US 5437993, 1995.

213. Pastinen, K. Visuri, and M. Leisola, "Xylitol purification by cross-linked glucose isomerase crystals," Biotechnology Techniques, vol. 12, no. 7, pp. 557–560, 1998.

214. B. Brandt, A. Hidalgo, and U. T. Bornscheuer, "Immobilization of enzymes in microtiter plate scale,"Biotechnology Journal, vol. 1, no. 5, pp. 582–587, 2006.

215. Cao, Carrier-bound immobilized enzymes: principles, applications and design, Wiley-VCH, Weinheim, Germany, 2005.

216. W. Pitcher, "Design and operation of immobilized enzyme reactors," in Enzymes for Industrial Reactors, R. Messing, Ed., pp. 151–199, Academic Press, New York, NY, USA, 1975.

Chapter 8

BIOENGINEERING RECOMBINANT DIACYLGLYCEROL ACYLTRANSFERASES

Heping Cao

U.S. Department of Agriculture, Agricultural Research Service Southern Regional Research Center U.S.A.

INTRODUCTION

The complete genomes of many organisms including human, mouse, Arabidopsis, and rice have been sequenced. However, the functions of the proteins encoded by a large percentage of the genes in these organisms have not been determined. The immediate challenge of the post-genomic biology is to determine the biological functions of proteins coded for by those unknown genes. Many endogenous proteins occur in extremely low abundance (such as the anti-inflammatory protein tristetraprolin, TTP) (Cao et al., 2004) and are labile (such as omega-3 fatty-acid desaturase, FAD3) (O'Quin et al., 2010), which are major problems inherent to characterization of those proteins. Recombinant proteins can be used as an alternative source to endogenous proteins. Production of active proteins in large quantities is necessary for the study of protein structure and function (Cao et al., 2003). Purified recombinant proteins are also important for the production of antibodies (Cao 2004; Cao et al., 2008; Cao et al., 2004) and pharmaceutical reagents. Unfortunately, a great number of proteins are difficult to express and purify. Those proteins include membrane proteins, lipid-associated proteins, and lowabundance proteins. The causes of the difficulties in protein expression and purification are various, among which are protein insolubility, protein degradation, and low-level protein expression (Cao 2010). Therefore, production of high-quality recombinant proteins requires optimization of protein expression and purification procedures in each case. Diacylglycerol acyltransferases (DGATs) catalyze the last and rate-limiting step of triacylglycerol (TAG) biosynthesis in eukaryotic organisms. DGAT genes have been isolated from many organisms. At least two forms of DGATs are present in mammals (Cases et al., 1998; Cases et al., 2001) and plants (Lardizabal et al., 2001; Shockey et al., 2006) with additional forms reported in burning bush (Euonymus alatus) (Durrett et al., 2010), peanut (Saha et al., 2006), and Arabidopsis (Rani et al., 2010). Plants and animals deficient in DGATs accumulate less TAG (Smith et al., 2000; Stone et al., 2004; Zou et al., 1999). Animals with reduced DGAT activity are resistant to diet-induced obesity (Chen et al., 2004; Smith et al., 2000) and lack milk production (Smith et al., 2000). Over-expression of DGAT enzymes increases TAG content in plants (Andrianov et al., 2010; Bouvier-Nave et al., 2000; Burgal et al., 2008; Durrett et al., 2010; Jako et al., 2001; Lardizabal et al., 2008; Xu et al., 2008), animals (Kamisaka et al., 2010; Liu et al., 2009; Liu et al., 2007; Roorda et al., 2005), and yeast (Kamisaka et al., 2007). DGATs have nonredundant functions in TAG biosynthesis in species

such as mice (Stone et al., 2004) and tung tree (Vernicia fordii) (Shockey et al., 2006). Mice deficient in DGAT1 are viable, have modest decreases in TAG, and are resistant to dietinduced obesity (Chen et al., 2002; Smith et al., 2000). In contrast, mice deficient in DGAT2 have severe reduction of TAG and die shortly after birth (Stone et al., 2004). The fact that DGAT1 is unable to compensate for the deficiency in DGAT2 indicates the nonredundant functions of each DGAT isoform in TAG biosynthesis during animal development. Therefore, understanding the roles of DGATs in plants and animals will have tremendous implications in creating new oilseed crops with value-added properties and in providing clues for therapeutic intervention in obesity and related diseases. Over-production of DGATs has been the subject of a number of studies, but progress has been slow in the characterization of the enzymes because DGATs are integral membrane proteins (Shockey et al., 2006; Stone et al., 2006) and difficult to express and purify (Cheng et al., 2001; Weselake et al., 2006). Information regarding the expression of DGAT genes in E. coli is limited. The expression of DGAT1 and DGAT2 as full-length proteins in E. coli had not been reported. We recently developed a reliable procedure for the expression and purification of tung DGATs in E. coli (Cao et al., 2010; Cao et al., 2011).

BIOENGINEERING RECOMBINANT DIACYLGLYCEROL ACYLTRANSFERASES

DGAT genes have been identified in a wide range of organisms

Database search identified at least 115 DGAT sequences from 69 organisms including plants (such as Arabidopsis, barley, caster bean, cauliflower, corn, rape, rice, sorghum, soybean, tobacco, tung tree), animals (such as bird, chimpanzee, cow, dog, fish, fly, frog, monkey, mosquito, mouse, pig, rabbit, rat, sheep, worm), fungi (such as yeast), and human. The names of organisms, the subfamilies of DGATs (DGAT1 and DGAT2) and the GenBank accession numbers are listed in Table 1. Although more than two isoforms of DGATs are found in some species, most of them could be classified into the DGAT1 or DGAT2 subfamily according to their sequence similarities and phylogenetic analysis (data not shown). However, DGAT3 (Saha et al., 2006) and DGAT4 (Rani et al., 2010) were reported recently which have very different sequences with those of DGAT1 and DGAT2. DGAT1 and DGAT2 subfamilies have many conserved residues among the diverse species. However, addition of DGAT3 and DGAT4 from Arabidopsis (GenBank accession number: AAN31909.1), caster bean (GenBank accession number: XP_002519339.1), peanut (GenBank accession number: AY875644.1), and yeast (GenBank accession number: DG315417.1) to the multiple sequence alignment completely destroyed all the conserved residues (data not shown), which is contrary to the general belief that the active sites of the enzymes should have certain degree of conservation during the evolution because all are supposed to catalyze the same/similar biochemical reaction.

Table 1: DGAT1 and DGAT2 sequence information (DGAT3 and DGAT4 are not included in the Table because of their divergent sequences). A: animal, F: fungus, P: plant

No.	Organism	DGAT	GenBank accession number	No.	Organism	DGAT	GenBank accession number
1	Aedes aegypti (A)	1	XP_001658299	59	Medicago truncatula (P)	2	ACJ84867.1
2	Ajellomyces capsulatus (F)	1	EGC41804.1	60	Nicotiana tabacum (P, tobacco)	1	AAF19345.1

3	*Anolis carolinensis* (A)	2	XP_003225477.1	61	*Nematostella vectensis* (A, worm)	2a	XP_0016304 35.1
4	*Ashbya gossypii* (F)	2	NP_983542.1	62	*Nematostella vectensis* (A, worm)	2b	XP_0016333 22.1
5	*Arthroderma otae* (F)	1	EEQ31683.1	63	*Nematostella vectensis* (A, worm)	2c	XP_0016355 48.1
6	*Arabidopsis thaliana* (P)	1	NP_179535.1	64	*Ovis aries* (A, sheep)	1	NP_0011036 34.1
7	*Arabidopsis thaliana* (P)	2	NP_566952	65	*Ovis aries* (A, sheep)	2	XP_0015188 99.1
8	*Bubalus bubalis* (A, buffalo)	1	AAZ22403.1	66	*Oryctolagus cuniculus* (A, rabbit)	1	XP_0027244 27.1
9	*Brassica juncea* (P)	1a	AAY40784.1	67	*Olea europaea* (P, tree)	1	AAS01606.1
10	*Brassica juncea* (P)	1b	AAY40785.1	68	*Olea europaea* (P, tree)	2	ADG22608.1
11	*Brassica napus* (P)	1a	AAD45536.1	69	*Oryza sativa* (P, rice)	1	NP_0010548 69.2
12	*Brassica napus* (P)	1b	AAD40881.1	70	*Oryza sativa* (P, rice)	2a	NP_0010479 17
13	*Brassica napus* (P)	2	ACO90187	71	*Oryza sativa* (P, rice)	2b	NP_0010575 30
14	*Brassica napus* (P)	2	ACO90188	72	*Ostreococcus tauri* (algae)	2	XP_0030835 39.1
15	*Bos taurus* (A, cow)	1	NP_777118.2	73	*Pongo abelii* (A)	2	XP_0028223 04.1
16	*Bos taurus* (A, cow)	2a	DAA21853.1	74	*Paracoccidioides brasiliensis* (F)	1	EEH17170.1
17	*Bos taurus* (A, cow)	2b	XP_875499.3	75	*Perilla frutescens* (P)	1	AAG23696.1
18	*Bos taurus* (A, cow)	2c	XP_002683800.1	76	*Polysphondyliu m pallidum* (F)	1	EFA85004.1
19	*Caenorhabditis elegans* (A, worm)	2a	NP_505413.1	77	*Polysphondyliu m pallidum* (F)	2	EFA83646.1
20	*Caenorhabditis elegans* (A, worm)	2b	NP_872180.1	78	*Physcomitrella patens* (P, moss)	1	XP_0017709 29.1
21	*Canis familiaris* (A, dog)	1b	XP_849176.1	79	*Physcomitrella patens* (P, moss)	1	XP_0017587 58.1
22	*Canis familiaris* (A, dog)	1c	XP_858062.1	80	*Physcomitrella patens* (P, moss)	2b	XP_0017777 26.1
23	*Capra hircus* (A, sheep)	1	ABD59375.1	81	*Picea sitchensis* (P, tree)	2	ABK26256.1

24	*Ciona intestinalis* (A)	2	XP_002120879 .1	82	*Pan troglodytes* (A, chimpanzee)	1	XP_520014.2
25	*Chlamydomonas reinhardtii* (algae)	2a	XP_001694904 .1	83	*Pan troglodytes* (A, chimpanzee)	2	XP_527842.2
26	*Chlamydomonas reinhardtii* (algae)	2b	XP_001693189 .1	84	*Phaeodactylum tricornutum* (F)	1	XP_0021777 53.1
27	*Chlorella variabilis* (algae)	1	EFN50697.1	85	*Populus trichocarpa* (P, tree)	1a	XP_0023082 78.1
28	*Chlorella variabilis* (algae)	2	EFN51306.1	86	*Populus trichocarpa* (P, tree)	1b	XP_0023305 10.1
29	*Dictyostelium discoideum* (mold)	1	XP_645633.2	87	*Populus trichocarpa* (P, tree)	2	XP_0023176 35.1
30	*Dictyostelium discoideum* (mold)	2	XP_635762.1	88	*Ricinus communis* (P, castor bean)	1	XP_0025141 32.1
31	*Drosophila melanogaster* (A, fly)	1a	NP_609813.1	89	*Ricinus communis* (P, castor bean)	1	XP_0025285 31.1
32	*Drosophila melanogaster* (A, fly)	1d	NP_995724.1	90	*Rattus norvegicus* (A, rat)	1	NP_445889.1
33	*Danio rerio* (A, zebrafish)	1a	NP_956024.1	91	*Rattus norvegicus* (A, rat)	2	NP_0010123 45.1
34	*Danio rerio* (A, zebrafish)	1b	NP_00100245 8.1	92	*Sorghum bicolor* (P, sorghum)	1a	XP_0024371 65.1
35	*Danio rerio* (A, zebrafish)	2	NP_00102536 7.1	93	*Sorghum bicolor* (P, sorghum)	1b	XP_0024394 19.1
36	*Euonymus alatus* (P)	1	AAV31083.1	94	*Sorghum bicolor* (P, sorghum)	2	XP_0024526 52.1
37	*Euonymus alatus* (P)	2	ADF57328.1	95	*Saccharomyces cerevisiae* (F, yeast)	2	NP_014888.1
38	*Elaeis oleifera* (P)	2	ACO35365.1	96	*Saccoglossus kowalevskii* (A, worm)	1	XP_0027361 60.1
39	*Echium pitardii* (P)	1	ACO55635.1	97	*Selaginella moellendorffii* (P)	1	XP_0029641 65.1
40	*Glycine max* (P, soybean)	1a	AAS78662.1	98	*Selaginella moellendorffii* (P)	2	XP_0029720 54.1

41	Glycine max (P, soybean)	1b	BAE93461.1	99	Spirodela polyrhiza (P)	2	AAQ89590.1
42	Glycine max (P, soybean)	2	ACU20344.1	100	Schizosaccharo myces pombe (F, yeast)	2	XP_0017131 60.1
43	Helianthus annuus (P)	2	ABU50328.1	101	Sus scrofa (A, pig)	1	NP_999216.1
44	Homo sapiens (human)	1	NP_036211.2	102	Tribolium castaneum (A)	1	XP_975142.1
45	Homo sapiens (human)	2a	AAQ88896.1	103	Tribolium castaneum (A)	2	XP_975146.1
46	Homo sapiens (human)	2b	NP_835470.1	104	Toxoplasma gondii (A)	1	AAP94209.1
47	Hordeum vulgare (P, barley)	2	BAJ85730.1	105	Taeniopygia guttata (A, bird)	2	XP_0021876 43.1
48	Ictalurus punctatus (A, catfish)	2b	NP_00118800 5.1	106	Tropaeolum majus (P)	1	AAM03340. 2
49	Jatropha curcas (P)	1	ABB84383.1	107	Vernicia fordii (P, tung tree)	1	DQ356680.1
50	Lotus japonicas (P)	1	AAW51456.1	108	Vernicia fordii (P, tung tree)	2	DQ356682
51	Metarhizium acridum (F)	1a	EFY86774.1	109	Vernonia galamensis (P)	1	ABV21945.1
52	Metarhizium anisopliae (F)	1b	EFY97444.1	110	Vernonia galamensis (P)	2	ACV40232.1
53	Monodelphis domestica (A)	1	XP_001371565 .1	111	Vitis vinifera (P, grape)	1	XP_0022793 45.1
54	Monodelphis domestica (A)	2	XP_001365685 .1	112	Vitis vinifera (P, grape)	2	XP_0022636 26
55	Mus musculus (A, mouse)	1	NP_034176.1	113	Xenopus tropicalis (A, frog)	2	NP_989372.1
56	Mus musculus (A, mouse)	2	NP_080660.1	114	Zea mays (P, corn)	1b	EU039830
57	Macaca mulatta (A, monkey)	1	XP_001090134 .1	115	Zea mays (P, corn)	2	NP_0011501 74.1
58	Medicago truncatula (P)	1	ABN09107.1				

Literature survey of DGAT expression

A literature survey was performed to find out how many publications related to DGATs have been collected by the two most popular databases, PubMed and Scopus. The data in Table 2 indicate that approximately 1000 papers had been collected by the two databases during the past 28 years when using DGAT and diacylglycerol acyltransferase as search terms in title/abstracts/keywords. Approximately four times of publications were obtained when using the full name of the enzyme "diacylglycerol acyltransferase" as a search term instead of using the abbreviation "DGAT" in the database search.

More than half of the publications were from animals and approximately one quarter of the publications were from plants. Less than half of those publications dealt with expression of DGATs at the RNA and protein levels.

Some of the publications reported of using more than one organism in the same paper, resulting in the total number of publications less than the number of publications from plants, animals, and human adding together (Table 2). Similarly, the total expression papers are less than the combination because more than one expression methods were used in the same paper. Approximately 5% of the publications were related to heterologous expression. However, only a few papers were from E. coli expression system.

Table 2: Literature survey of publications related to DGAT expression in PubMed and Scopus databases (1982-2010)

Database	PubMed	PubMed	Scopus	Scopus
Search terms in title/abstracts/keywords	DGAT	diacylglycerol acyltransferase	DGAT	diacylglycerol acyltransferase
Total publications	216	817	255	1102
Plant	57	118	60	137
Human	74	203	72	316
Animal	138	588	164	760
Total expression papers	90	225	122	322
Plant expression	31	50	34	62
Human expression	31	85	42	131
Animal expression	53	144	78	220
E. coli expression	4	8	1	6
Yeast expression	17	32	17	33
Insect expression	5	12	7	15

Recombinant DGAT expression update

Expression and purification of recombinant DGATs from any source represents a challenge because DGATs are integral membrane proteins (Hobbs et al., 1999; Siloto et al., 2008; Weselake et al., 2006). In addition, more than 40% of the total amino acid residues are hydrophobic (Table 3). Yeast was the preferred host for DGAT expression (Bouvier-Nave et al., 2000; Burgal et al., 2008; Cao et al., 2010; He et al., 2004; Kalscheuer et al., 2004; Kalscheuer & Steinbuchel 2003; Kroon et al., 2006; Liu et al., 2011; Liu et al., 2010; ManasFernandez et al., 2009; Mavraganis et al., 2010; Milcamps et al., 2005; Nykiforuk et al., 2002; Quittnat et al., 2004; Shockey et al., 2006; Siloto et al., 2009; Wagner et al., 2010; Xu et al., 2008; Yu et al., 2008) followed by insect cells (Buszczak et al., 2002; Cases et al., 1998; Cases et al., 2001; Lardizabal et al., 2001). A limited number of reports used other host cells including E. coli (Saha et al., 2006; Siloto et al., 2008; Weselake et al., 2006) and human cells (Cheng et al., 2001). The great majority of the yeast and insect cell expression studies were designed to confirm the functions of full-length cloned genes. A few studies were directly related to the expression and purification of recombinant DGATs using E. coli expression system for functional and structural studies. The recombinant N-terminal region of Brassica napus DGAT1 was purified from E. coli with a predicted molecular mass of 13,278 Da which was confirmed by MALDI-TOF mass spectrometry. However, the apparent molecular mass on SDS-PAGE was doubled and the native size was four times of the size of the monomer

due to self-association (Weselake et al., 2006). The N-terminal region of mouse DGAT1 was also studied in a similar way (Siloto et al., 2008). Full-length DGAT1 or DGAT2 from any organism was, however, not successfully expressed in E. coli (Hobbs et al., 1999; Weselake et al., 2006). The exceptional case was that expression of soluble peanut DGAT (DGAT3) in E. coli resulted in high levels of DGAT activity and the formation of labeled TAG (Saha et al., 2006), although its sequence is very different from those of DGAT1 or DGAT2.

Table 3: Tung DGATs properties and amino acid composition

	Tung tree DGAT1	Tung tree DGAT2	DGAT1 - DGAT2
Length (aa)	526	322	204
Molecular weight	59773.84	36726.20	23047.64
Isoelectric point (PI)	8.91	9.24	- 0.33
Charge at pH 7	11.78	8.44	3.34
Charged (RKHYCDE) (%)	27.00	23.60	3.40
Acidic (DE) (%)	7.98	7.14	0.84
Basic (KR) (%)	10.08	9.63	0.45
Polar (NCQSTY) (%)	25.86	21.74	4.12
Hydrophobic (AILFWV) (%)	41.06	43.48	-2.42

Bioengineering recombinant DGAT for expression in bacteria

We recently described a procedure for over-expression of recombinant full-length DGAT1 and DGAT2 in a bacterial expression system (Cao et al., 2010; Cao et al., 2011). DGAT1 is much larger than DGAT2, although they are similar in other properties and amino acid composition (on % of frequency basis) (Table 3). The two DGAT isoforms have only limited sequence identity and similarity (Figure 1). We were able to express both proteins in E. coli as full-length recombinant proteins. In our study, we engineered a maltose binding protein (MBP) tag at the amino terminus and 6 histidine residues (His-tag) at the carboxyl terminus of full-length tung DGATs (Table 4).

Table 4: Primers for PCR-amplification of the full-length DGAT1 and DGAT2 insert sequences

Primer	Sequence (5′ to 3′)	Comments
DGAT1 forward	AATATT<u>GGTACC</u>CTGTTTCAGGGTCC GACAATCCTTGAAACGCCG	*Kpn*I site underlined Codons for PreScission protease site Colored
DGAT1 reverse	CGATTA<u>ACTAGT</u>AGCTAGCTCAATG ATGATGATGATGATGTCTTGATTCGG TAGTCCC	*Spe*I site underlined Codons for 6 His Colored
DGAT2 forward	AATATT<u>GGTACC</u>CTGTTTCAGGGTCC GGGGATGGTGGAAGTTAAG	*Kpn*I site underlined Codons for PreScission protease site Colored
DGAT2 reverse	CGATTA<u>ACTAGT</u>AGCTAGCTCAATG ATGATGATGATGATGAAAAATTTCA AGTTTAAG	*Spe*I site underlined Codons for 6 His Colored

We engineered plasmids pMBP-DGAT1-His and pMBP-DGAT2-His for expressing the fulllength tung tree type 1 and type 2 diacylglycerol acyltransferases (DGAT1 and DGAT2, GenBank Accession No. DQ356680 and DQ356682, respectively (Shockey et al., 2006) as fusion proteins in E. coli. The recombinant proteins contained MBP at the amino terminus and His-tag at the carboxyl terminus. The cloning vector pMBP-hTTP (Figure 2) was reported previously (Cao et al., 2003).

lasmids pMBP-DGAT1-His (Figure 3) and pMBPDGAT2-His (Figure 4) were constructed by replacing the hTTP fragment in plasmid pMBPhTTP (Figure 2) with the PCR-amplified DGAT1 and DGAT2 fragments at the KpnI and SpeI sites (Table 4). Existing DGAT plasmid DNAs were used as the templates for PCRamplification of the DGAT DNA open reading frames (Shockey et al., 2006). DGAT forward primers contained DNA sequence for a KpnI/Asp718I restriction enzyme recognition site followed by a PreScission protease cleavage site (5′-CTGTTTCAGGGTCCG-3′) (Cao et al., 2003) which codes for 5 amino acid residues (LFQGP) between MBP and DGAT protein sequences (Table 4). DGAT reverse primers contained sequence for a His-tag (5′- ATGATGATGATGATGATG-3′) coding for 6 histidine residues at the carboxyl terminus of DGATs (Table 4).

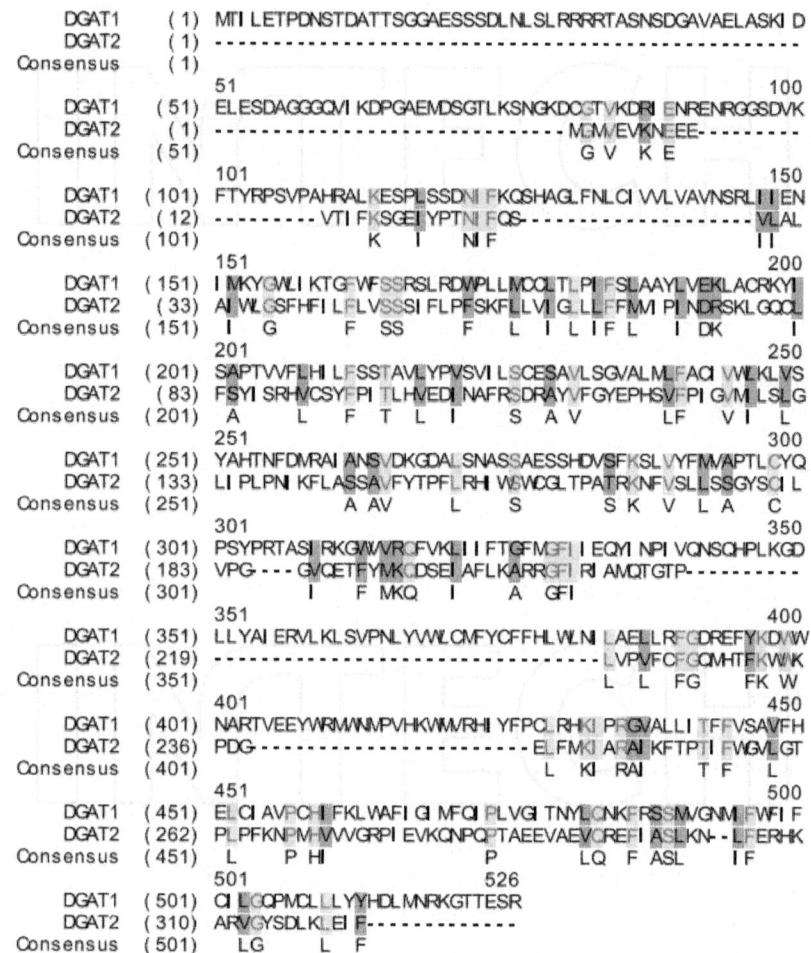

Figure. 1: Alignment of tung tree DGAT1 and DGAT2 amino acid sequences.

PThe successful expression of full-length recombinant DGATs was probably due to the fusion to MBP, which was shown to increase the solubility of target proteins such as human and mouse TTP (Cao et al., 2003; Cao et al., 2008; Kapust & Waugh 1999). Although we engineered double affinity tags for facilitating purification of recombinant DGAT from E. coli, recombinant DGATs were only partially purified from the extract by either type of affinity beads [amylose resin and nickel-nitrilotriacetic agarose (Ni-NTA) beads] or both kinds of beads in tandem. Our data, together with the various published reports cited in the previous section, underline the tremendous challenges that exist for the purification of recombinant full-length DGAT proteins.

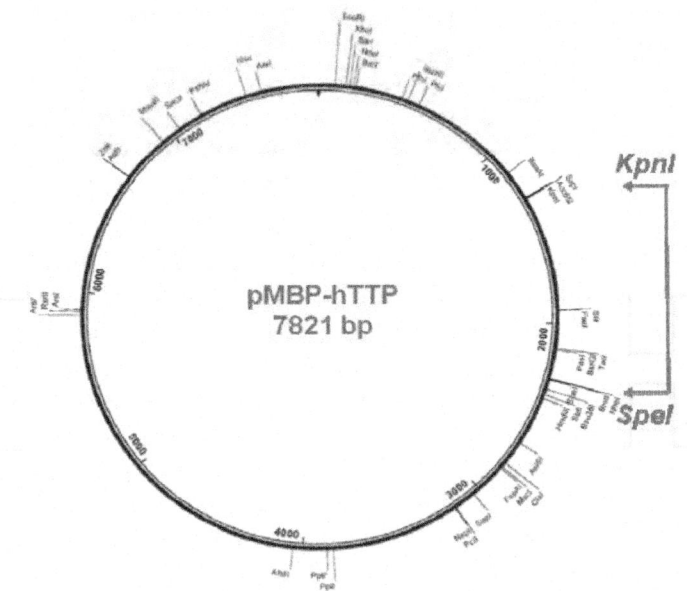

Figure. 2: Plasmid map of E. coli expression vector pMBP-hTTP.

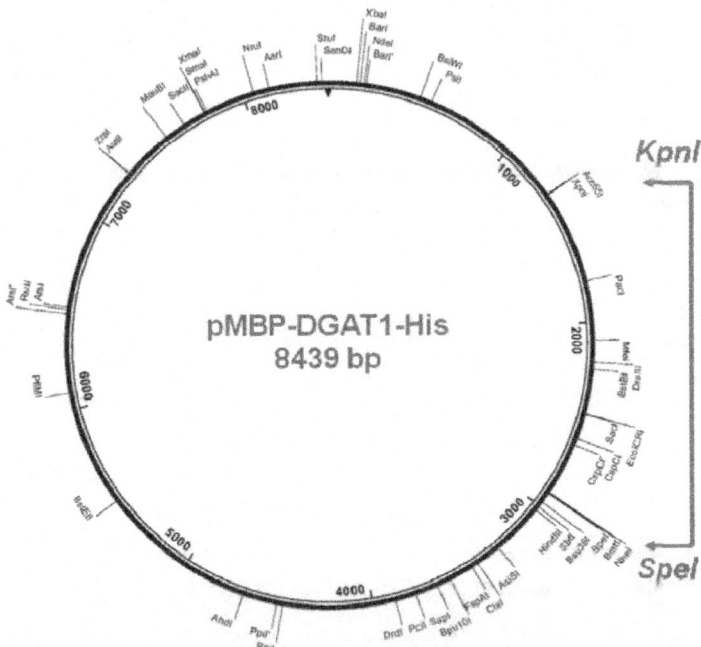

Figure. 3: Plasmid map of E. coli expression vector pMBP-DGAT1-His.

Conclusion Diacylglycerol acyltransferases (DGATs) catalyze the last and rate-limiting step of triacylglycerol (TAG) biosynthesis in eukaryotic organisms. At least 115 DGAT sequences are identified from 69 organisms in the GenBank databases. Only a few papers have been published in the last 28 years on the expression of the recombinant DGAT proteins in a bacterial expression system. None of the full-length DGAT1 or DGAT2 had been expressed in E. coli expression system. The difficulties in DGAT expression and purification are due to the nature of these proteins being integral membrane proteins with more than 40% of the total amino acid residues being hydrophobic. Therefore, progress in characterization of the enzymes has been slow. We recently developed a procedure for full-length DGAT expression in E. coli. Expression plasmids were engineered to express tung DGATs fused to maltose binding protein and poly-histidine. The development of the technique should help to purify full-length DGATs for further studies such as raising high-titer antibodies and studying the structure-function relationship. Understanding the roles of DGATs in plant oil biosynthesis will help to create new oilseed crops with value-added properties.

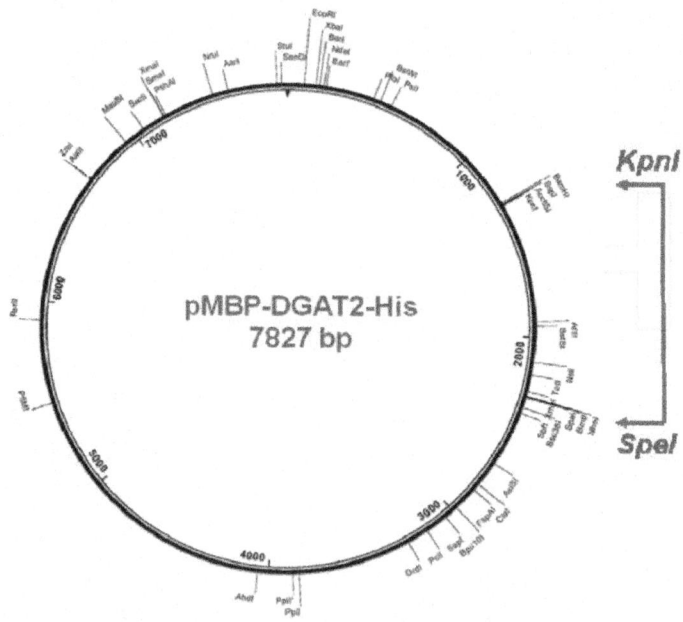

Figure. 4: Plasmid map of E. coli expression vector pMBP-DGAT2-His.

The elucidation of the precise roles of DGATs in animal and human fat synthesis and deposition may provide clues for nutritional and therapeutic intervention in obesity and related diseases.

ABBREVIATIONS

DGAT, diacylglycerol acyltransferase; FAD3, omega-3 fatty-acid desaturase; His, poly histidine; MBP, maltose binding protein; Ni-NTA, nickel-nitrilotriacetic agarose; TAG, triacylglycerol; TTP; tristetraprolin.

REFERENCES

1. Andrianov, V.; Borisjuk, N.; Pogrebnyak, N.; Brinker, A.; Dixon, J.; Spitsin, S.; Flynn, J.; Matyszczuk, P.; Andryszak, K.; Laurelli, M.; Golovkin, M., & Koprowski, H. (2010) Tobacco as a production platform for biofuel: overexpression of Arabidopsis DGAT and LEC2 genes increases accumulation and shifts the composition of lipids in green biomass. Plant Biotechnol J 8: 277-287

2. Bouvier-Nave, P.; Benveniste, P.; Oelkers, P.; Sturley, S. L., & Schaller, H. (2000)Expression in yeast and tobacco of plant cDNAs encoding acylCoA:diacylglycerol acyltransferase. Eur J Biochem 267: 85-96

3. Burgal, J.; Shockey, J.; Lu, C.; Dyer, J.; Larson, T.; Graham, I., & Browse, J. (2008) Metabolicengineering of hydroxy fatty acid production in plants: RcDGAT2 drivesdramatic increases in ricinoleate levels in seed oil. Plant Biotechnol J 6: 819-831

4. Buszczak, M.; Lu, X.; Segraves, W. A.; Chang, T. Y., & Cooley, L. (2002) Mutations in themidway gene disrupt a Drosophila acyl coenzyme A: diacylglycerolacyltransferase. Genetics 160: 1511-1518

5. Cao, H. (2004) Expression, purification, and biochemical characterization of theantiinflammatory tristetraprolin: a zinc-dependent mRNA binding proteinaffected by posttranslational modifications. Biochemistry 43: 13724-13738

6. Cao, H. (2010) Recombinant protein production technology [Review]. J Jiangxi Agric Univ32: 1018-1031

7. Cao, H.; Chapital, D. C.; Shockey, J. M., & Klasson, T. K. (2011). Expression of tung treediacylglycerol acyltransferase 1 in E. coli. BMC Biotechnol, 11:72.Cao, H.; Chapital, D. C.; Howard, O. D.; Jiang, X. N.; Shockey, J. M., & Klasson, K. T.(2011) Purification of recombinant tung tree diacylglycerol acyltransferases fromE. coli. The FASEB Journal 25: 765.8

8. Cao, H.; Dzineku, F., & Blackshear, P. J. (2003) Expression and purification of recombinanttristetraprolin that can bind to tumor necrosis factor-alpha mRNA and serve as asubstrate for mitogen-activated protein kinases. Arch Biochem Biophys 412: 106-120

9. Cao, H.; Lin, R.; Ghosh, S.; Anderson, R. A., & Urban, J. F., Jr. (2008) Production andcharacterization of ZFP36L1 antiserum against recombinant protein fromEscherichia coli. Biotechnol Prog 24: 326-333

10. Cao, H.; Tuttle, J. S., & Blackshear, P. J. (2004) Immunological characterization of tristetraprolin as a low abundance, inducible, stable cytosolic protein. J Biol Chem 279: 21489-21499

11. Cases, S.; Smith, S. J.; Zheng, Y. W.; Myers, H. M.; Lear, S. R.; Sande, E.; Novak, S.; Collins,C.; Welch, C. B.; Lusis, A. J.; Erickson, S. K., & Farese, R. V., Jr. (1998)

12. Identification of a gene encoding an acyl CoA:diacylglycerol acyltransferase, akey enzyme in triacylglycerol synthesis. Proc Natl Acad Sci U S A 95: 13018-13023

13. Cases, S.; Stone, S. J.; Zhou, P.; Yen, E.; Tow, B.; Lardizabal, K. D.; Voelker, T., & Farese, R.V., Jr. (2001) Cloning of DGAT2, a second mammalian diacylglycerolacyltransferase, and related family members. J Biol Chem 276: 38870-38876

14. Chen, H. C.; Rao, M.; Sajan, M. P.; Standaert, M.; Kanoh, Y.; Miura, A.; Farese, R. V., Jr., &Farese, R. V. (2004) Role of adipocyte-derived factors in enhancing insulinsignaling in skeletal muscle and white adipose tissue of mice lacking AcylCoA:diacylglycerol acyltransferase 1. Diabetes 53: 1445-1451

15. Chen, H. C.; Smith, S. J.; Ladha, Z.; Jensen, D. R.; Ferreira, L. D.; Pulawa, L. K.; McGuire, J.G.; Pitas, R. E.; Eckel, R. H., & Farese, R. V., Jr. (2002) Increased insulin and leptin sensitivity in mice lacking acyl CoA:diacylglycerol acyltransferase 1. J Clin Invest 109: 1049-1055

16. Cheng, D.; Meegalla, R. L.; He, B.; Cromley, D. A.; Billheimer, J. T., & Young, P. R. (2001)Human acyl-CoA:diacylglycerol acyltransferase is a tetrameric protein. Biochem J359: 707-714

17. Durrett, T. P.; McClosky, D. D.; Tumaney, A. W.; Elzinga, D. A.; Ohlrogge, J., & Pollard,M. (2010) A distinct DGAT with sn-3 acetyltransferase activity that synthesizesunusual, reduced-viscosity oils in Euonymus and transgenic seeds. Proc Natl AcadSci U S A 107: 9464-9469

18. He, X.; Turner, C.; Chen, G. Q.; Lin, J. T., & McKeon, T. A. (2004) Cloning andcharacterization of a cDNA encoding diacylglycerol acyltransferase from castorbean. Lipids 39: 311-318

19. Hobbs, D. H.; Lu, C., & Hills, M. J. (1999) Cloning of a cDNA encoding diacylglycerolacyltransferase from Arabidopsis thaliana and its functional expression. FEBSLett 452: 145-149

20. Jako, C.; Kumar, A.; Wei, Y.; Zou, J.; Barton, D. L.; Giblin, E. M.; Covello, P. S., & Taylor,D. C. (2001) Seed-specific over-expression of an Arabidopsis cDNA encoding adiacylglycerol acyltransferase enhances seed oil content and seed weight. PlantPhysiol 126: 861-874

21. Kalscheuer, R.; Luftmann, H., & Steinbuchel, A. (2004) Synthesis of novel lipids inSaccharomyces cerevisiae by heterologous expression of an unspecific bacterialacyltransferase. Appl Environ Microbiol 70: 7119-7125

22. Kalscheuer, R. & Steinbuchel, A. (2003) A novel bifunctional wax ester synthase/acylCoA:diacylglycerolacyltransferase mediates wax ester and triacylglycerolbiosynthesis in Acinetobacter calcoaceticus ADP1. J Biol Chem 278: 8075-8082

23. Kamisaka, Y.; Kimura, K.; Uemura, H., & Shibakami, M. (2010) Activation ofdiacylglycerol acyltransferase expressed in Saccharomyces cerevisiae:overexpression of Dga1p lacking the N-terminal region in the Deltasnf2disruptant produces a significant increase in its enzyme activity. Appl MicrobiolBiotechnol 88: 105-115Kamisaka, Y.; Tomita, N.; Kimura, K.; Kainou, K., & Uemura, H. (2007) DGA1(diacylglycerol acyltransferase gene)overexpression and leucine biosynthesisissignificantly increase lipid accumulation in the Deltasnf2 disruptant ofSaccharomyces cerevisiae. Biochem J 408: 61-68

24. Kapust, R. B. & Waugh, D. S. (1999) Escherichia coli maltose-binding protein isuncommonly effective at promoting the solubility of polypeptides to which it isfused. Protein Sci 8: 1668-1674

25. Kroon, J. T.; Wei, W.; Simon, W. J., & Slabas, A. R. (2006) Identification and functionalexpression of a type 2 acyl-CoA:diacylglycerol acyltransferase (DGAT2) indeveloping castor bean seeds which has high homology to the major triglyceridebiosynthetic enzyme of fungi and animals. Phytochemistry 67: 2541-2549

26. Lardizabal, K.; Effertz, R.; Levering, C.; Mai, J.; Pedroso, M. C.; Jury, T.; Aasen, E.; Gruys,K., & Bennett, K. (2008) Expression of Umbelopsis ramanniana DGAT2A in seedincreases oil in soybean. Plant Physiol 148: 89-96Lardizabal, K. D.; Mai, J. T.; Wagner, N. W.; Wyrick, A.; Voelker, T., & Hawkins, D. J.(2001) DGAT2 is a new diacylglycerol acyltransferase gene family: purification,cloning, and expression in insect cells of two polypeptides from Mortierellaramanniana with diacylglycerol acyltransferase activity. J Biol Chem 276: 38862-38869Liu, L.; Shi, X.; Bharadwaj, K. G.; Ikeda, S.; Yamashita, H.; Yagyu, H.; Schaffer, J. E.; Yu, Y.H., & Goldberg, I. J. (2009) DGAT1 expression increases heart

triglyceride contentbut ameliorates lipotoxicity. J Biol Chem 284: 36312-36323

27. Liu, L.; Zhang, Y.; Chen, N.; Shi, X.; Tsang, B., & Yu, Y. H. (2007) Upregulation ofmyocellular DGAT1 augments triglyceride synthesis in skeletal muscle andprotects against fat-induced insulin resistance. J Clin Invest 117: 1679-1689

28. Liu, Q.; Siloto, R. M.; Snyder, C. L., & Weselake, R. J. (2011) Functional and topologicalanalysis of yeast acyl-coa:diacylglycerol acyltransferase 2, an endoplasmicreticulum enzyme essential for triacylglycerol biosynthesis. J Biol Chem 286:13115-13126

29. Liu, Q.; Siloto, R. M., & Weselake, R. J. (2010) Role of cysteine residues in thiolmodification of acyl-CoA:diacylglycerol acyltransferase 2 from yeast. Biochemistry49: 3237-3245

30. Manas-Fernandez, A.; Vilches-Ferron, M.; Garrido-Cardenas, J. A.; Belarbi, E. H.; Alonso,D. L., & Garcia-Maroto, F. (2009) Cloning and molecular characterization oftheacyl-CoA: diacylglycerol acyltransferase 1 (DGAT1) gene from Echium. Lipids 44:555-568

31. Mavraganis, I.; Meesapyodsuk, D.; Vrinten, P.; Smith, M., & Qiu, X. (2010) Type IIdiacylglycerol acyltransferase from Claviceps purpurea with ricinoleic acid, ahydroxyl fatty acid of industrial importance, as preferred substrate. Appl EnvironMicrobiol 76: 1135-1142

32. Milcamps, A.; Tumaney, A. W.; Paddock, T.; Pan, D. A.; Ohlrogge, J., & Pollard, M. (2005)Isolation of a gene encoding a 1,2-diacylglycerol-sn-acetyl-CoA acetyltransferasefrom developing seeds of Euonymus alatus. J Biol Chem 280: 5370-5377

33. Nykiforuk, C. L.; Furukawa-Stoffer, T. L.; Huff, P. W.; Sarna, M.; Laroche, A.; Moloney, M.M., & Weselake, R. J. (2002) Characterization of cDNAs encoding diacylglycerolacyltransferase from cultures of Brassica napus and sucrose-mediated inductionof enzyme biosynthesis. Biochim Biophys Acta 1580: 95-109

34. O'Quin, J. B.; Bourassa, L.; Zhang, D.; Shockey, J. M.; Gidda, S. K.; Fosnot, S.; Chapman, K.D.; Mullen, R. T., & Dyer, J. M. (2010) Temperature-sensitive post-translationalregulation of plant omega-3 fatty-acid desaturases is mediated by theendoplasmic reticulum-associated degradation pathway. J Biol Chem 285: 21781-21796

35. Quittnat, F.; Nishikawa, Y.; Stedman, T. T.; Voelker, D. R.; Choi, J. Y.; Zahn, M. M.;Murphy, R. C.; Barkley, R. M.; Pypaert, M.; Joiner, K. A., & Coppens, I. (2004) On

36. the biogenesis of lipid bodies in ancient eukaryotes: synthesis of triacylglycerolsby a Toxoplasma DGAT1-related enzyme. Mol Biochem Parasitol 138: 107-122Rani, S. H.; Krishna, T. H.; Saha, S.; Negi, A. S., & Rajasekharan, R. (2010) Defective incuticular ridges (DCR) of Arabidopsis thaliana, a gene associated with surfacecutin formation, encodes a soluble diacylglycerol acyltransferase. J Biol Chem 285:38337-38347

37. Roorda, B. D.; Hesselink, M. K.; Schaart, G.; Moonen-Kornips, E.; Martinez-Martinez, P.;Losen, M.; De Baets, M. H.; Mensink, R. P., & Schrauwen, P. (2005) DGAT1overexpression in muscle by in vivo DNA electroporation increasesintramyocellular lipid content. J Lipid Res 46: 230-236

38. Saha, S.; Enugutti, B.; Rajakumari, S., & Rajasekharan, R. (2006) Cytosolic triacylglycerolbiosynthetic pathway in oilseeds. Molecular cloning and expression of peanutcytosolic diacylglycerol acyltransferase. Plant Physiol 141: 1533-1543

39. Shockey, J. M.; Gidda, S. K.; Chapital, D. C.; Kuan, J. C.; Dhanoa, P. K.; Bland, J. M.;Rothstein, S. J.; Mullen, R. T., & Dyer, J. M. (2006) Tung tree DGAT1 and DGAT2have nonredundant functions in triacylglycerol biosynthesis and are localized todifferent subdomains of the endoplasmic reticulum. Plant Cell 18: 2294-2313Siloto, R. M.; Madhavji, M.; Wiehler, W. B.; Burton, T. L.; Boora, P. S.; Laroche, A., &Weselake, R. J. (2008) An N-terminal fragment of mouse DGAT1 binds differentacyl-CoAs with varying affinity. Biochem Biophys Res Commun 373: 350-354

40. Siloto, R. M.; Truksa, M.; Brownfield, D.; Good, A. G., & Weselake, R. J. (2009) Directedevolution of acyl-CoA:diacylglycerol acyltransferase: Development andcharacterization of Brassica napus DGAT1 mutagenized libraries. Plant PhysiolBiochem 47: 456-461

41. Smith, S. J.; Cases, S.; Jensen, D. R.; Chen, H. C.; Sande, E.; Tow, B.; Sanan, D. A.; Raber, J.;Eckel, R. H., & Farese, R. V., Jr. (2000) Obesity resistance and multiplemechanisms of triglyceride synthesis in mice lacking Dgat. Nat Genet 25: 87-90

42. Stone, S. J.; Levin, M. C., & Farese, R. V., Jr. (2006) Membrane topology and identificationof key functional amino acid residues of murine acyl-CoA:diacylglycerolacyltransferase-2. J Biol Chem 281: 40273-40282

43. Stone, S. J.; Myers, H. M.; Watkins, S. M.; Brown, B. E.; Feingold, K. R.; Elias, P. M., &Farese, R. V., Jr. (2004) Lipopenia and skin barrier abnormalities in DGAT2-deficient mice. J Biol Chem 279: 11767-11776

44. Wagner, M.; Hoppe, K.; Czabany, T.; Heilmann, M.; Daum, G.; Feussner, I., & Fulda, M.(2010) Identification and characterization of an acyl-Co A:diacylglycerolacyltransferase 2 (DGAT2) gene from the microalga O. tauri. Plant Physiol Biochem

45. 48: 407-416Weselake, R. J.; Madhavji, M.; Szarka, S. J.; Patterson, N. A.; Wiehler, W. B.; Nykiforuk, C.L.; Burton, T. L.; Boora, P. S.; Mosimann, S. C.; Foroud, N. A.; Thibault, B. J.;Moloney, M. M.; Laroche, A., & Furukawa-Stoffer, T. L. (2006)

46. Acyl-CoA-bindingand self-associating properties of a recombinant 13.3 kDa N-terminal fragment ofdiacylglycerol acyltransferase-1 from oilseed rape. BMC Biochem 7: 24

47. Xu, J.; Francis, T.; Mietkiewska, E.; Giblin, E. M.; Barton, D. L.; Zhang, Y.; Zhang, M., &Taylor, D. C. (2008) Cloning and characterization of an acyl-CoA-dependentdiacylglycerol acyltransferase 1 (DGAT1) gene from Tropaeolum majus, and astudy of the functional motifs of the DGAT protein using site-directedmutagenesis to modify enzyme activity and oil content. Plant Biotechnol J 6: 799-88

48. Yu, K.; Li, R.; Hatanaka, T., & Hildebrand, D. (2008) Cloning and functional analysis oftwo type 1 diacylglycerol acyltransferases from Vernonia galamensis.Phytochemistry 69: 1119-1127

49. Zou, J.; Wei, Y.; Jako, C.; Kumar, A.; Selvaraj, G., & Taylor, D. C. (1999) The Arabidopsisthaliana TAG1 mutant has a mutation in a diacylglycerol acyltransferase gene.Plant J 19: 645-653

Chapter 9

BIOENGINEERING OF VASCULAR CONDUITS

A. Pontini[1], M.M. Sfriso[2], M.I. Buompensiere[2],V.Vindigni[1], and F. Bas-setto[1]

[1]Department of Neurosensorial Specialties, Institute of Plastic Reconstructive Surgery and Burn Unit, Padova University Hospital, Padova, Italy

[2]Department of Molecular Medicine, Human Anatomy Section, Padova University Hospi- tal, Padova, Italy

INTRODUCTION

Tissue engineering research was applied in the last few years to vascular conduit field, aiming to obtain a suitable and ready to use substitute for vessel replacement. Based on the possibilities to obtain in vitro a biocompatible structure it is established that is theoretically and experi- mentally possible to provide vessels that can be employed to replace both diseased than damaged native blood vessels overcoming the massive worldwide clinical need and the poor supply of natural graft and, the same time, offering a better long term performance than the artificial conduits. The challenges reported in literature about the approach of tissue engi- neering for replacing blood vessels are continuously increasing. It has been reported natural vessel like structure, with similar elastic wall properties that are necessary for the cyclic blood flow loading with similar native vascular diameter to allow a perfect match with the host vessel. Fundamental are also the result obtained in term of antithrombotic lumen [1, 2].

Particularly important achievement were reached for cardiovascular application but also the potential range of application could easily been expanded to all microsurgical and vascular applications [3]. Tissue engineering has been projected as an alternative treatment to these problems by replacing the damaged tissue or organ function with constructs which are bio fabricated based on the required tissue or organ features [4]. In particular, cardiovascular tissue engineering is more valuable and relevant compared to other fields of tissue engineering mainly because it increases life expectancy, preserve the extremities, and provide solutions to a large number of disease [5]. Tissue engineering could at least been see as an interdisciplinary field that applies the principles of engineering and life sciences towards the development of functional substitutes for damaged tissues. It is strictly related to the fundamental concept of utilizing the body's natural biological response to tissue damage in conjunction with engineering principles [6, 7]. Besides, tissue engineering is planned to produce biomimetic constructs, which resemble normal tissues.

Moreover, the main objective of tissue engineering is the restoration of function through the delivery of living elements which become integrated in the patient [8].Tissue engineering strategies have three basic components: firstly, the cells or source which must express the appropriate genes and maintain the appropriate phenotype in order to preserve the specific function of the tissue, secondly, the bioactive agents or signals that induce cells to function, and thirdly, the scaffolds that house the cells and act as substitute for the damaged tissue [9, 10]. The source may either be embryonic stem cells (ESC) or adult stem cells(ASC) in origin, the scaffolds may be categorized as synthetic, biological, or composite, and the signals may include growth factors/ cytokines, adhesion factors, and bioreactors [11]. Develop a bioengi- neered tissue is due to a precise clinical needs and different ways were searched to obtain the answer (Figure 1).

So, joint to the ideal construct is also the research of the best tissue engi- neering approaches in terms of biocompatibility, feasibility and costs. So from 1999 till now several studies, both in vitro than in vivo, were presented, particularly inherent on vessel scaffold, which a large number of information about cell sources, technology application, outcomes and future perspectives [12]. Until nowadays the most encouraging result are limited in basic and experimental research but the rapidly increasing of optimal result permit to access to numerously clinical trials. Biomaterials technologies for vascular replacement must obtain an ideal graft that could overcome the needing of autologous vessel, that often it's not applicable, but, at the same time, providing similar properties. In fact the ideal bioengineered vascular substitute must be not thrombogenic, overall in small caliber vessel, and also when a long graft is needed. Thrombosis mechanism into vascular substitutes, especially in artificial one, is the main cause of obliteration and subsequent failure of most microvascular prostheses.

Autologous native vessel, are the most currently used material for small-diameter arterial replacement. Immune acceptance is a major advantage offered by this technique but the time of dissecting, harvesting and preparing autologous graft limited the microvascular emergency surgery and, in elective surgery, it could be possible that no one suitable vessel could be harvested. For that reason, the tissue engineering was applied to improved prosthetic performance at the blood-biomaterial interface. Different approach were described to optimize vascular bioengi- neered conduits as completely bio-resorbeable vascular prostheses with the capacity for induce regeneration and growth of a new vascular segment, biologic scaffold enhanced by stem cell seeding, decelularizzed native vessel with or without cell enhancement. In vitro ed in vivo study of all these different approaches shown the possibility to overcome the limitations of the artificial prostheses that are nonviable and based on allogenic materials lacking the capacity of growth, repair, and remodeling. The use of bio engineered vascular conduit are fundamental in small caliber vessel where the artificial replacement is affected by a very low patency rate meanwhile the possibility to obtain bioengineered large vessel replacement is actually less important due to the satisfactory result and still less expensive use of artificial or homograft conduit. Synthetic prostheses offer to microsurgery a possible solution for microvascular need of a ready to use and simple to manage small diameter vessel.

Availability in multiple different diameters and lengths, uncomplicated storage associated with easy handling are some of main advantages of such grafts; nevertheless, inherent thrombogenicity and compliance mismatch could represent their drawbacks. The research aim is to obtain and ideal prostheses, particu- larly in term of biocompatibility and satisfactory patency rate in long period of time. In fact, similar outcomes as large vessel replacement were not achieved in microvascular surgery. Best performance was obtained when the blood flow is high and the resistance is low, because those conditions allowed to overcome possible thrombogenic events that occur in large part in small-diameter prostheses. Multiple strategies were studied to overcome these limitations applying tissue engineering techniques. Ideally artificial conduit ready to use should be composed of viable tissue, able to contract in response to hemodynamic forces and chemical stimulation, and secrete physiological blood vessel substances. Anastomoses using artificial prostheses should also allow complete healing without immunologic reaction, remodeling according to surroundings environment, and even have the ability to grow when placed in children.

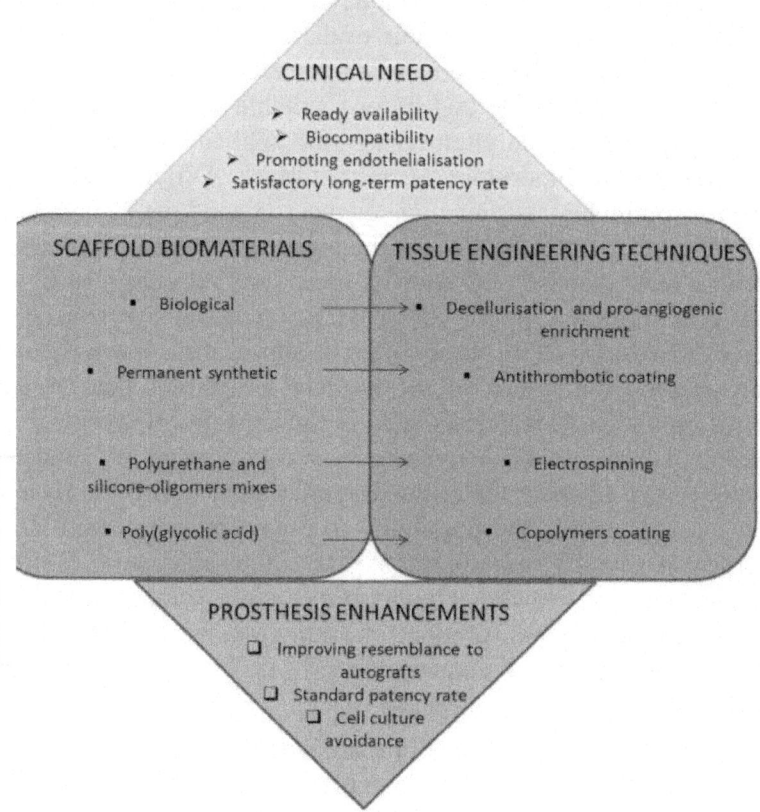

Figure 1: Clinical need and bioengineering of vascular conduit. Adapted from: Pontini et al. "Alternative Conduits for Microvascular Anastomoses" Surg Innov. 2013 Aug 20;21(3):277-282.

BLOOD FLOW AND FUNCTIONAL IMPLICATION FOR VASCULAR REPLACEMENT

Blood vessels are the conduit that allows the transportation in the human body to the organs and tissues of blood, oxygen, nutrients and catabolities removal. Small caliber arteries (< 6 mm), account for most of deaths in the United States every year [1] Atherosclerosis, due to numerously damage mechanism of media and intima tunica in vessel lumen, cause a progressive occlusion that could lead to a severe blood flow impairment and organ failure. Most common vascular disease based on critic vessel damage could be the cardiac infarction subsequent to a coronary artery occlusion, claudication or chronic ulcer because peripheral arterial disease, or stroke due to occlusion of carotid or cerebral arteries. Arterial replacement is widely consid- ered a common treatment for vascular disease, accounting for more than 1.4 million arterial bypass operations performed annually in the United States [2] The "gold standard" treatment is based on autologous vessels employment such saphenous veins and mammary arteries but one-third of patients are lacking for suitable veins, often for general vascular disease or past harvesting in vascular procedures. Nevertheless autologous vein graft have shown high patency rate failure in the long term period principally due to intimal hyperplasia [13]. Need for vascular grafts are also important in reconstructive surgery, vascular trauma, organ transplantation, so a large number of vascular conduit are needed in clinical daily practice. Besides are must to be considered that a significant morbidity and high economical costs are associated with autologous vessel preparation. Multiple factor are at least involved in a widely recognized need for an efficient, readily available and simple to manage small-diameter vascular graft. The first step on not autolo- gous vessel replacement was constituted by artificial vessel based on different permanent material as polyurethane, polyethylene terephthalate and polytetrafluoroethylene (ePTFE). All these prosthetic materials have proved to be inferior to autologous conduits, especially for small caliber. Low patency rate outcome with important thrombosis risk, infection and low performance at anastomosis site have determined the progressive discharged of artificial conduits. [14] The biological approach provided by tissue engineering was thought to allow a better performance, compatibility and host matching. [15]

BIOENGINEERED VESSEL: DEFINITION AND DEVELOPMENT

Vessel replacement is a clinical need in many vascular field. Blood vessel diseases, such as atherosclerosis and arteritis as reported from Ross in 1993 and Wilcox in 1996. The Chronic Venous Insufficiency (CVI) as described from Moriyama in 2011 and thrombosis remain globally the major vascular problems. Therapies for such diseases often require replacement of those vessels with vascular grafts. Autologous arteries or veins are the best substitutes for small-diameter (internal diameter < 6 mm) vessels as shown also from Tu and colleagues in 1997. However, in some cases of acute vascular disease, amputation or previous harvest, the implant of autologous vessels could be limited of the biodisponibility of patient. The most important described application were the arterial occlusive disease such coronorapathy and ischemic periferical condition treated with artery by-pass [16]. Furthermore were described use also in thoracic surgery and for dialysis access [17, 18]. Literature review of the historical approach in vascular bioengineering shown multiple and different approach worldwide to obtain the best biocompatible and long lasting replacement. The research start from modification of synthetic material and was develop to biologic material assessment. There have been so many attempts to develop a small-diameter vascular graft made of synthetic or natural polymers. The synthetic polymeric materials include polyethylene terephthalate and expanded polytetrafluoroethylene (ePTFE) as described from Teebken in 2002. Although these polymeric vascular grafts have been successfully employed to replace blood vessels above 6 mm in ID, these polymeric grafts cannot be used for treatment of smalldiameter vascular diseases due to thrombus formation as demonstrated from Veith in 1986 and then from Chard in 1987. Coating of the intimal side with antithrombogenic materials, such as heparin was the approach for example of Devine in 2001, polyethylene oxide was the attempt of Kidane in 1999, or, previously, with endothelial cells as described from James in 1998, has been applied to solve this problem as we also reported below. Unfortunately these approaches still remaining doubtfull in vivo and in long-term and are considered unsuccessful. For that reason tissue engineered blood vessels (TEBV) arising as a promising approach to address the shortcoming of such problems. Many design criteria have been proposed for the development of blood vessels scaffold as it's possible to read in the works of Conte et al. in 1998; Mitchell et al in 2003 and Teebken et al. also in 2003. Scaffold must be biocompatible, i.e. non thrombogenic, non-immunogenic, and resistant to infection, all of which are associated with a confluent, quiescent, non-activated endothelium. Furthermore, it must induce an acceptable healing response that does not result in inflammation, hyperplasia, or fibrous capsule formation, and, ideally, leads to the integration of the graft into the body such that it eventually becomes indistinguishable from a native vessel. It must possess appropriate mechanical properties, which include physiological compliance, the ability to withstand long-term hemodynamic stress without failure, and no susceptibility to permanent creep that can lead to aneurysm formation. Scaffold must have an appropriate permeability to water, solutes, and cells and must exhibit physiological properties, such as vasoconstriction/relaxation responses. Finally, easy handling and suturability are crucial for such vessels to be viable from a surgical standpoint. These design criteria are quite challenging given the demanding mechanical environment of the cardiovascular system. Although different approaches attempt to meet these criteria in different ways, it is widely held that 3 components are necessary for these criteria to be met:

- a biocompatible component with high tensile strength to provide mechanical support (collagen fibers or their analogue);
- a biocompatible elastic component to provide recoil and prevent aneurysm formation (elastin fibers or their analogue);

• a non-activated, confluent endothelium to prevent thrombosis.

In 1986, Weinberg and Bell generated what was widely regarded as the first tissue-engineered blood vessel substitute, consisting of cultures of bovine endothelial cells (ECs), smooth muscle cells (SMCs) and fibroblasts embedded in a collagen gel. However, the graft lacked sufficient strength and was unsuitable for implantation. This construct was evaluated in vivo as an arterial implant only after reinforcement with Dacron® as shown from Matsuda in 1995. Various methods of improving the mechanical properties of collagen gels (e.g., crosslinking agents such as glutaraldehyde) have been investigated, but none has proven to yield a structurally stable tissue-engineered vascular grafts (TEVG) as reported from Charulatha in 2003. As an alternative to collagen for natural ECM-based scaffolds, fibrin holds particular promise because of its ability to induce collagen and elastin synthesis and improved mechan- ical properties as shown from Swartz in 2005. Furthermore, encouraging result, even if in larger diameter vessel, was achieved by combining fibrin gels with biodegradable polymeric scaffolds followed by seeding of autologous arterial-derived cells, from the group of Tschoeke in 2009 as also endothelialized vessels have been successfully implanted in the carotid arteries of sheep from Koch and his team in 2010. Decellularized tissue, often in the form of a Xenogenic, can serve as a naturally available scaffold. Examples of such scaffolds were realized by Lantz in 1993, who used the small intestinal submucosa (SIS) as a vascular implant. The SIS was decellularized and then im- planted in aorta, carotid and femoral arteries of dogs. The grafts resulted completely endo- thelialized at 28 days post-implantation. At 90 days, the grafts were histologically similar to normal arteries and veins and contained a smooth muscle media and a dense fibrous connec- tive tissue adventitia. Follow-up periods of up to 5 years found no evidence of infection, intimal hyperplasia, or aneurysmal dilation. One infection-challenge study suggested that SIS may be infection resistant, possibly because of early capillary penetration of the SIS (2 to 4 days after implantation) and delivery of body defenses to the local site. Kaushal, in 2001, has employed a decellularized porcine iliac arteries, seeded them with endothelial progenitor cells (EPCs), and implanted the constructs into ovine carotid arteries. These TEVG constructs remained patent out to 130 days and were remodeled into neovessel, whereas the unseeded control group occluded within 15 days. These results indicate that decellularized vascular scaffolds are susceptible to early failure unless first undergoing endothelialisation or additional modification. In fact, Simon in 2003, shown as elements of the ECM are exposed to physical and chemical stresses during the process of decellularization, which can adversely affect the biomechanical properties of the ECM. This deterioration might ultimately lead to degenerative structural graft failure. Additional drawbacks of decellularized materials included the inability to modify the ECM content and architecture, the variability among donor sources, and the risk of viral transmission from animal tissue.

In 2011 Quint has developed a unique method of developing decellularized tissue for smalldiameter arterial grafts using biodegradable polymers. They developed a different approach to arterial tissue engineering that can substantially reduce the waiting time for a graft. Tissueengineered vessels (TEVs) were grown from banked porcine smooth muscle cells that were allogenic to the intended recipient, using a biomimetic perfusion system. The engineered vessels were then decellularized, leaving behind the mechanically robust extracellular matrix of the graft wall. The acellular grafts were then seeded with cells that were derived from the intended recipient, EPC or EC, on the graft lumen. TEV were then implanted as end-to-side grafts in the porcine carotid artery, which is a rigorous test-bed due to its tendency for graft occlusion. The EPC-and EC-seeded TEV all remained patent for 3D in this study, whereas the contralateral control vein grafts were patent in only 3/8 implants. Going along with the improved patency, the cell-seeded TEV demonstrated less neointimal hyperplasia and fewer proliferating cells than did the vein grafts. Proteins in the mammalian target of rapamycin signaling pathway tended to be decreased in TEV compared with vein grafts, implicating this pathway in the TEV's resistance to occlusion from intimal hyperplasia. These results indicate that a readily available, decellularized tissue-engineered vessel can be seeded with autologous endothelial progenitor cells to provide a biological vascular graft that resists both clotting and intimal hyperplasia. Decellularized xenografts have been identified as potential scaffolds for small-diameter vascular substitutes. Xiong, for example, in 2013 shown a work that has aimed to develop and investigate a biomechanically functional and biocompatible acellular conduit using decellu- larized porcine saphenous arteries (DPSAs), through a modified decellularization process using Triton X-100 solution and serum-containing medium. Histological and biochemical analysis indicated a high degree of cellular removal and preservation of the extracellular matrix. Bursting pressure tests showed that the DPSAs could withstand a pressure of 1854 ± 164 mm Hg. Assessment of in vitro cell adhesion and biocompatibility showed that porcine pulmonary artery endothelial cells were able to adhere and proliferate on DPSAs in static and rotational culture. After interposition into rabbit carotid arteries in vivo, DPSAs showed patency rates of 60% at 1 month and 50% at 3 months. No aneurysm and intimal hyperplasia were observed in any DPSAs. All patent grafts showed regeneration of vascular elements, and thrombotic occlusion was found to be the main cause of graft failure, probably due to remain- ing xenoantigens. The purpose of this work was to evaluate the effects of using a decellularization protocol in samples of rabbit and human arteries and veins, involving mechanical processes and enzy- matic reactions in order to obtain a scaffold suitable for the implantation in an organism recipient. Subsequently, a further purpose of this thesis is to obtain a new type of scaffold derived from skeletal muscle decellularization. Resuming all the mentioned aspect we can confirm that autologous vessels and vascular allografts are the most reliable vessel source, but often their supply are insufficient for their widespread application [19]. For that reason we saw the use of synthetic prosthetic vessel, before tissue engineering onset. The principal vessel typology needed in a large number of vascular substitution is represented by small caliber vessel (< 2 mm) and synthetic material failed because a low patency rate due to thrombosis [20]. As listed before, the thrombogenic surface of many engineered materials becomes particularly relevant in microvascular grafts. Patency properties of vascular grafts could be considered the key point to obtain a conduit with a relevant chance of stable replacement of damaged vessel. In fact, thrombosis is the main mechanism of obliteration and subsequent failure of most microvascular anastomoses using artificial conduits. Various methods have been recorded to avoid it, such as coatings with antithrombotic drugs, as heparin, hirudin, aspirin, or tissue factor pathway inhibitor [21]. There have been attempts to emulate the endothelial cellular surface which, coated with heparan sulphate proteoglycan, produces a negative surface charge which helps to prevent platelet adherence. Some prostheses are therefore coated internally with heparin sulphate, which is quickly degraded, and some materials with an electronegative surface have been created, with uncertain results [22]. So far, many researchers have described seeding endo- thelial cells in conduits. Laube et al. reported a patency rate of 90% in 27 months for ePTFE prostheses used in coronary bypass, after additional incubation with endothelial cells which allowed them to adhere to the material [23]. The major limitation of this method is the need for cell cultures and withdrawal of tissue from the patient, and in any case it remains a two-step procedure. A tubular structure of ePTFE is also left in place, with the risk of later infection. Constructs composed entirely of cells (Tissue Engineered Blood Vessels: TEBV) have been studied to overcome these compli- cations [24]. Although the method promises amazing results, it is time-consuming and very expensive (Fig. 2). To avoid the cost of cell cultures, many researchers have tried to improve endothelial coverage of prostheses by coating them with endothelial-friendly compounds with good haemocom- patibility. E-PTFE prostheses have been coated with perlecan [6] and endothelial-specific adhesion proteins such as fibrin–and hirudin [25, 26].

Fibronectin coating seems to be a successful method, apart from loss of lining at high flow rates. This is why a functional ligand for fibronectin was used, with covalent binding of short peptide sequences (Arg-Gly-Asp, RGD) to improve cell adhesion. Instead of coating prostheses with the above substances, another possibility would be to use absorbable, already biocom- patible biomaterials, to make entire prostheses. In spite of all these experiments, endothelialisation in various types of vascular prostheses has been shown in animals but never satisfactorily in humans. The type of material is not the only essential point to allow the endothelialisation but it's fundamentally correlate with the physiopathology of endothelialisation, which takes place in three main ways: trans-anastomotic endothelialisation; transmural, and due to 'fall-out' of circulating pluripotent cells. Therefore, trying to enhance endothelialisation means acting on each of these three modalites of cell growth.

Trans-anastomotic endothelialisation (TAE) appears to be very difficult in humans. Early studies on synthetic prostheses report that they cannot be longer than 0.5 cm, even after prolonged implantation. In spite of a long period of observation, internal endothelialisation has not been observed in humans, except in sites of anastomosis [27]. Several factors have been observed to influence this, such as species, senescence, anatomic dimensions of the vessel, and prosthetic materials, but even in animals TAE is limited [28]. Study of endothelial cells, both human and canine, compared in vitro, suggest that human cells have a greater potential for migration but a lower capacity for adhesion, which may explain the lack of re-endothelialisation in vivo, when blood flow may obstruct cell adhesion [29]. Instead, the transmural pathway seems to enhance rapid endothelialisation, according to recent studies on materials with sufficient porosity. Pore size takes on importance in these studies, since the prosthesis must be sufficiently large to allow cell growth, but not too large to cause loss of intercellular adhesion [30]. Materials with differently sized pores inside and outside the conduit have even been experi- mented, in order to obtain an biocompatible surface internally and a colonizable one externally. Pore size also alters the haemocompatibility of biomaterials, as well as their compliance and degradation time. An optimal pore size for vascular engineering has been hypothesized, ranging from 30 to 50 microns. It appears that smaller pores would not allow growth of endothelial cells, and larger ones would cause excessive leakage of blood. Pores in the walls of prosthetic materials can also avoid intimal hyperplasia. It has been hypothesized that a thrombus initially deposited on the walls of the prothesis later organizes itself into muscle-like tissue, which then gives rise to intimal hyperplasia. The precocious growth of endothelial tissue would avoid thrombosis and thus the consequent cascade of events leading to intimal hyperplasia. Increased pore size causes increased failure compliance of the material. Several studies have shown that vascular implants with fibers organized in a circular fashion do not cause dilation. Even surgical technique could determine an influence in local hyper- plasia at the anastomotic site [31]. Anyway, this is local, since cells undergo mechanical stress and are thus conditioned in their spatial orientation. "Fall-out healing" leads to the formation of endothelial islands, with no connection with the formation of trans-anastomotic or transmural tissue. This is a late phenomenon that appears to be the mechanism for repairing small vascular lesions however, recent studies show how this mechanism may be enhanced, by attracting EPC cells to partic- ipate [32] One of the last procedures presented to improve the patency rate is constituted from nano modification of the vascular lumen with heparin addition to obtain a very low trombo- genic vessel The employment and the wide possibility offered in the vascular field from the use of nanotechnology was also investigated [33].

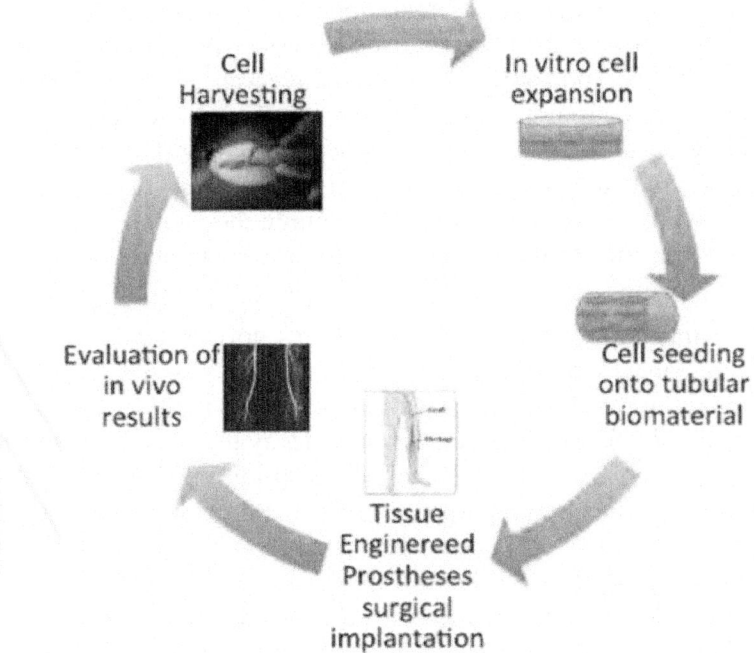

Figure 2: The in vivo –in vitro production of a tissue engineered prostheses. Adapted from: Pontini et al. "Alternative Conduits for Microvascular Anastomoses" Surg In-nov. 2013 Aug 20;21(3):277-282.

METHODS AND RESULT IN VASCULAR TISSUE ENGINEERING

Many research groups have approached the problem of developing the ideal prosthesis in a variety of ways. Their main activities ranges around the triad scaffold-cells-growth factor. Scaffolds are ideal biomaterials for conduits, and cells can be seeded and cultivated on them, after preconditioning with various growth factors. To date, despite numerous scaffolds that have been manufactured trough varied forms of tissue engineering techniques, the construc- tion of an entirely biomimetic blood vessels is still underway. To achieve a successful clinical application of tissue-engineered blood vessels, the bio fabrication of vascular grafts necessi- tates a vigorous yet time-efficient biotechnological process [34]. Several tissue engineering strategies have emerged to address biological flaws at the bloodmaterial interface of the synthetic scaffolds, hence, paving the way to vascular cell seeding and design of bioactive polymers for in situ regeneration. Moreover, advances in biomaterial design have been directed towards the generation of suitable materials that does not only mimic the native vascular tissue's mechanical properties but also promote cell growth, inhibit thrombogenicity, and facilitate extracellularr matrix production. In addition, an important characteristic of artificial scaffolds in advanced bioma- terial vessels substitutes is not just the tolerance of the cells but the capacity to mimic the natural ECM in order to regulate extent and strength of cell adhesion, growth activity, cel ldifferen-tiation, and maturation to the desired phenotype [35, 36].

The extracellular matrix proteins such as collagen, elastin, fibronectin, vitronectin, and laminins which mediate cell-material adhesion have been thoroughly assessed in an earlier review [37].Materials for vascular replacements should be biomimetic in such a way that they should be resistant not only to thrombosis, but also to inflammation, and neointimal prolifer-ation, and for all intents and purposes, they should resemble the native vessels [37]. For these reasons, it is necessary to investigate the physical, chemical, and biological properties and modifications of materials to further understand the molecular mechanism of the cell material interaction [37]. The lack of endothelial cells on the luminal surface of the artificial grafts contributes to synthetic graft thrombogenicity and promotes intimal proliferation within the graft. Endothe- lial cell (EC) seeding on the synthetic grafts has been attempted to mitigate these problems. The first group to perform endothelial cell (EC) isolation and their subsequent transplantation into vascular graft were Herring and his group [38]. Current researches indicate the signifi- cance of such process in vascular tissue engineering. The polymer surfaces which have been formerly investigated for endothelial attachment, proliferation, and function had been listed in an earlier review [39]. On the other hand, the synthetic polymers for reconstructing blood vessels for clinical practice which are based on polyethylene terephthalate (PET) or polytetra- fluoroethylene (PTFE) had been previously reviewed [37]. Furthermore, blood vessel stem cells have been studied in combination with recent and alternative types of scaffolds/polymers. Parallel to this, in scaffold-based blood vessel engineering, bioreactors and pulsatile flow systems, designed by many scientists, have been found to progress the mechanical property of the engineered blood vessels by augmenting the deposition and remodeling of extracellular matrix as well as the maturation and differentiation of self-assembled micro tissues [40]. Bioreactors, which were originally designed for industrial use, have high degree of reprodu- cibility, control, and automation for specific experimental bioprocesses and these have been the reasons for their transfer to large scale applications including vascular tissue engineering. The bioreactors allow scientists to manipulate the environment and the parameters such as pH, temperature, pressure, nutrient supply, and waste removal in order mimic the in vivo physiological condition and allow biological or biochemical processes to occur and subse- quently develop the desired tissue [41]. Taken together, the formation of a microvasculature within a tissue-engineered organ or tissue will depend on multiple factors: the biochemical environment, EC type, the micro-architecture presented by the scaffold material, and mechanical signals [42]. Due to the goal of developing biomimetic blood vessel scaffolds, many groups have designed such biomaterials. The polymers used in scaffold fabrication for tissue engineered blood vessels started from polyglycolic (PGA) to varied types such as polyglycolic acid-poly-L-lactic acid (PGAPLLA), Collagen/Elastin, chitosan, Poly (glycerol sebacate) (PGS), and very recently polyglycolide knitted fiber, and an L-lactide and ε-caprolactone copolymer sponge crosslinked to amniotic fluid. Furthermore, amniotic membranes have been used as scaffolds which signify that scaffold based tissue-engineered blood vessels can be fabricated from autologous cells at a reduced manufacturing period. Resuming the most important that have been used as scaffolds, we could generally subdivided them into four main categories: permanent material included polyurethane material, resorbe- able material and biological ones as allografts, xenografts, and derived products.

Permanent materials

Synthetic polymers such as e-PTFE and Dacron have not provided a satisfactory results in small diameter vessels (

Polyurethane materials

These polymers are biocompatible and highly versatile, since their tensile strength and radial compliance vary according to segment composition, stiff segments being responsible for tensile strength and soft segments for elasticity [31] Originally produced as permanent biomaterials, they do deteriorate in vivo, due to oxidation and enzymatic and cell-mediated degradation, with the result that their biostability is under revision. The differing composition of PU segments may lead to products with various degrees of biostability. PU have been combined with highly crystalline segments such as polycarbonates and silicon oligomers to increase their stability [34]. The most relevant development of small diameter vascular prostheses composed by PU revealed a total of 22 articles on polyurethanes, 14 in vitro, 4 on production of material and its medical properties, and only 4 in vivo. The cellular compatibility of several PU (associated with other substances) has also been studied according to method of preparation, e.g., the use of porous structures. Electrospinning has been applied to other materials in the field of vascular engineering, and produces small diameter fibers with good tensile strength on the final material.

As regards chemical composition, PU has been combined with silk fibroin, showing better histocompatiblity of pure PU after implant in rat muscular tissue [44]. Many experiments have also been made on the mixed-composition PU PDMS (polydimethylsiloxane), a silicon-based polymer. In this case, PDMS not only increased biostability but also increased haemocompat- ibility and immunocompatibility. In vivo studies show encouraging long-term viability: in one, a PEUU/PDMS polymer was created with the spray phase inversion technique in a tubular form with two-phase porosity [45]. It showed good re-endothelialisation 24 months after implant. Another in vivo study in this series used poly(ester urethane)urea (PEUU) combined with a thrombogenic polymer not similar to a phospholipid, poly(2-methacryloyloxyethyl phosphorylcholine-co-methacryloy- loxyethyl butylurethane) (PMBU), to create a fibrillar scaffold by electrospinning, with good tensile strength and compliance. The association with PMBU made the PU less prone to platelet deposition and hypertrophy of muscle cells. The in vivo patency of 1.3-mm conduits implanted in rat aorta after 8 weeks varied from 40% for pure PU to 67% for PU PMBU [46].

Bioresorbable materials

These materials may be synthetic or biopolymers already constituting the extracellular matrix. The most common absorbable biomaterials are polyesters. This category contains poly(α- hydroxylester poly(L-lactic acid) (PLLA), poly(-glycolic acid) (PGA), polylactone polyor- thoesters (POE) and polycarbonates. When these materials are implanted in vivo, their polymeric structure is subject to a hydrolysis process and metabolization of the resulting products, such as lactic and glycolic acids. Their safety and biocompatibility are now established. However, the mechanical resistance of these products does not reach the desired levels – an anticipated outcome, as PGA was originally in the form of a non-woven fabric, and thus does not have measurable tensile strength [28]. Most studies have therefore concentrated on preconditioning methods to increase resistance, e.g., use of pulsatile flow bireactors, alternative techniques of cell culture, and administration of various growth factors [47.]Another substantial problem with PGA is its stiffness, which does not confer the elastic properties of native arterial tissues. In this case too, the use of copolymers has improved results. A fibrillar scaffold based on polyglycolic or polylactic acid coated with a 50:50 L-lactate or L-caprolactone (PCLA/PGA or PCLA/PLA) copolymer has been specifically tested for vascular repair, resulting in compliance closer to that of the original vessel with better surgical handling [48]. One study showed how PGA-based matrices have greater cellularity and production of proteins of extracellular matrices based on PHAV and P4HB. The authors explained this phenomenon as due to the higher porosity of PGA (> 90%), yielding a contact surface greater than that of cells. To support the remodeling process in vivo, a biomaterial that functions only as a temporary absorbable guide, similar to an in vivo "Artery-Bioregeneration Assist Tube" (ABAT), which can promote the sequential and complete regeneration of vascular structures at the implanta- tionsite,entirelymade of Hyaluronic Acid was used in differentin vivo experimental model [49].

Other example of bioresorbeable material used for vascular purposed could be represented from collagen, which the author have experienced in small caliber vessel replacement in vivo experimental studies (Fig. 3)

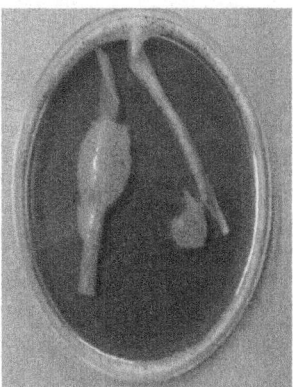

Figure 3: Example of pre-in vivo experimental implant of a biocompatible artificial collagen based artery scaffold. Pro- tocol study of the author.

Biological materials

These biomaterials are widely available and they are of course excellent substrates for cell adhesion. In addition, the processing method can retain all their advantageous mechanical properties (tensile strength, elasticity) [50]. As the main disadvantages are possible residual antigenicity and infection after implant, techniques for their decellularization and sterilization have been refined [51]. Most recent article published on biological graft mainly deal with materials already naturally present as tubular structures in the body (arteries, veins, urethers) and submitted to decellularization. They are often studied as allo-or xenografts, and enriched with cells, bFGF, heparin and VEGF to improve patency in the long term. Of special interest for the physiopathology of tissue healing after implant is one study reporting trends after implants of decellularised porcine arteries in rat, concluding that the initial inflammation due to integration in tissues does not interfere with long-term modelling. One in vitro study examines the creation of a biotube produced by reaction to a foreign body1-3. These studies show the good mechanical properties of this biomaterial, but also the poor long-term patency of conduits [52, 53]. Decellularization have so represented one of the most reliable procedure to obtain an ideal scaffold for vascular replacement, in particular for its peculiar property to retain native ECM that is the fundamental aspect for cell seeding and cell host colonization (Fig.4)

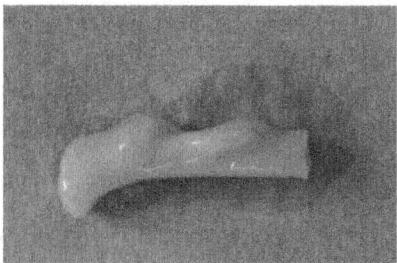

Figure 4: Example of pre-in vivo experimental implant of a decellularized artery. Protocol study of the authors.

There is not a unique decellularization protocol but every tissue need specific reactions and solutions. Protocols usually require a combination of physical, chemical and enzyme processes: the first phases are dedicated to rupture the cell membranes in order to release the intracellular components, which are then separated, dispersed, and degraded by enzymatic detergents and solutions. The last phases are directed to the elimination of cellular debris, that remained within the matrix, and of the reagents, that could interfere with the subsequent recellulariza- tion of scaffold or cause adverse reactions into recipient organism [54]. Some examples of physical treatment are the fast cycle of freeze/thaw that induce effective rupture of cell membranes, or the application of mechanical forces or the agitation of the material in combination with chemical detergents to promote the solubilization of homoge- nous membranous components [55]. As regards chemical treatments, the commonly solutions are usually composed by acid or alkaline solutions which are used for decellularization of thin layers of tissue samples as the bladder submucosa [56]. This type of treatment is effective in removing cell debris and, at the same time, has an antibacterial and antifungal activity [57]. Despite this, it has been demon- strated that the use of chemical solvents for prolonged times may alter the structure of the matrix, causing the detachment of glycosaminoglycans (GAGs) from collagen fibers [58]. Ionic detergents are another category of chemical agents usually employed during decellularization and they are involvede, with effectiveness, in removing the nuclear and cytoplasmic frag- ments. Their disadvantage could be the possible denaturation of ECM proteins. In the decellularization protocols are also used chelating agents such as EDTA, whose function is to bind and ions, which are physiologicaly essential for covalent bonds be- tween cells and matrix components, such as collagen fibers and fibronectin. In this way their use facilitate the loss of interconnections of the cells, resulting in disintegration and removal of cellular material [59]. To improve the effectiveness of the decellularization process, it is necessary the use of enzyme solutions, mainly consisting of trypsin, which specifically cleaves the protein bonds and exerts its maximum activity at the temperature of 37°C and pH 8. In addition to trypsin, is also frequently the use of the endonucleases, which include the deoxyribonuclease and ribonu clease. These enzymes catalyze the hydrolysis of covalent bonds of, respectively, DNA and RNA. However, it is recommended to avoid the exceeding exposure of samples with the action of the enzyme solutions because they could damage in the structure of the extracellular matrix. All decellularization solutions are usually implemented with antibiotic or antifungal agents to prevent any bacterial or fungal contamination. To obtain an effectively decellularized scaffolds, it is important to remove the antigenic components, such as cell surface receptors, cytoplasmic proteins and nucleic acids that, due to their immunogenic properties, could trigger a defensive reaction of the immune system of the recipient. In contrast, components of the extracellular matrix, such as collagen and elastic fibers, are widely conserved among individuals of the same or different species and, therefore, they usually do not evoke the immunogenic reaction of the host. A low intensity inflammatory response of the receiving was observed, at least, at histological level and there have been reports of significant rejection reactions, as described by Kasimir and colleagues, after implant of porcine valve prosthesis in pediatric patients [60].

Clinical relevance and rationale for the use of ECM as a biologic scaffold

The use of ECM derived from decellularized tissue is increasingly frequent in regenerative medicine and tissue engineering strategies, with recent applications including the use of threedimensional ECM scaffolds prepared by whole organ decellularization [61, 62]. Clinical products such as surgical mesh materials composed of ECM are harvested from a variety of allogenic or xenogeneic tissue sources, including dermis, urinary bladder, small intestine, mesothelium, pericardium, and heart valves, and from several different species. The potential advantage of tissue specificity for maintaining selected cell functions and phenotype has been suggested by studies of cells and ECM isolated from tissues and organs such as the liver [63], respiratory tract, nerve [64], adipose [65], and mammary gland [66]. The ECM has been shown to influence cell mitogenesis and chemotaxis, direct cell differentiation [67-71], and induce constructive host tissue remodelling responses [72-74]. It is likely that the three-dimensional ultrastructure, surface topology, and composition of the ECM all contribute to these effects. There is also evidence that residual cellular material attenuates or fully negates the constructive tissue remodelling advantages of biologic scaffold materials in vivo [75]. Therefore, tissue processing methods, including decellularization, are critical determinants of clinical success [76]. It should be understood that every cell removal agent and method will alter ECM composition and cause some degree of ultrastructure disruption. Minimization of these undesirable effects rather than complete avoidance is the objective of decellularization

Scaffold Free Method

An alternative method based on a scaffold free technique was also described and provide the advancement in the cell technology field. The application of such procedure depends from the necessity to improve the bioengineered vessel in terms to overcome process like the chronic inflammation, thrombosis, rejection, and poor mechanical properties of allogeneic or xenoge- neic and synthetic vessels that as previously reported have impaired their clinical applications [77]. In addition it has emerged due to the failure cell to cell interaction and the assembly and alignment of ECM components, and the complex host response to scaffolds, [78]. In scaffoldfree tissue engineering approach, the fabrication of the tissue construct is anchored in the crucial capability of the cells to manufacture their own extracellular matrix [39]. In 1998, the first scaffold-free tissue-engineered human blood vessel was established by L' Heureux and has been replicated for further preclinical evaluation using rat and mice models in 2006 [77]. Years later, groups of scientists reported a fully biological self-assembly approaches by implementing rapid prototyping bioprinting method and stimulation via bioreactors for scaffold-free small diameter vascular reconstruction [80]. Similar to scaffold-based technique, in tissue engineering for scaffold-free blood vessels, the bioreactors were also used to provide specific biochemical and physical signals to regulate cell differentiation, ECM production, and tissue assembly by using chemical, mechanical, or electromagnetic stimulation techniques to produce de novo tissue with properties comparable to the damaged or desired tissues [81, 82]. There are many types of launched bioreactors however, in engineering the vascular tissues, designs of various bioreactors have been based on the expansion and recoil properties of blood vessels, and so the combinations of stress, strain, and perfusion stimulation in biomimetic bioreactors have successfully developed vascular tissues [83, 84]. In case of cell senescence problem, lifespan extension via telomerase expression in vascular cells (smooth muscle cells and endothelial cells) from elderly patients has been found as an effective strategy for engineering autologous blood vessels and eventu- ally provides bypass conduit for atherosclerotic diseases. Human telomerase, composed of an RNA component and a reverse transcriptase (hTERT), maintains the telomere length at the ends of the chromosomes [85, 86].Absence of hTERT expression in mature somatic cells induces lack of telomerase activity thus its ectopic expression has been shown to restore telomerase activity, arrest telomere shortening and senescence in some cells [87]. While high cell popula- tion is essential in cell-based vessel biofabrication and the expansion process is lengthy, cellbased therapies are more promising in terms of efficacy despite the fact that they are more complex and costly than scaffold-based techniques. Therefore, many researchers have focused on this approach and the representative studies are presented in. Among the well-studied scaffold-free techniques are the coculture system, sheet-based engineering, decellularization, direct cell injection, bioprinting, and bio fabrication in a bioreactor system.

CONSIDERATION ON COST/EFFECTIVENESS

Nevertheless systematic studies about tissue engineering costs and clinical benefit are not still available, due to many differences in procedure, techniques and materials, is well known that is a medicine promising but expensive tool. In fact, since now most of the result are due to high cost basic research with low effect on daily clinical practice and, in most cases, not widely available. The scientific surgeon community agree on necessity to use the best achievement in cell tecnologies to obtain an ideal and easy way to replace damaged tissue but also the cost to obtain it must be controlled. If the industry employ large sum of money to obtain a product its cost could be not widely available and so a novel promising technology could have a low impact worldwide

More than 70 tissue engineering related start-up companies spent more than $600 million/year, with only two FDA-approved tissue-engineered products [88]. Given the modest performance in clinically approved organs, tissue engineering still remain a promising field. Often is a lacking in experimental model that avoid a perfect matching with human clinical situation. The community of bioengineering technology is advocating the application of clinically driven methodologies in large animal models enabling clinical translation. The employment of sophisticated tecnologies in cell treatment as in decellularization process, cell isolation and in vitro expansion, costs of growth factor and bioreactor and so on are the most principles obstacle to a low cost wide available bioengineered tissue. The huge clinical needing and the necessity of the low risk application of engineered substitute still represent a limitation in prevision of clinical application about most of the promising experimental result. At the same time such encouraging result provide the base for further development and spending limitation. Since now all in vitro advancement in vessel replacement technology must consider its success not only in term of ideal vessel production but also in terms of feasibility, economic and time saving procedure.

CONCLUSION

Critical reading of researches in the field of microvascular tissue engineering gave the general impression of progress in the search for an ideal replacement for small diameter vessels but the goal its aimed to is still lacking. In fact, even if several studies seems to be promising they must be completely proved in vivo in human clinical situation and in long term period as also they must obtain a therapeutic result within an acceptable cost for the community. Tissue Engineering in the context of Regenerative Medicine has been hailed for many years as one of the most important topics in medicine in the twenty-first century. While the first clinically relevant efforts were mainly concerned with the generation of bioengineered skin substitutes, subsequently tissue engineering applications have been continuously extended to a wide variety of tissues and organs. The advent of either embryonic or mesenchymal adult stem-cell technology has fostered many of the efforts to combine this promising tool with tissue engineering approaches and has merged the field into the term Regenerative Medicine. As a typical example in translational medicine, the discovery of a new type of cells called telocytes that have been described in many organs and have been detected by electron microscopy opens another gate to regenerative medicine. Besides cell-therapy strategies, the application of gene therapy combined with tissue engi- neering has been investigated to generate tissues and organs. The vascularization of constructs plays a crucial role besides the matrix and cell substitutes.

Therefore, novel in vivo models of vascularization have evolved allowing axial vascularization with subsequent transplantation of constructs. This article is intended to give an overview over some of the most recent developments and possible applications in regenerative medicine through the perspective of tissue engineering achievements and cellular research. The synthesis of bioengineering with innovative methods of molecular biology and stem-cell technology appears to be very promising. Most studies indicate the use of absorbable biomaterials, in view of their good integration, with the hope of developing autogenous vessels to replace prostheses. However, not one of these products has yet been approved for clinical experimentation. Degradability is one of the characteristics which tend to dissuade surgeons at the crucial moment of implant. In addition, synthetic not degradable material could not offer adequate surface to maintain and adequate patency in long period. There are many gaps in the examined articles. The first problem, already examined by many authors, is variability in animal models, which hinders direct comparison of results. Homogeneous studies on mechanical studies are also lacking, since so many of them focus on tensile strength, and neglect compliance, which is an essential feature of vessels. An effective model of an artificial vessel is very far from being achieved, particularly consid- ering the field of microvascular graft. So its development must take into account the context in which it could be applied. Experimental models have already been super-ceded, if we think that the application of a bio-absorbable prosthesis means that cells must be able to reconstruct a new artery and that, in clinical microsurgery practice, this must be achieved in already damaged arteries However, the procedures are time-consuming and very expensive, requiring dedicated laboratories able to guarantee sterility and suitability for in vivo re-implantation of cell cultures. As regards urgent procedures, such as revascularisation of all types, the cell culture step should be avoided. The ideal choice would be ready-to-use materials, that actually are needing their improvement.

REFERENCES

1. Tu JV, Pashos CL, Naylor CD, et al. Use of cardiac procedures and outcomes in elder- ly patients with myocardial infarction in the United States and Canada. N Engl J Med 1997;336:1500–1505.

2. McKee JA, Banik SS, Boyer MJ, et al. Human arteries engineered in vitro. EMBO Rep 2003;4:633–638.

3. Wang X, Lin P, Yao Q, Chen C. Development of small-diameter vascular graft. World J Surg. 2007 Apr;31(4):682-9

4. J. L. Platt, "Preface: future approaches to replacement of organs, " American Journal of Transplantation, 2004 vol. 4, no. 6, pp.5–6.

5. B. Ogle, M. Cascalho, and J. L. Platt, "Fusion of approaches to the treatment of organ failure, " American Journal of Transplantation 2004vol. 4, supplement 6, pp. 74–77.

6. J. Yang, M. Yamato, C. Kohno et al., "Cell sheet engineering: recreating tissues with- out biodegradable scaffolds, " Biomaterials, 2005 vol. 26, no. 33, pp. 6415–6422.

7. N. L'Heureux, N. Dusserre, A. Marini, S. Garrido, L. de la Fuente, and T. McAllister, "Technology insight: the evolution of tissue-engineered vascular grafts—from re- search to clinical practice, " Nature Clinical Practice Cardiovascular Medicine, 2007 vol. 4, no. 7, pp. 389–395.

8. J. R. Porter, T. T. Ruckh, and K. C. Popat, "Bone tissue engineering: a review in bone biomimetics and drug delivery strategies, 2009 Biotechnology Progress, vol. 25, no. 6, pp. 1539– 1560.

9. W. Ji, Y. Sun, F. Yang et al., "Bioactive electrospun scaffoldsdelivering growth factors and genes for tissue engineering applications, " Pharmaceutical Research, 2011 vol. 28, no. 6, pp. 1259–1272.

10. J. P. Vacanti and R. Langer, "Tissue engineering: the designand fabrication of living replacement devices for surgical reconstruction and transplantation, " 1999 The Lan- cet, vol. 354, supplement 1, pp. S32–S34,

11. B. S. Kim and D. J. Mooney, "Development of biocompatible synthetic extracellular matrices for tissue engineering, " Trends in Biotechnology, 1998 vol. 16, no. 5, pp. 224– 230

12. Judee Grace Nemeno-Guanzon, Soojung Lee, Johan Robert Berg, 1, 2 Yong Hwa Jo, Jee Eun Yeo,, 3 BoMi Nam, Yong-Gon Koh, and Jeong Ik Lee Trends in Tissue Engi- neering for Blood Vessels J Biomed Biotechnol. 2012;2012:956345.

13. McKee JA, Banik SS, Boyer MJ, et al. Human arteries engineered in vitro. EMBO Rep 2003;4:633–638.

14. Guidoin R, Chakfé N, Maurel S, How T, Batt M, Marois M, Gosselin C. (1993). Ex- panded polytetrafluoroethylene arterial prostheses in humans: histopathological study of 298 surgically excised grafts. 1993 Biomaterials;14(9):678-93

15. Bordenave L, Fernandez P, Rémy-Zolghadri M, Villars S et al. In vitro endothelial- ized ePTFE prostheses: clinical update 20 years after the first realization. Clin Hem- orheol Microcirc; 2005 33(3):227-34

16. Cooper GJ, Underwood MJ, Deverall PB Arterial and venous conduits for coronary artery bypass. A current review. Eur J Cardiothorac Surg. 1996;10(2):129-40

17. Klopsch C, Steinhoff G. Tissue-engineered devices in cardiovascular surgery. Eur Surg Res. 2012;49(1):44-52

18. Tillman BW, Yazdani SK, Neff LP, Corriere MA, Christ GJ, Soker S, Atala A, Geary RL, Yoo JJ. Bioengineered vascular access maintains structural integrity in response to arteriovenous flow and repeated needle puncture. J Vasc Surg. 2012 Sep;56(3): 783-93.

19. Schmedlen RH, Elbjeirami WM, Gobin AS, West JL. Tissue engineered small-diame- ter vascular grafts. Clin Plast Surg. 2003 Oct;30(4):507-17.

20. Mooney DJ Mazzoni CL, Breuer C, McNamara K, Hern D, Vacanti JP, Langer R. Sta- bilized polyglycolic acid fibre-based tubes for tissue engineering. Biomaterials; 1996 17(2):115-24

21. Kim BS, Mooney DJ Engineering smooth muscle tissue with a predefined structure. J Biomed Mater Res; 1998, 41(2): 322-32.

22. Yao Y, Wang J, Cui Y, Xu R, Wang Z, Zhang J, Wang KLi Y, Zhao Q, Kong D. Effect of sustained heparin release from PCL/chitosan hybrid small-diameter vascular grafts on anti-thrombogenic property and endothelialization. Acta Biomater. 2014 Jun;10(6):2739

23. Laube HR, Duwe J, Rutsch W, Konertz W Clinical experience with autologous endo- thelial cell-seeded polytetrafluoroethylene coronary artery bypass grafts. J Thorac Cardiovasc Surg; 2000, 120(1):134-41.

24. Kumar TR, Krishnan LK A stable matrix for generation of tissue-engineered non- thrombogenic vascular grafts. Tissue Eng. 2002, Oct;8(5):763-70

25. Hang Z, Wang Z, Liu S, Kodama M. (2004). Pore size, tissue ingrowth, and endothe- lialization of small-diameter microporous polyurethane vascular prostheses. Bioma- terials; 2004. 25(1):177-87

26. Matsuda T, Nakayama Y. (1996). Surface microarchitectural design in biomedical ap- plications: in vitro transmural endothelialization on microporous segmented polyur- ethane films fabricated using an excimer laser. J Biomed Mater Res; 1996 31(2):235-42.

27. Berger K, Sauvage LR, Rao AM, Wood SJ (1972). Healing of arterial prostheses in man: its incompleteness. Ann Surg; 1972 175(1):118-2728

28. Zilla P, Bezuidenhout D, Human P (2007). Human, Prosthetic vascular grafts: wrong models, wrong questions and no healing. Biomaterials;

2007 28(34):5009-2.

29. Dixit P, Hern-Anderson D, Ranieri J, Schmidt CE. (2001). Vascular graft endotheliali- zation: comparative analysis of canine and human endothelial cell migration on nat- ural biomaterials. J Biomed Mater Res;2001, 56(4):545-55.

30. Kidane AG, Salacinski H, Tiwari A, Bruckdorfer KR, Seifalian AM. Anticoagulant and antiplatelet agents: their clinical and device application(s) together with usages to engineer surfaces. Biomacromolecules; 2004 5(3):798-813.

31. Tiwari A, Cheng KS, Salacinski H, Hamilton G, Seifalian AM.. Improving the paten- cy of vascular bypass grafts: the role of suture materials and surgical techniques on reducing anastomotic compliance mismatch. Eur J Vasc Endovasc Surg; 200325(4): 287-95.

32. Avci-Adali M, Ziemer G, Wendel HP. Induction of EPC homing on biofunctionalized vascular graftsfor rapid in vivo self-endothelialization- -a review of current strategies. Biotechnol Adv; 201028(1):119-29

33. Rathore A, Cleary M, Naito Y, Rocco K, Breuer C. Development of tissue engineered vascular grafts and application of nanomedicine Wiley Interdiscip Rev Nanomed Nanobiotechnol. 2012 May-Jun;4(3):257-72

34. Atala AL et al. Principles of Regenerative Medicine. Elsevier: Burlington, Massachu- setts. First ed ed. 2008

35. L. Bacakova, E. Filova, F. Rypacek, V. Svorcik, and V. Stary, "Cell adhesion on artifi- cial materials for tissue engineering, " Physiological Research, 2004 vol. 53, supplement 1, pp. S35–S45.

36. L. Bacakova, E. Filova, D. Kubies et al., "Adhesion and growth of vascular smooth muscle cells in cultures on bioactive RGD peptide- carrying polylactides, " Journal of Materials Science 2007 vol. 18, no. 7, pp. 1317–1323.

37. M. Parizek, K. Novotna, and L. Bacakova, "The role of smoothmuscle cells in vessel wall pathophysiology and reconstruction using bioactive synthetic polymers, " Phys- iological Research, 2011 vol. 60, no. 3, pp. 419–437

38. M. Herring, A. Gardner, and J. Glover, "A single staged technique for seeding vascu- lar grafts with autogenous endothelium, " Surgery, 1978, vol. 84, no. 4, pp. 498–504.

39. H.M. Nugent and E. R. Edelman, "Tissue engineering therapy for cardiovascular dis- ease, " Circulation Research, 2003, vol. 92, no.10, pp. 1068–1078

40. L. Buttafoco, P. Engbers-Buijtenhuijs, A. A. Poot, P. J. Dijkstra, I. Vermes,

and J. Fei- jen, "Physical characterization of vascular grafts cultured in a bioreactor, " Biomateri- als, 2006 vol. 27, no. 11, pp. 2380–2389.

41. Biotechnology, 2004vol. 22, no. 2, pp. 80–86.

42. M. C. Peters, P. J. Polverini, and D. J. Mooney, "Engineeringvascular networks in po- rous polymer matrices, " Journal of Biomedical Materials Research, 2002 vol. 60, no. 4, pp. 668–678.

43. Kidane AG, Salacinski H, Tiwari A, Bruckdorfer KR, Seifalian AM. Anticoagulant and antiplatelet agents: their clinical and device application(s) together with usages to engineer surfaces. Biomacromolecules;2004, 5(3):798-813.

44. Wang W, Jin B, Ouyang C, Li Yet al. Acute phase reaction of different macromolecule vascular grafts healing in rat muscle. Sheng Wu Gong Cheng Xue Bao ;2010, 26(1): 79-84.

45. Khorasani MT, Shorgashti S. (2006). Fabrication of microporous polyurethane by spray phase inversion method as small diameter vascular grafts material. J Biomed Mater Res A;2006, 77(2):253-60.

46. Hong Y, Ye SH, Nieponice A, Soletti L et al. A small diameter, fibrous vascular con- duit generated from a poly(ester urethane)urea and phospholipid polymer blend. Bi- omaterials;2009, 30(13):2457-67.

47. Sodian R, Hoerstrup SP, Sperling JS, Martin DP et al. Evaluation of biodegradable, three-dimensional matrices for tissue engineering of heart valves. ASAIO J;2000, 46(1):107-10.

48. Watanabe M, Shin'oka T, Tohyama S, Hibino N et al. Tissue-engineered vascular au- tograft: inferior vena cava replacement in a dog model. Tissue Eng;2001, 20017(4): 429-39.

49. Pandis L, Zavan B, Abatangelo G, Lepidi S, Cortivo R, Vindigni V.. Hyaluronanbased scaffold for in vivo regeneration of the rat vena cava: Preliminary results in an animal model. J Biomed Mater Res A2010, ;93(4):1289-96.

50. Schmidt CE, Baier JM. Acellular vascular tissues: natural biomaterials for tissue re- pair and tissue engineering. Biomaterials;2000, 21(22):2215-31.

51. Chlupác J, Filová E, Bacáková L. Blood vessel replacement: 50 years of development and tissue engineering paradigms in vascular surgery. Physiol Res; 2009, 58 Suppl 2:S119-39.

52. Hinds MT, Rowe RC, Ren Z, Teach J, Wu PC et al. Development of a reinforced por- cine elastin composite vascular scaffold. J Biomed Mater Res A;2006, 77(3):458-69.

53. Pavcnik D, Obermiller J, Uchida BT, Van Alstine W et al. Angiographic evaluation of carotid artery grafting with prefabricated small-diameter, small-intestinal submucosa grafts in sheep. Cardiovasc Intervent Radiol. 2009, 32(1):106-13.

54. Gilbert TW, Sellaro TL, Badylak SF. Decellularization of tissues and organs Biomate- rials. 2006 Jul;27(19):3675-83

55. Schenke-Layland K, Vasilevski O, Opitz F, König K, Riemann I, Halbhuber KJ, Wah- lers T, Stock UA. Impact of decellularization of xenogeneic tissue on extracellular matrix integrity for tissue engineering of heart valves. J Struct Biol. 2003 Sep;143(3): 201-8

56. Wilshaw SP, Kearney JN, Fisher J, Ingham E. Production of an acellular amniotic membrane matrix for use in tissue engineering Tissue Eng. 2006 Aug;12(8):2117-29.

57. Petersen TH, Calle EA, Colehour MB, Niklason LE. Matrix composition and mechan- ics of decellularized lung scaffolds Cells Tissues Organs. 2012;195(3):222-31

58. Teebken OE1 , Bader A, Steinhoff G, Haverich A.Tissue engineering of vascular grafts: human cell seeding of decellularised porcine matrix Eur J Vasc Endovasc Surg. 2000 Apr;19(4):381-6

59. Grauss RW, Hazekamp MG, Oppenhuizen F, van Munsteren CJ, Gittenberger-de

60. Groot AC, DeRuiter MC. Histological evaluation of decellularised porcine aortic valves: matrix changes due to different decellularisation methods. Eur J Cardiothorac Surg. 2005 Apr;27(4):566-71

61. Kasimir MT, Rieder E, Seebacher G, Silberhumer G, Wolner E, Weigel G, Simon PComparison of different decellularization procedures of porcine heart valves Int J Artif Organs. 2003 May;26(5):421-7

62. K.Weinzierl, A. Hemprich, and B. Frerich, "Bone engineeringwith adipose tisssue de- rived stromal cells, " Journal of Cranio-Maxillofacial Surgery 2006, vol. 34, no. 8, pp. 466–471.

63. Y. Zhu, T. Liu, K. Song, X. Fan, X. Ma, and Z. Cui, "Adiposederived stem cell: a bet- ter stem cell than BMSC, " Cell Biochemistry and Function 2008, vol. 26, no. 6, pp. 664– 675.

64. S. H. Bhang, S. W. Cho, J. M. Lim et al., "Locally delivered growth factor enhances the angiogenic efficacy of adiposederived stromal cells transplanted to ischemic limbs, " Stem Cells, 2009 vol. 27, no. 8, pp. 1976–1986.

65. K. Rubina, N. Kalinina, A. Efimenko et al., "Adipose stromal cells

stimulate angio- genesis via promoting progenitor cell differentiation, secretion of angiogenic factors, and enhancing vessel maturation, " Tissue Engineering A, 2009 vol. 15, no. 8, pp. 2039– 2050.

66. 65T. J. Lee, S. H. Bhang, H. S. Yang et al., "Enhancement of longterm angiogenic effica- cy of adipose stem cells by delivery of FGF2, " Microvascular Research, 2012vol. 84, no. 1, pp. 1–8.

67. Sterodimas, J. de Faria, B.Nicaretta, and I. Pitanguy, "Tissue engineering with adi- pose-derived stem cells (ADSCs): current and future applications, " Journal of Plastic, Reconstructive and Aesthetic Surgery, 2010 vol. 63, no. 11, pp. 1886–1892.

68. S. Levenberg, J. S. Golub, M. Amit, J. Itskovitz-Eldor, and R. Langer, "Endothelial cells derived from human embryonic stem cells, " Proceedings of the National Academy of Sciences of the United States of America, 2002 vol. 99, no. 7, pp. 4391–4396.

69. M. Hristov, W. Erl, and P. C. Weber, "Endothelial progenitor cells: mobilization, dif- ferentiation, and homing, " Arteriosclerosis, Thrombosis, and Vascular Biology, 2003 vol. 23, no. 7, pp. 1185– 1189.

70. M. T. Hinds, M. Ma, N. Tran et al., "Potential of baboon endothelial progenitor cells for tissue engineered vascular grafts, " Journal of Biomedical Materials Research A, 2008 vol. 86, no.3, pp. 804–812.

71. X. Wu, E. Rabkin-Aikawa, K. J. Guleserian et al., "Tissueengineered microvessels on three-dimensional biodegradable scaffolds using human endothelial progenitor cells, " American Journal of Physiology, 2004 vol. 287, no. 2, pp. H480–H487.

72. J. M. Hill, G. Zalos, J. P. J. Halcox et al., "Circulating endothelial progenitor cells, vas- cular function, and cardiovascular risk, " The New England Journal of Medicine, 2003 vol. 348, no. 7, pp. 593–600.

73. Kawamoto, T. Asahara, and D. W. Losordo, "Transplantation of endothelial progeni- tor cells for therapeutic neovascularization, " Cardiovascular RadiationMedicine, 2002 vol. 3, no. 3-4, pp. 221–225.

74. T. Shirota, H. He, H. Yasui, and T. Matsuda, "Human endothelial progenitor cellseeded hybrid graft: proliferative and antithrombogenic potentials in vitro and fabri- cation processing, " Tissue Engineering, 2003, vol. 9, no. 1, pp. 127–136.

75. S. Kaushal, G. E. Amiel, K. J. Guleserian et al., "Functionalsmall-diameter neovessels created using endothelial progenitor cells expanded ex vivo, " Nature Medicine, 2001, vol. 7, no. 9, pp. 1035–1040.

76. A. Kocher, M. D. Schuster, M. J. Szabolcs et al., "Neovascularization

of ischemic myo- cardium by human bone-marrowderived angioblasts prevents cardiomyocyte apop- tosis, reduces remodeling and improves cardiac function, " Nature Medicine, 2001 vol. 7, no. 4, pp. 430–436.

77.   Assmus, V. Sch°achinger, C. Teupe et al., "Transplantation of progenitor cells and re- generation enhancement in acute myocardial infarction (TOPCARE-AMI), " Circula- tion, 2002, vol. 106, no. 24, pp. 3009–3017.

78.   N. L'Heureux, S. Pˆaquet, R. Labb´e, L. Germain, and F. A. Auger, "A completely bio- logical tissue-engineered human blood vessel, " The FASEB Journal, 1998 vol. 12, no. 1, pp. 47–56.

79.   L. Germain, M. Remy-Zolghadri, and F. Auger, "Tissue engineering of the vascular system: from capillaries to larger blood vessels, " Medical and Biological Engineering and Computing, 2000, vol. 38, no. 2, pp. 232–240.

80.   79H. Ozaki and H. Karaki, "Organ culture as a usefulmethod for studying the biology of blood vessels and other smoothmuscle tissues, " Japanese Journal of Pharmacology, 2002, vol. 89, no. 2, pp. 93–100.

81.   N. L'Heureux, N. Dusserre, G. Konig et al., "Human tissue engineered blood vessels for adult arterial revascularization, " Nature Medicine, 2006, vol. 12, no. 3, pp. 361–365.

82.   C. Norotte, F. S. Marga, L. E. Niklason, and G. Forgacs, "Scaffold-free vascular tissue engineering using bioprinting, " Biomaterials, 2009, vol. 30, no. 30, pp. 5910–5917

83.   S. Chaterji, K. Park, and A. Panitch, "Scaffold-free in vitro arterial mimetics: the im- portance of smooth muscle-endothelium contact, " Tissue Engineering A, 2010 vol. 16, no. 6, pp. 1901– 1912.

84.   Z. H. Syedain, L. A. Meier, J. W. Bjork, A. Lee, and R. T. Tranquillo, "Implantable ar- terial grafts from human fibroblasts and fibrin using a multi-graft pulsed flowstretch bioreactor with noninvasive strength monitoring, " Biomaterials, 2011, vol. 32, no. 3, pp. 714–722.

85.   J. Zhao, L. Liu, J.Wei et al., "A novel strategy to engineer smalldiameter vascular grafts from marrow-derived mesenchymal stem cells, " Artificial Organs, 2012, vol. 36, no. 1, pp. 93–101.

86.   L. Bacakova, E. Filova, F. Rypacek, V. Svorcik, and V. Stary, "Cell adhesion on artifi- cial materials for tissue engineering, " Physiological Research, 2004 vol. 53, supplement 1, pp. S35–S45.

87.   E. Oragui, M. Nannaparaju, and W. S. Khan, "The role of bioreactors in tissue engi- neering for musculoskeletal applications, " The Open

Orthopaedics Journal, 2011, vol. 5, supplement 2, pp. 267–270.

88. N. Plunkett and F. J. O'Brien, "IV.3. bioreactors in tissue engineering, " Studies in Health Technology and Informatics, 2010 vol. 152, pp. 214–230.

89. T. M. Nakamura, G. B. Morin, K. B. Chapman et al., "Telomerase catalytic subunit homologs from fission yeast and human, " Science, vol. 277, 1997 no. 5328, pp. 955– 959.

90. Othman SF, Xu H, Mao JJ. Future role of MR elastography in tissue engineering and regenerative medicine. J Tissue Eng Regen Med. 2013

Chapter 10

EVALUATION OF LOCALLY ESTABLISHED REFERENCE INTERVALS FOR HEMATOLOGY AND BIOCHEMISTRY PARAMETERS IN WESTERN KENYA

Collins Odhiambo[1], Boaz Oyaro[1], Richard Odipo[1], Fredrick Otieno[1], George Alemnji[2], John Williamson[3], Clement Zeh[3]

[1] Kenya Medical Research Institute, Kisumu, Kenya

[2] U.S. Centers for Disease Control and Prevention (CDC), Bridgetown, Barbados

[3] U.S. Centers for Disease Control and Prevention (CDC), Kisumu, Kenya

ABSTRACT

Background

Important differences have been demonstrated in laboratory parameters from healthy persons in different geographical regions and populations, mostly driven by a combination of genetic, demographic, nutritional, and environmental factors. Despite this, European and North American derived laboratory reference intervals are used in African countries for patient management, clinical trial eligibility, and toxicity determination; which can result in misclassification of healthy persons as having laboratory abnormalities.

Methods

An observational prospective cohort study known as the Kisumu Incidence Cohort Study (KICoS) was conducted to estimate the incidence of HIV seroconversion and identify determinants of successful recruitment and retention in preparation for an HIV vaccine/prevention trial among young adults and adolescents in western Kenya. Laboratory values generated from the KICoS were compared to published region-specific reference intervals and the 2004 NIH DAIDS toxicity tables used for the trial.

Results

About 1106 participants were screened for the KICoS between January 2007 and June 2010. Nine hundred and fifty-three participants aged 16 to 34 years, HIV-seronegative, clinically healthy, and non-pregnant were selected for this analysis. Median and 95% reference intervals were calculated for hematological and biochemistry parameters. When compared with both published region-specific reference values and the 2004 NIH DAIDS toxicity table, it was shown that the use of locally established reference intervals would have resulted in fewer participants classified as having abnormal hematological or biochemistry values compared to US derived reference intervals from DAIDS (10% classified as abnormal by local parameters vs. >40% by US DAIDS). Blood urea nitrogen was most often out of range if US based intervals were used: <10% abnormal by local intervals compared to >83% by US based reference intervals.

Conclusion

Differences in reference intervals for hematological and biochemical parameters between western and African populations highlight importance of developing local reference intervals for clinical care and trials in Africa.

Introduction

The burden of diseases such as HIV/AIDS, tuberculosis, and malaria is heaviest in sub-Saharan Africa compared to the rest of the world [1, 2]. For example, sub-Saharan Africa has the highest prevalence and incidence of HIV infection globally. As such, a major of many recent HIV prevention, care and treatment initiatives are being conducted within the region [3,4]. including most phase I /IIB HIV-1 vaccine trials [5].

With increasing clinical trials in sub-Saharan Africa to combat these diseases, there is a need for accurate clinical laboratory reference intervals for appropriate participant screening, disease progression monitoring and evaluation of possible clinical trial-associated toxicity and adverse events [6]. Reference intervals are important for guiding patient treatment and management as well as identifying abnormal hematologic values [7]. For example, the complete blood count and CD4 determination are important laboratory tests in HIV-endemic regions [8]. The level of hemoglobin concentration has utility as a prognostic indicator while CD4 is used to make decisions regarding initiation of antiretroviral drugs and to monitor disease progression. These tests require accurate reference intervals for correct interpretation of laboratory results. However, currently used reference intervals in many countries in sub-Saharan

Africa are derived from populations in Europe and North America [6, 9]. Since hematologic parameters are affected not only by individual factors such as age, sex and lifestyle, but also by population and ecological factors such as ethnic background, climate, exposure to pathogens and altitude, they vary not only between individuals but also between populations [10]. Thus, there is not a universal definition of 'normal' hence it is important to define reference intervals that are suited to the particular population of interest [10]. A few studies conducted in Africa over the last decade have highlighted differences in hematologic parameters between the local population and Caucasian populations in Europe and North America [11–15]. More recently, a study highlighted differences in hematological and biochemistry values between adolescent and young adult males [15]. Despite these recorded inter-population differences in reference values for different geographical regions, few data exist in Africa to provide locally-derived values [7, 12, 13, 16].

Despite these recorded differences, the Division of AIDS (DAIDS) National Institute of Health toxicity tables [17], are still used for grading the severity of adult and pediatric adverse events, whether or not they are considered to be related to the study intervention. This leads to unnecessary exclusion of would be participants misclassified as having abnormal hematologic parameters thereby escalating operational costs especially in phase I safety trials where there may not be a control group [18–20]. This may also lead to improper patient management through misclassification of adverse events. Due to these differences, there is a need to develop and test locally-derived age specific reference intervals within African populations.

While it is desirable to generate reference intervals for different populations, the procedure remains a challenge due to the prohibitive cost involved in performing these studies and the limitation in identifying suitable healthy reference individual. Thus, the recommendation by the Clinical and Laboratory Standards Institute (CLSI) that all diagnostic laboratories must determine and maintain their own reference interval for each laboratory parameter is impractical. CLSI further recommends that if it is not possible to establish the detailed reference studies, then validation of published reference intervals can be performed using own methodology for the population served by the laboratory. Zeh et al have recently established reference intervals for use in western Kenya [15]. These intervals were generated from a study conducted on 13–34 year old, clinically healthy, HIV-seronegative, non-pregnant residents of western Kenya. Because the established reference intervals were from a population in Siaya County in western Kenya, our aim was to validate these established reference intervals for use in Kisumu County of western Kenya. We also retrospectively determined the proportion of participants in an observational prospective cohort study known as the Kisumu Incidence Cohort Study (KICoS), who would be misclassified as having abnormal hematological parameters using the established reference intervals and compared our findings to those obtained using the 2004 NIH DAIDS toxicity tables.

MATERIALS AND METHODS

Study population

This analysis utilized 953 samples obtained from 1106 participants screened in the KICoS conducted between January 2007 and June 2010 at the KEMRI/ CDC Clinical Research Center (CRC) within New Nyanza Provincial General Hospital, Kisumu. The laboratory where the study was conducted is accredited by the South African National Accreditation System [21].

KiCoS was an observational prospective cohort study designed to estimate the incidence of HIV seroconversion and to identify determinants of successful recruitment and retention in preparation for an HIV vaccine or prevention trial among young adults and adolescents in Kisumu, western Kenya. Healthy adolescent (16–17 years) and young adult (18–34 years) residents of Kisumu who reported having sexual intercourse at least once in the past three months were eligible for the study. The study was conducted in the catchment area of Kisumu, a city of approximate population of 578,865 as projected by central bureau of statistics by 2006 in western Kenya [22]. All participants underwent screening for HIV-1 and HSV-2 among other sexually transmitted infections. Signs and symptoms were collected both in a self administered Audio Computer Assisted Self Interview (ACASI) (for STI symptoms) and a clinician administered Computer Assisted Personal Interview (CAPI) for all other symptoms. Blood samples were collected for complete blood count, HIV and HSV-2 testing with laboratory results.

Ethical Approval

Ethical approval for the study was obtained from KEMRI and CDC ethics review committee/institutional review board. Written informed consent was obtained from each participant prior to study initiation. Minors (<18 years of age) were classified as "mature" or "non-mature" using legal definitions [23]. Mature minors could consent to study participation as they would for HIV counseling and testing in Kenya. Non-mature minors went through a two-step written consent process involving consent from the parent or guardian followed by written individual assent from the minor.

Blood Collection And Hiv Serology

Whole blood was collected in EDTA vacutainer tubes (Becton Dickinson, Franklin Lakes, NJ) and transported to the KEMRI/CDC HIV-research laboratory for processing and analysis within six hours of specimen collection. HIV status was determined from whole blood using HIV rapid test kits as follows: Determine (Abbot Laboratories, Tokyo, Japan), and Unigold (Trinity Biotech Plc, Bray, Ireland), with Bioline (Standard Diagnostics Inc., Korea) as a tie breaker.

Pregnancy Testing

A urine pregnancy test was administered to all females who were not visibly pregnant, using First Sign HCG One Step (UNIMED International, Inc., South San Francisco, CA, USA).

Hematological Analysis

Absolute white blood cell counts and percentages for leukocytes (WBC) with differentials (neutrophils, lymphocytes, monocytes, eosinophils, and basophils), erythrocytes (RBC) with parameters (hemoglobin (Hb), hematocrit (Hct), MCV, and MCH), and platelet counts were determined from whole blood using a Coulter ACT 5Diff CP analyzer (Beckman Coulter, France). This was performed within 24 hours of sample collection as recommended by the manufacturer.

Biochemistry analysis

Clinical chemistries were analyzed from serum obtained from serum separation tubes (Becton Dickinson, Franklin Lakes, NJ). Samples were analyzed for alanine aminotransferase (ALT), creatinine (Cr), and blood urea nitrogen (BUN) using the Cobas Integra 400 plus biochemistry analyzer (Roche, Germany) per the manufacturer's instructions.

Quality Control

Quality control protocols included running known standards each day before testing samples. In addition, the laboratory is enrolled in external quality assurance testing programs with the College of American Pathologists (lymphocyte immunophenotyping, hematology, and clinical chemistry) and the United Kingdom National External Quality Assurance Service (lymphocyte immunophenotyping). The laboratory has satisfactory performance in UK NEQAS (Lymphocyte Immunophenotyping) and CAP Clinical Chemistry as well as CAP Hematology over the past three years.

Statistical Analysis

Data were collected on optical character recognition (OCR) enabled forms and entered with scanners. Cross-checking and data cleaning was performed regularly We followed the guidelines of the Clinical Laboratory Standards Institute (CLSI, Wayne, PA, USA) for reference interval determination [10]. While these guidelines are meant for establishing new reference intervals, the basic principles also apply to validation of reference intervals [8]. The median and the 2.5 and 97.5 percentiles were calculated for each hematological parameter. Study participants were partitioned into two age groups: those 17 and younger (adolescents) and those 18 and older (young adults) and analyzed using SAS v9.1 (Cary, NC, USA). The Wilcoxon test was used to compare hematological parameters between the two age groups separately for males and females and to compare males and females separately by age group. A two-sided P value of ≤ 0.05 was considered significant.

We compared our data against reference intervals from the established reference intervals for western Kenya (Table 1), the Massachusetts General Hospital (MGH), USA reference intervals and the U.S. NIH Division of AIDS (DAIDS) toxicity tables, to determine the number (and percentage) of study participants who had values outside the established reference intervals (2.5 to 97.5 percentiles) or who had any adverse events as graded by the DAIDS criteria.

Table 1: Hematological, immunologic and biochemistry reference intervals (median and 95th-percentile) stratified by age and gender from a 13–34 years old cohort in rural western Kenya (2003–2005) [15]

Parameter	Age 13–17 years Male	Female	Age 18–34 years Male	Female
RBC (10^6 Cells/µl)	4.9 (4.1–5.8)	4.7 (3.3–5.4)	5.3 (4.3–6.5)	4.5 (3.4–5.7)
Hb (g/dL)	13.1(10.6–15.6)	12.2 (8.1–14.2)	14.2 (11.4–16.9)	12.1 (8.0–14.2)
HCT (%)	38.8 (29.3–48.1)	35.6 (24.8–43.1)	41.7 (32.6–51.5)	35.8 (23.2–44.3)
MCV (fL)	79 (62–92)	78 (57–91)	80 (55–98)	79 (60–94)
PLT (10^3 cells/µl)	224 (103–386)	233 (134–439)	201 (102–307)	220 (88–439)
WBC(10^3 cells/µl)	5.6 (3.3–8.3)	5.2 (3.9–10.2)	5.3 (2.5–7.4)	5.6 (3.3–9.7)
Ne (10^3 cells/µl)	1.9 (0.8–5.0)	2.0 (1.1–3.1)	2.0 (0.8–3.9)	2.3 (1.3–3.8)
Ly (10^3 cells/µl)	2.2 (1.0–4.2)	2.2 (1.1–3.1)	2.2 (1.0–3.5)	2.2 (1.3–3.8)
Mo (10^3 cells/µl)	0.5 (0.2–0.7)	0.4 (0.2–0.7)	0.5 (0.2–0.9)	0.5 (0.3–0.8)
Eo (10^3 cells/µl)	0.4 (0.1–1.8)	0.4 (0.1–2.2)	0.5 (0.1–1.7)	0.4 (0.1–1.3)
Ba (10^3 cells/µl)	0.04 (0.02–0.30)	0.04 (0–0.10)	0.04 (0.01–0.19)	0.04 (0–0.20)
CD4 (10^3 cells/mm³)	874 (367–1571)	934 (465–1553)	811 (462–1306)	866 (440–1602)
CD8(10^3 cells/mm³)	468 (196–988)	505 (195–1068)	486 (201–1104)	472 (262–1167)
CD4%	42 (32–56)	44 (30–56)	41 (29–54)	44 (32–55)
CD8%	23.1(12.4–36.4)	23.5 (17.0–34.8)	24.6 (14.9–44.0)	24.3 (17.5–35.0)
CD4:CD8 ratio	1.8 (1.0–3.1)	1.8 (0.9–3.2)	1.6 (0.8–2.8)	1.8 (0.8–2.8)
ALT (µ/L)	20.5 (4.9–42.4)	17.4 (4.2–65.3)	22.4 (12.0–80.6)	18.9 (10.7–61.3)
AST (µ/L)	26.9 (17.0–59.2)	22.6 (12.0–43.1)	26.7 (12.5–69.3)	22.2 (13.5–48.5)
T-Bil (µmol/L)	13.9 (5.7–62.6)	9.7 (3.7–38.5)	13.8 (5.3–50.7)	11.5 (5.8–36.1)
Creatinine (µmol/L)	66.3 (49.6–103.7)	64.5 (48.0–87.6)	83.1 (54.2–137.8)	70.7 (52.4–96.8)
Glu (mmol/L)	3.8 (2.2–6.6)	3.8 (2.0–7.0)	3.7 (2.1–9.0)	3.8 (2.1–6.0)
BUN (mmol/L)	2.5 (1.7–4.1)	2.3 (1.2–4.8)	3.0 (1.8–5.3)	2.8 (1.4–4.5)

doi:10.1371/journal.pone.0123140.t001

RESULTS

Sample Collection Results

Out of 1106 participants screened for eligibility, 534 (48.3%) were males while 572 (51.7%) were females. Following screening, a total of 153 (13.8%) participants was excluded of which 125 (81.7%) were HIV-1 infected, 20 (13.1%) were pregnant and 8 (5.2%) both HIV-1 infected and pregnant. Thus, 499 (93.4%) male and 454 (79.4%) clinically healthy female participants were selected for this analysis. Of the male participants, 22.0% (110) were adolescents and 78.0% (389) were young adults while adolescents and young adults constituted 29.1% (132) and 70.9% (322) of the female participants respectively. The number of participants tested for each parameter was within the sample size (N = 120) recommended by the CLSI for the establishment of reference intervals except the male adolescent group which had 110 participants. However all gender and age groups had sample size above the number required for reference interval transference (N = 60) [10].

Hematology and Chemistry Reference Intervals

Tables 2 and 3 summarizes the calculated median and 95th percentile reference interval for hematological and biochemistry parameters for adolescents and young adults respectively obtained from this study. The reference intervals were generally comparable although our upper reference limit for some parameters was slightly higher than those of the established reference intervals. There were significant differences in Hb, RBC, Hct, creatinine, ALT and BUN between male and female participants in both adolescent and young adult cohorts with males having higher values but these differences were not clinically relevant (Table 4). We also observed significant differences in the hematological indices among males by age, with the young adults having a higher median as compared to adolescents in Hb (15.1 g/dL versus 14.2 g/dL), Hct (45.4% versus 42.6%), RBC (5.4X10^6/ μL versus 5.2 X10^6/ μL), ALT (17.4 μ/L versus 16.4 μ/L), creatinine (93 μmol/L versus 65 μmol/L) and neutrophils (2.6X10^3/ μL versus 2.2X10^3/ μL). Compared to the males, there was little variation in these parameters among female adolescent and adult participants except creatinine. However, females had significantly higher PLT, lymphocytes and WBC than males in both adolescent and young adult cohorts. There were significant differences in neutrophil counts between male and female adolescents, with the females having higher counts than males. There were no gender or age differences in absolute basophil, eosinophil and monocytes counts.

Table 2: Adolescent hematological and biochemistry reference values (median and 95th-percentile) comparison between locally-established reference intervals for western Kenya versus reference values established from the Kisumu Incidence cohort study in western Kenya (2007–2010)

Parameter	Local interval (Age 13–17 years) [15] Male	Female	This study (Age 16–17 years) Male	Female
RBC (10⁶ Cells/µl)	4.9 (4.1–5.8)	4.7 (3.3–5.4)	5.2 (4.3–6.4)	4.9 (3.7–6.0)
Hb (g/dL)	13.1(10.6–15.6)	12.2 (8.1–14.2)	14.2 (11.1–16.7)	12.7 (7.5–14.8)
HCT (%)	38.8 (29.3–48.1)	35.6 (24.8–43.1)	42.6 (33.7–49.7)	38.0 (24.2–43.7)
MCV (fL)	79 (62–92)	78 (57–91)	80.9 (66.2–91.5)	79.0 (52.5–88.5)
PLT (10³ cells/µl)	224 (103–386)	233 (134–439)	215 (112–474)	264 (126–448)
WBC(10³ cells/µl)	5.6 (3.3–8.3)	5.2 (3.9–10.2)	5.2 (3.6–9.1)	6.0 (3.6–9.5)
Ne (10³ cells/µl)	1.9 (0.8–5.0)	2.0 (1.1–3.1)	2.2 (0.9–6.7)	2.7 (1.3–5.8)
Ly (10³ cells/µl)	2.2 (1.0–4.2)	2.2 (1.1–3.1)	2.2 (1.4–3.4)	2.5 (1.2–3.9)
Mo (10³ cells/µl)	0.5 (0.2–0.7)	0.4 (0.2–0.7)	0.4 (0.2–0.8)	0.5 (0.2–1.1)
Eo (10³ cells/µl)	0.4 (0.1–1.8)	0.4 (0.1–2.2)	0.22 (0.03–1.64)	0.24 (0.05–1.4)
Ba (10³ cells/µl)	0.04 (0.02–0.30)	0.04 (0–0.10)	0.04 (0.02–0.09)	0.04 (0.02–0.11)
ALT (µ/L)	20.5 (4.9–42.4)	17.4 (4.2–65.3)	16.4 (7.8–33.9)	14.1 (5.7–32.5)
Creatinine (µmol/L)	66.3 (49.6–103.7)	64.5 (48.0–87.6)	65 (39–89)	51 (40–69)
BUN (mmol/L)	2.5 (1.7–4.1)	2.3 (1.2–4.8)	2.7 (1.2–4.5)	2.4 (1.2–4.2)

doi:10.1371/journal.pone.0123140.t002

Table 3: Adult hematological and biochemistry reference values (median and 95th-percentile) comparison between locally-established reference intervals versus reference values from the Kisumu Incidence cohort study in western Kenya (2007–2010)

Parameter	Local interval (Age 18–34 years) [15] Male (n = 110)	Female (n = 132)	This study (Age 18–34 years) Male (n = 389)	Female (n = 322)
RBC (10⁶ Cells/µl)	5.3 (4.3–6.5)	4.5 (3.4–5.7)	5.4 (4.6–6.6)	4.8 (4.0–5.8)
Hb (g/dL)	14.2 (11.4–16.9)	12.1 (8.0–14.2)	15.1 (12.6–17.2)	12.8 (9.0–14.9)
HCT (%)	41.7 (32.6–51.5)	35.8 (23.2–44.3)	45.4 (38.1–51.6)	38.6 (28.6–44.2)
MCV (fL)	80 (55–98)	79 (60–94)	84.0 (67.4–93.6)	80.4 (59.3–93.2)
PLT (10³ cells/µl)	201 (102–307)	220 (88–439)	227 (126–356)	270 (147–454)
WBC(10³ cells/µl)	5.3 (2.5–7.4)	5.6 (3.3–9.7)	5.6 (3.3–9.6)	5.9 (3.7–9.1)
Ne (10³ cells/µl)	2.0 (0.8–3.9)	2.3 (1.3–3.8)	2.6 (1.3–5.2)	2.7 (1.3–5.0)
Ly (10³ cells/µl)	2.2 (1.0–3.5)	2.2 (1.3–3.8)	2.1 (1.2–3.4)	2.3 (1.4–3.8)
Mo (10³ cells/µl)	0.5 (0.2–0.9)	0.5 (0.3–0.8)	0.4 (0.2–0.7)	0.4 (0.2–0.8)
Eo (10³ cells/µl)	0.5 (0.1–1.7)	0.4 (0.1–1.3)	0.23 (0.04–1.6)	0.21 (0.04–1.2)
Ba (10³ cells/µl)	0.04 (0.01–0.19)	0.04 (0–0.20)	0.04 (0.01–0.14)	0.04 (0.02–0.09)
ALT (µ/L)	22.4 (12.0–80.6)	18.9 (10.7–61.3)	17.4 (8.4–54.7)	13.5 (7.2–34.1)
Creatinine (µmol/L)	83.1 (54.2–137.8)	70.7 (52.4–96.8)	93 (69–123)	78 (57–100)
BUN (mmol/L)	3.0 (1.8–5.3)	2.8 (1.4–4.5)	2.8 (1.5–5.0)	2.4 (1.2–4.1)

doi:10.1371/journal.pone.0123140.t003

Table 4: Test of difference in hematologic and clinical chemistry parameters between gender and age-groups from the 16–34 years old cohort in Kisumu Kenya (2007–2010)

Parameter	Gender	n	Age 16–17 years median	p-value (gender)	n	Age 18–34 years median	p-value (gender)	P-value (age)
Hemoglobin (g/dL)	Female	132	12.7 (7.5–14.8)	<0.0001	322	12.8 (9.0–14.9)	<0.0001	0.3143
	Male	110	14.2 (11.1–16.7)		389	15.1 (12.6–17.2)		<0.0001
Hematocrit (%)	Female	132	38.0 (24.2–43.7)	<0.0001	322	36.6 (28.6–44.2)	<0.0001	0.2242
	Male	110	42.6 (33.7–49.7)		389	45.4 (38.1–51.6)		<0.0001
WBC (x1000)	Female	132	6.0 (3.6–9.5)	0.0025	322	5.9 (3.7–9.1)	0.0002	0.4387
	Male	110	5.2 (3.6–9.1)		389	5.6 (3.3–9.6)		0.5766
RBC (x10¹²/L)	Female	132	4.9 (3.7–6.0)	<0.0001	322	4.8 (4.0–5.8)	<0.0001	0.4424
	Male	110	5.2 (4.3–6.4)		389	5.4 (4.6–6.6)		<0.0001
Lymphocytes (x10⁹/L)	Female	132	2.5 (1.2–3.9)	0.0112	322	2.3 (1.4–3.8)	<0.0001	0.0789
	Male	110	2.2 (1.4–3.4)		389	2.1 (1.2–3.4)		0.0261
Neutrophiles (x10⁹/L)	Female	132	2.7 (1.3–5.8)	0.0112	322	2.7 (1.3–5.0)	0.0538	0.4565
	Male	110	2.2 (0.9–6.7)		389	2.6 (1.3–5.2)		0.0169
PLT(x10⁹/L)	Female	132	264 (126–448)	<0.0001	322	270 (147–454)	<0.0001	0.3589
	Male	110	215 (112–474)		389	227 (126–356)		0.1218
ALT (µ/L)	Female	132	14.1 (5.7–32.5)	0.0030	322	13.5 (7.2–34.1)	<0.0001	0.2750
	Male	110	16.4 (7.8–33.9)		388	17.4 (8.4–54.7)		0.0417
BUN (mmol/L)	Female	132	2.4 (1.2–4.2)	0.0241	322	2.4 (1.2–4.1)	<0.0001	0.7883
	Male	110	2.7 (1.2–4.5)		388	2.8 (1.5–5.0)		0.0604
Creatinine (µmol/L)	Female	132	51 (40–69)	<0.0001	322	78 (57–100)	<0.0001	<0.0001
	Male	110	65 (39–89)		388	93 (69–123)		<0.0001

doi:10.1371/journal.pone.0123140.t004

Comparison with locally established reference intervals, US MGH and NIH-DAIDS toxicity tables

Using the US-based MGH values, most of the KiCoS participants would have been misclassified as out of range with the highest misclassification in BUN parameter which would result in over 80% of participants excluded (Tables 5 and 6). However, using the locally established reference intervals, very few of the KiCoS participants would have been misclassified as out of range with the highest misclassification (<10%) being BUN in both adult and adolescent cohorts except males in the later (14.5%). Using the US-based MGH values, about a quarter (26.4%) of our adult female and 7.2% of adult male participants would have been misclassified as having out of range Hb levels (Table 5). In contrast, using the established reference intervals for western Kenya, only 1.2% and 1.0% of adult female and male participants would have been misclassified as having out of range Hb levels. This observation was similar for other red cell indices including Hct, MCV and RBC count with higher proportion of female participants misclassified.

Table 5: Out of range and frequency of adverse events in the Kisumu Adult cohort obtained from comparison with values from locally-established reference intervals, DAIDS and North American derived MGH values

Parameter	Gender	This Study 95% reference interval	n	Local reference [15] 95% reference interval	n	%	Out of range comparison MGH-USA [9] 95% reference interval	n	%	2004 DAIDS [17] Cut-off	N	%
Hemoglobin (g/dL)	Female	9.0–14.9	322	8.0–14.2	4	1.2	12–16	85	26.4	>10.9	40	12.4
	Male	12.6–17.2	389	11.4–16.9	4	1.0	13.5–17.5	28	7.2	>10.9	3	0.8
Hematocrit (%)	Female	28.6–44.2	322	23.2–44.3	2	0.6	36–46	83	25.8			
	Male	38.1–51.6	389	32.6–51.5	3	0.8	41–53	37	9.5			
MCV (%)	Female	59.3–93.2	322	60–94	12	3.7	80–100	153	47.5			
	Male	67.4–93.6	389	55–98	3	0.8	80–100	103	26.5			
WBC (x10⁹/L)	Female	3.7–9.1	322	3.3–9.7	8	2.5	4.5–11.0	37	11.5	>2.5	1	0.3
	Male	3.3–9.6	389	2.5–7.4	0	0	4.5–11.0	82	21.1	>2.5	0	0
RBC (x10¹²/L)	Female	4.0–5.8	322	3.4–5.7	1	0.3	4.0–5.2	7	2.2			
	Male	4.6–6.6	389	4.3–6.5	2	0.5	4.2–6.3	2	0.5			
Lymphocytes (x10⁹/L)	Female	1.4–3.8	322	1.3–3.8	7	2.2	1.0–4.8	0	0			
	Male	1.2–3.4	389	1.0–3.5	4	1.0	1.0–4.8	4	1.0			
Neutrophils (x10⁹/L)	Female	1.3–5.0	322	1.3–3.8	10	3.1	1.8–7.7	0	0	>1.3	10	3.3
	Male	1.3–5.2	389	0.8–3.9	0	0	1.8–7.7	4	1.0	>1.3	13	4.0
PLT (x10⁹/L)	Female	147–454	322	88–439	1	0.3	150–350	12	3.7	≥125	4	1.2
	Male	126–356	389	102–307	4	1.0	150–350	30	7.7	≥125	8	2.1
Eosinophils (10³ cells/µl)	Female	0.04–1.2	322	0–1.3	7	2.2	0–0.5	52	16.1			
	Male	0.04–1.6	389	0–1.7	5	1.3	0–0.5	77	19.8			
ALT (µ/L)	Female	7.2–34.1	322	0–61.3	1	0.3	0–35	7	2.2	<76.6	0	0
	Male	8.4–54.7	389	0–80.6	4	1.0	0–35	39	10.0	<100.8	2	0.5
BUN (mmol/L)	Female	1.2–4.1	322	1.4–4.5	23	7.1	3.6–7.1	305	94.7			
	Male	1.5–5.0	389	1.8–5.3	32	8.2	3.6–7.1	324	83.3			
Creatinine (µmol/L)	Female	57–100	322	0–96.8	18	5.6	0–133	0	0	<106.5	4	1.2
	Male	69–123	389	0–137.8	2	0.5	0–133	2	0.5	<151.6	1	0.3

doi:10.1371/journal.pone.0123140.t005

Table 6: Out of range and frequency of adverse events in the Kisumu Adolescent cohort obtained from comparison with values from locally-established reference intervals and DAIDS values

Parameter	Gender	This Study 95% reference interval	n	Local reference [15] 95% reference interval	n	%	Out of range comparison MGH-USA [9] 95% reference interval	n	%	2004 DAIDS [17] Cut-off	N	%
Hemoglobin (g/dL)	Female	7.5–14.8	132	8.1–14.2	7	5.3	12–16	37	28.0	>10.9	18	13.6
	Male	11.1–16.7	110	10.6–15.6	1	0.9	13.5–17.5	36	32.7	>10.9	2	1.8
Hematocrit (%)	Female	24.2–43.7	132	24.8–43.1	2	1.5	36–46	30	22.7			
	Male	33.7–49.7	110	29.3–48.1	2	1.8	41–53	41	37.3			
MCV (%)	Female	59.3–93.2	132	57–91	6	4.5	80–100	76	57.6			
	Male	67.4–93.6	110	62–92	1	0.9	80–100	48	43.6			
WBC (x10⁹/L)	Female	3.6–9.5	132	3.9–10.2	3	2.3	4.5–11.0	12	9.1	>2.5	3	2.3
	Male	3.6–9.1	110	3.3–8.3	0	0	4.5–11.0	27	24.5	>2.5	0	0
RBC (x10¹²/L)	Female	3.7–6.0	132	3.3–5.4	1	0.8	4.0–5.2	4	3.0			
	Male	4.3–6.4	110	4.1–5.8	2	1.8	4.2–6.3	2	1.8			
Lymphocytes (x10⁹/L)	Female	1.2–3.9	132	1.1–3.1	4	3.0	1.0–4.8	2	1.5			
	Male	1.4–3.4	110	1.0–4.2	0	0	1.0–4.8	0	0			
Neutrophiles (x10⁹/L)	Female	1.3–5.8	132	1.1–3.1	3	2.3	1.8–7.7	20	15.2	>1.3	3	2.3
	Male	0.9–6.7	110	0.8–5.0	0	0	1.8–7.7	27	24.5	>1.3	9	8.2
PLT (x10⁹/L)	Female	126–448	132	134–439	2	1.5	150–350	6	4.5	≥125	3	2.3
	Male	112–474	110	103–386	1	0.9	150–350	14	12.7	≥125	8	7.3
Eosinophils (10³ cells/µl)	Female	0.04–1.2	132	0–2.2	1	0.8	0–0.5	24	18.2			
	Male	0.04–1.6	110	0–1.8	0	0	0–0.5	22	20.0			
ALT (µ/L)	Female	5.7–32.5	132	0–65.3	1	0.8	0–35	3	2.3	<76.6	0	0
	Male	7.8–33.9	110	0–42.4	1	0.9	0–35	2	1.8	<100.8	0	0
BUN (mmol/L)	Female	1.2–4.2	132	1.2–4.8	4	3.0	3.6–7.1	116	87.9			
	Male	1.2–4.5	110	1.7–4.1	16	14.5	3.6–7.1	96	87.3			
Creatinine (µmol/L)	Female	40–69	132	0–87.6	0	0	0–133	0	0	<106.5	0	0
	Male	39–89	110	0–103.7	0	0	0–133	0	0	<151.6	0	0

doi:10.1371/journal.pone.0123140.t006

Using the 2004 NIH DAIDS toxicity grading to select participants eligible for the study (Tables5 and 6), 12.8% (n = 58) of female participants and 1.0% (n = 5) of male participants would have been classified as having an abnormal Hb level. However, only 2.4% (n = 11) of the female participants and 1.0% (n = 5) of male participants would have been classified as having out of range values using the locally established reference intervals. Similarly, 3.4% (n = 17) male and 1.5% (n = 7) female participants would have been classified as having an abnormal platelet count using the 2004 NIH DAIDS toxicity grading while only 1.0% (n = 5) male and 0.7% (n = 3) female participants would have been classified as out of range using the established reference interval for western Kenya.

DISCUSSION

With increasing clinical trials in Africa in an effort to combat tropical diseases [24], a need arises to consider the health status of the likely participants in such studies [12]. In this regard, several African studies have generated reference intervals for use in the respective regions [13, 15, 24, 25]. While it is important to develop locally derived reference intervals that ensure proper assessment of volunteers in clinical trials, monitoring of laboratory-based adverse events and prevention of unnecessary exclusion, it is important to evaluate their use within the local population. To our knowledge, this is one of the first evaluations of established reference intervals reported in sub-Saharan Africa. In this study, we evaluate the use of hematological and biochemistry reference intervals established for western Kenya using specimen drawn from participants in a HIV incidence cohort study in Kisumu. Our values were comparable to those of the established reference intervals for most parameters although our median values were slightly higher for most hematological parameters. This may be so given that the samples for this evaluation were drawn from an urban population that may have had access to better healthcare, clean water and nutrition than the rural population from where the established reference intervals for western Kenya [15] were obtained. Moreover, using the US MGH reference interval, the overall out of range MCV values constituted 54.0% [15] of the study population in rural western Kenya but only 27.6% in this study. Low MCV is an indirect marker of iron deficiency [26]. This is further corroborated by the low eosinophil counts observed in this study. Our eosinophil counts are comparable to those obtained from an urban population of blood donors in Uganda [18] in contrast to higher counts in a similar study in a rural population in the same country [27]. Our values for ALT and creatinine were also lower than those of Zeh et al [15]. Similarly, a study in Cameroon designed to establish reference intervals for biochemical parameters reported statistical differences in biochemistry reference parameters between participants from urban and rural geographic regions [28]. While this might not necessitate the need to establish separate intervals, consideration should be made when applying such intervals within specific populations.

Using the US-based MGH reference intervals to hypothetically select participants in a trial based on Hb, WBC counts, neutrophil counts, eosinophil counts and platelets, 51.5% (n = 491) of the total participants would have been excluded from participating in the study (Table 7). However, using the established reference intervals for western Kenya, only 6.7% (n = 64) of participants would have been excluded from participating in the study. Including BUN in the selection criteria would result in exclusion of over 80% of participants. This was similar in other African studies [7, 13, 15, 27] suggesting that this may result from a common environmental or genetic factor [12, 13]. Thus use of locally established reference intervals would reduce the overall screening to enrollment ratio in this case. This reduces the overall cost of screening and theoretically would reduce the time period for screening by reaching the study target within a shorter time period. Eller et al. have documented similar findings in a study of healthy adult Ugandan blood donors [18]. It is not surprising to see that the adolescent cohort resulted in the most out of range values hence yielded the least enrolled participants using the western-derived reference intervals and toxicity tables. Thus, partitioning of male adolescents needs to be considered in future trials.

Table 7: Hypothetical enrollment using local reference intervals compared to US-derived reference intervals and the DAIDS toxicity tables

Age category	Sex	No of participants	Number (%) Enrolled Local intervals [15]	US MGH [9]	*DAIDS [17]
Adolescents	Male	110	106 (96.4%)	27 (24.5%)	70 (63.6%)
	Female	132	113 (85.6%)	67 (50.8%)	81 (61.4%)
Adults	Male	389	374 (96.1%)	196 (50.4%)	289 (74.3%)
	Female	322	296 (92.0%)	172 (53.4%)	218 (67.7%)
	Total	953	889 (93.3%)	462 (48.5%)	658 (69.0%)

* eosinophil count grading using US derived values for adults.

doi:10.1371/journal.pone.0123140.t007

Similarly, using the 2004 NIH US DAIDS toxicity grading for screening, 31.0% (n = 295) of participants would have been excluded from the study. Although the table has been revised for some parameters including neutropenia [29], a large proportion of our study participants would still have been excluded based on Hb levels. Moreover, the toxicity table does not take into account the significant difference in red blood cell parameters between males and females, thus, a majority of those excluded would constitute female participants.

The CLSI guidelines recommend the collection of specimen from healthy volunteers for use in validating reference intervals. Thus, a limitation of this study was the failure to screen for possible asymptomatic parasitic infections like malaria and helminthes which are endemic within the study region. However, our Hb values were much higher than those from a study within the same region that screened out malaria infected participants [13]. Moreover, our eosinophil counts were much lower than the two studies within the region which screened participants from a rural population [13, 15]. A second limitation may have been that this was a self-selected population of participants willing to participate in a cohort study. However, this represents a similar population that would be willing to participate in a clinical trial thus provides a good sample to evaluate the use of the locally established reference intervals. Moreover, the HIV prevalence of the study population (12.0%) is comparable to the prevalence within the general population [30].

Given that the number of clinical trials and persons receiving clinical services is expected to increase substantially in sub-Saharan Africa, there is a need for the establishment and evaluation of locally derived clinical laboratory reference values to ensure appropriate general health assessment, treatment monitoring, and efficient implementation of clinical trials. Even more important is the need for the establishment of toxicity grading tables for application in clinical care among Africans based on the documented differences between laboratory reference intervals from African and Caucasian populations. This study confirms that the hematological and biochemistry reference intervals established by Zeh et al. are valid for use in participant recruitment in western Kenya.

ACKNOWLEDGMENTS

Disclaimer: The findings and conclusions in this article are those of the authors and do not necessarily represent the views of the CDC. Use of trade names is for identification purposes only and does not constitute endorsement by the CDC or the Department of Health and Human Services.

We are grateful to the study participants, the Kisumu Incidence Cohort study team, the HIV research laboratory, Kenya Medical Research Institute (KEMRI) and Kenya Ministry of Health whose participation made this study possible. This paper is published with the permission of the Director of KEMRI.

AUTHOR CONTRIBUTIONS

Conceived and designed the experiments: CO CZ. Performed the experiments: BO RO. Analyzed the data: CO JW CZ. Contributed reagents/materials/ analysis tools: GA FO CZ. Wrote the paper: CO JW GA CZ. Reviewed final manuscript: CO BO RO FO GA JW CZ.

REFERENCES

1. The Global Fund to Fight AIDS, TB and Malaria;"Global FundARVFactSheet. (01-June-2009). Available: http://www. theglobalfund.org/en/publications/annualreports/

2. UNAIDS Global Report: UNAIDS Report on the Global AIDS Epidemic 2013 WHO press. Geneva, Switzerland. Available:.http://www.unaids. org/en/resources/documents/2013/20130923_UNAIDS_Global_ Report_2013. Accessed 07 March 2015.

3. PEPFAR. The United States President's Emergency Plan for AIDS relief. Seventh Annual Report to Congress on PEPFAR; 2011. Available:http:// www.pepfar.gov/press/seventhannualreport/. Accessed 07 March 2015.

4. UNAIDS (2010) Report on the global AIDS epidemic. Geneva, Switzerland, WHO press. UNAIDS.

5. Esparza J, Osmanov S. HIV vaccines: a global perspective. Curr Mol Med. 2003;3:183–193. pmid:12699356 doi: 10.2174/1566524033479825

6. Jaoko W, Nakwagala FN, Anzala O, Manyonyi GO, Birungi J, Nanvubya A, et al. Safety and immunogenicity of recombinant low-dosage HIV-1 A vaccine candidates vectored by plasmid pTHr DNA or modified vaccinia

virus Ankara (MVA) in humans in East Africa. Vaccine 2008; 26: 2788–2795. doi: 10.1016/j.vaccine.2008.02.071. pmid:18440674

7. Saathoff E, Schneider P, Kleinfeldt V, Geis S, Haule D, Maboko L, et al. Laboratory reference values for healthy adults from southern Tanzania. Trop Med Int Health. 2008; 13: 612–625. doi: 10.1111/j.1365-3156.2008.02047.x. pmid:18331386

8. Lawrie D, Coetzee LM, Becker P, Mahlangu J, Stevens W, Glencross DK. Local reference ranges for full blood count and CD4 lymphocyte count testing. S Afr Med J. 2009; 99: 243–248. pmid:19588777

9. Kratz A, Ferraro M, Sluss PM, Lewandrowski KB. Case records of the Massachusetts General Hospital. Weekly clinicopathological exercises. Laboratory reference values. N Engl J Med. 2004; 351: 1548–1563. pmid:15470219 doi: 10.1056/nejmcpc049016

10. National Committee for Clinical laboratory Standards. How to define and determine reference intervals in the clinical laboratory; approved guideline-second-edition. Wayne, PA, USA: NCCLS C28-A2, vol 20 (13); 2000.

11. Buchanan AM, Muro FJ, Gratz J, Crump JA, Musyoka AM, Sichangi MW, et al. Establishment of haematological and immunological reference values for healthy Tanzanian children in Kilimanjaro Region. Trop Med Int Health. 2010; 15: 1011–1021. doi: 10.1111/j.1365-3156.2010.02585.x. pmid:20636301

12. Karita E, Ketter N, Price MA, Kayitenkore K, Kaleebu P, Nanvubya A, et al. CLSI derived hematology and biochemistry reference intervals for healthy adults in eastern and southern Africa. PLOS One. 2009;4: e4401. doi: 10.1371/journal.pone.0004401. pmid:19197365

13. Kibaya RS, Bautista CT, Sawe FK, Shaffer DN, Sateren WB, Scott PT, et al. Reference ranges for the clinical laboratory derived from a rural population in Kericho, Kenya. PLOS One. 2008; 3:e3327. doi: 10.1371/journal.pone.0003327. pmid:18833329

14. Menard D, Mandeng MJ, Tothy MB, Kelembho EK, Gresenguet G, Talarmin A. Immunohematological reference ranges for adults from the Central African Republic. Clin Diagn Lab Immunol. 2003; 10: 443–445. pmid:12738646 doi: 10.1128/cdli.10.3.443-445.2003

15. Zeh C, Amornkul PN, Inzaule S, Ondoa P, Oyaro B, Mwaengo DM, et al. Population-based biochemistry, immunologic and hematological reference values for adolescents and young adults in a rural population in Western Kenya. PLOS One. 2011; 6:e21040. doi: 10.1371/journal.pone.0021040. pmid:21713038

16. Tsegaye A, Messele T, Tilahun T, Hailu E, Sahlu T, Doorly R, et al. Immunohematological reference ranges for adult Ethiopians. Clin Diagn Lab Immunol. 1999; 6: 410–414. pmid:10225845

17. DAIDS. Division of AIDS Table for Grading the Severity of Adult and Pediatric Adverse Events. Bethseda, MD, USA. DAIDS: 2004

18. Eller LA, Eller MA, Ouma B, Kataaha P, Kyabaggu D, Tumusiime R, et al. Reference intervals in healthy adult Ugandan blood donors and their impact on conducting international vaccine trials. PLOS One. 2008: 3: e3919. doi: 10.1371/journal.pone.0003919. pmid:19079547

19. 19.Lubega IR, Fowler MG, Musoke PM, Elbireer A, Bagenda D, Kafulafula G, et al. Considerations in using US-based laboratory toxicity tables to evaluate laboratory toxicities among healthy malawian and Ugandan infants. J Acquir Immune Defic Syndr. 2010; 55: 58–64. doi: 10.1097/QAI.0b013e3181db059d. pmid:20588184

20. Omosa-Manyonyi GS, Jaoko W, Anzala O, Ogutu H, Wakasiaka S, Malogo R, et al. Reasons for ineligibility in in phase 1 and 2A HIV vaccine clinical trials at Kenya AIDS vaccine initiative (KAVI), Kenya. PLOS One. 2011; 6: e14580. doi: 10.1371/journal.pone.0014580. pmid:21283743

21. SANAS South African National Accreditation System. Available:http://home.sanas.co.za/. Accessed 07 March 2015.

22. Central Bureau of Statistics. Ministry of Planning and National Development. Population distribution by administrative areas and urban centres, Kenya 1999 Population and Housing Census. Vol. 1. Nairobi, Kenya Central Bureau of Statistics; 1999.

23. National AIDS and STD Control Programme, Ministry of Public Health and Sanitation, Kenya. Guidelines for HIV Testing and Counselling in Kenya. NASCOP; 2008. Available: http://www.google.com/url?sa=t&r ct=j&q=&esrc=s&source=web&cd=1&ved=0CB0QFjAA&url=http% 3A%2F%2Fwww.who.int%2Fhiv%2Ftopics%2Fvct%2Fpolicy%2FKe nyaGuidelines_Final2009.pdf&ei=2h37VOyGMbLe7AaQroCICA&us g=AFQjCNEyh5-DfPUxq6eJZyvZPu6o-Og7zQ&bvm=bv.87611401,d. ZWU. Accessed 07 March 2015.

24. Excler J. AIDS vaccine efficacy trials: expand capacity and prioritize. 'Throughout Africa, Asia and Latin America state-of-the-art clinics and laboratories...exist where, 4 years ago, there were none'. Expert Rev Vaccines. 2006; 5: 167–170. pmid:16608417 doi: 10.1586/14760584.5.2.167

25. Kueviakoe IM, Segbena AY, Jouault H, Vovor A, Imbert M. Hematological reference values for healthy adults in Togo. ISRN Hematol. 2011: e3919. doi: 10.5402/2011/736062

26. Tefferi A. Anemia in adults: a contemporary approach to diagnosis. Mayo Clin Proc. 2003; 78: 1274–1280. pmid:14531486 doi: 10.4065/78.10.1274

27. Lugada ES, Mermin J, Kaharuza F, Ulvestad E, Were W, Langeland N, et al. Population-based hematologic and immunologic reference values for a healthy Ugandan population. Clin Diagn Lab Immunol. 2004; 11: 29–34. pmid:14715541 doi: 10.1128/cdli.11.1.29-34.2004

28. Alemnji GA, Mbuagbaw J, Folefac E, Teto G, Nkengafac S, Atems N, et al. Reference Physiological Ranges for Serum Biochemical Parameters among Healthy Cameroonians to Support HIV Vaccine and Related Clinical Trials. Afr J Health Sci. 2010; 17: 75–82. doi: 10.2174/1874241601003010066

29. Wells J, Shetty AK, Stranix L, Falkovitz-Halpern MS, Chipato T, Nyoni N, et al. Range of normal neutrophil counts in healthy Zimbabwean infants: implications for monitoring antiretroviral drug toxicity. J Acquir Immune Defic Syndr. 2006; 42: 460–463. pmid:16810112 doi: 10.1097/01.qai.0000224975.45091.a5

30. National AIDS and STI Control Programme N. Kenya AIDS Indicator Survey 2007: Preliminary Report. Nairobi, Kenya: Ministry of Health Kenya; 2008. Available:www.nacc.or.ke/nacc%20downloads/official_kais_report_2009.pdf. Accessed 2015 Mar 7.

Chapter 11

THE BIOENGINEERING AND INDUSTRIAL APPLICATIONS OF BACTERIAL ALKALINE PROTEASES: THE CASE OF SAPB AND KERAB

Bassem Jaouadi[1], Badis Abdelmalek[2], Nedia Zaraî Jaouadi[1] and Samir Bejar[1]

[1] Laboratory of Microorganisms and Biomolecules, Centre de Biotechnologie de SfaxUniversity of Sfax, Road of Sidi Mansour Km 6, Tunisia

[2] Laboratory of Biochemistry and Industrial Microbiology, Department of Industrial Chemistry, University Saad Dahlab of Blida, Algeria

INTRODUCTION

Enzymes have long been used as alternatives to chemicals to improve the efficiency and cost-effectiveness of a wide range of industrial systems and processes. They are currently used in basic and applied arenas of research as well as in a wide range of product design and manufacturing processes, such as those pertaining to the food, beverage, pharmaceutical, detergent, leather processing, and peptide synthesis industries (Gupta et al., 2002). Of particular interest to the aims of the present work, proteases have often been reported to constitute a resourceful class of enzymes with promising industrial applications. According to recent estimates, these enzymes account for nearly 65% of total worldwide enzyme sales (Anonyme, 2007; Rao et al., 1998). They are widely distributed in nature and play a vital role in life processes. They are particularly known for their capacity to hydrolyze peptide bonds in aqueous environments and to synthesize peptide bonds in non-aqueous biocatalysis.

Proteases have been employed in a wide array of applications for many years with satisfactory results. They constitute a large family of enzymes present in a wide range of living organisms, such as plants, animals and microorganisms. In biotechnologically oriented systems and processes, however, proteases from microbial origins have often been reported to have distinct advantages when compared to plant or animal proteases, particularly because they possess almost all the characteristics desired for biotechnological applications. Among these biocatalysts, high-alkaline proteases, which alone account for about 40% of the total worldwide enzyme sales (Kirk et al., 2002), proved particularly suitable for industrial use. This is mainly due to their high stability and activity under harsh conditions.

Nowadays, the use of alkaline protease-based detergents is preferred over the conventional synthetic ones. This is partly because of their better cleaning properties, higher performance efficiency at lower washing temperature, and safer dirt removal conditions (Gupta et al., 2002). Typically, a detergent protease needs to be active, stable, and compatible with the alkaline environment encountered under harsh washing conditions: pH 9 - 11, temperature of 20 - 60°C, as well as high concentrations of salt, bleach, and surfactant. Some of the alkaline proteases that are particularly preferred in contemporary detergent formulations include Savinase™ (Subtilisin 309), Subtilisin Novo (BPN′), Alcalase™ (Subtilisin Carlsberg; SC), Maxacal™ (Novozymes A/S, Denmark), BLAP S[b] (Henkel, Germany) and Properase™ (Genecor Int. USA). They are often reported to be stable at conditions of elevated temperatures and pH. Most of them have, however, been criticized for their limited efficiency in the presence of liquid or solid laundry detergents wherein their stability decreases (Beg and Gupta, 2003;Maurer, 2004). Therefore, the search for and screening of alternative microorganisms that produce detergent-stable enzymes and preserve their high activity and stability at extreme conditions would be highly desired, particularly within the framework of the persistent aspirations that consumers, industrialists and, by extension, researchers, have towards improved laundry detergents with powerful, safe and healthy cleansing abilities.

Various alkaline proteases have been reported to constitute appropriate additives for a variety of detergent, laundry and cleansing supplies as well as other leather processing, dyeing, and finishing applications. Keratinases are a group of mostly extracellular serine-proteases that have often been reported for their excellent potency to degrade keratins, a group of fibrous, insoluble and abundant structural proteins that constitute the major components of structures growing from the skin of vertebrates, such as hair, wool, nails, hooves, horns and feather quills. In fact, due to their high degree of cross-linking to disulphide bonds, hydrogen bonds, and hydrophobic interactions, these proteins show high stability and resistance to proteolytic hydrolysis (Coulombe and Omary, 2002).

Large amounts of keratin containing wastes are discharged every year from poultry, leather and meat processing industries. Current estimates indicate that the global annual discharge of feather from the poultry processing industry alone reaches millions of tons (C.A.S.T., 1995; Freeman et al., 2009). This keratinous poultry waste is degraded very slowly in nature and is, therefore, considered hazardous to the environment. Seeing that keratinous waste represents a valuable source for proteins and amino acids, several steam pressure and chemical treatment processes have been developed to convert feathers into feather meal for animals (Hess and FitzGerald, 2007). These physico-chemical conversion methods have, nevertheless, been reported to involve costly treatments under harsh temperature and pressure conditions that result in the loss of essential amino acids (Onifade et al., 1998). Alternatively, feather biodegradation processes have been proposed as viable substitutes (Ignatova et al., 1999; Xie et al., 2010).

Keratinolytic microorganisms can be employed in the manufacture of nutritious, cost-effective, environmentally safe feather meal for poultry, as well as in the enhancement of drug delivery, hydrolysis of prions, construction of biodegradable films, and production of biofuels (Brandelli et al., 2010). Additionally, these keratinolytic enzymes have a variety of current and potential applications in a wide range of biotechnological processes that involve keratin hydrolysis, including the enzymatic dehairing and catalysis for leather and cosmetic industries, the breaking down of recalcitrant matter for the laundry and detergent industries, the slowing down of nitrogen release for fertilizer and pesticide industries, and the production of biohydrogen and rare amino acids for animal feed and foodstuff industries (Bertsch and Coello, 2005).

Several microorganisms that possess keratinolytic activity have been reported to accede to the biodegradation of keratin waste by secreting keratinolytic peptidases into the culture medium and to offer valuable tools for the development of efficient and cost-effective keratin waste bioconversion methods (Onifade et al., 1998). In this respect, various keratinases have been purified from different microorganisms, namely fungi, such as *Microsporum* (Essien et al., 2009) and *Chryseobacterium indologenes* TKU014 (Wang et al., 2008), and bacteria, such as *Bacillus* (Pillai and Archana, 2008; Radha and Gunasekaran, 2008) and *Streptomyces* (Syed et al., 2009; Tatineni et al., 2008). As corresponds to their habitat, these bacteria are nutritionally quite versatile, and most of them produce extracellular hydrolytic enzymes that permit the use of high-molecular-weight biopolymers, such as proteins, polysaccharides, fats, and a variety of other substrates (Gupta et al., 1995). Among these enzymes, several serine peptidases have so far been isolated, purified, and characterized from various species, such as *S. griseus* (Awad et al., 1972; Johnson and Smillie, 1974), *S. fradiae* (Kitadokoro et al., 1994), *S. thermoviolaceus* SD8 (Chitte et al., 1999), and *S. graminofaciens* (Szabo et al., 2000).

Despite this large flow of data on keratinases, however, little information has so far been reported on the characterization and purification of keratinases from *Streptomyces*. Moreover, and particularly due to the relatively poor levels of stability and catalytic activity obtained for the *Streptomyces* enzymes so far investigated under the specific operational conditions required by current industrial applications, namely high temperature and pH values, as well as the presence of detergents or non-aqueous solvent, their practical application still remained very limited. Accordingly, the isolation and screening of new keratinolytically active *Streptomyces* strains from natural habitats could open new pathways for the discovery and use of novel keratinases.

The present chapter aim to provide an overview on the current quest for novel natural bacterial alkaline proteases with special emphasis on the purification and characterization of two enzymes, namely SAPB and KERAB, from isolated alkaline proteinase and keratinase producing microbial strains, whose promising properties and attributes are likely to open new pathways in current and future research and new possibilities for the improvement of current detergent formulations and leather processing industries. In fact, both SAPB and KERAB showed valuable operational characteristics that made them strong potential candidates for future application as additives in biotechnological applications and processes, particularly in detergent formulations and in dehairing during leather processing. They also showed relatively high stability in the presence of organic solvents, a feature which is highly desired in applications involving the biocatalysis of non-aqueous peptides. Accordingly, this chapter intends to report on the screening, identification, and phylogenetic analysis of the *Bacillus pumilus* strain CBS producing SAPB and the *Streptomyces* sp. strain AB1 producing KERAB. It also aims to describe the laundry detergent compatibility and high dehairing capacity of both enzymes, and to report on the ability of each strain or enzyme (SAPB or KERAB) alone to accomplish the whole keratin-degradation process of various keratinacious biowastes.

SCREENING AND IDENTIFICATION OF ALKALINE PROTEINASE AND KERATINASE PRODUCING MICROBES

The isolation and screening of micro-organisms from naturally occurring alkaline habitats and keratinacious biowaste is likely to help identify potential microbial strains capable of producing active and stable enzymes that can resist the aforementioned harsh substances and conditions present in detergent formulations and leather dehairing processes.

Screening of Alkaline Protease and Keratinase Producing Strains

A recent work by the authors (Jaouadi et al., 2009; Badis et al., 2009) involved the screening of about 125 bacterial strains (Bacilli and Actinomyces), originating from a collection of bacterial strains at the CBS and other strains that were previously isolated from surface soil samples at the Mitidja plain, North of Algeria (Badis et al., 2010), for protease and keratinase activities. Based on the ratio of the diameter of the clear zone (onto skimmed milk or keratin-containing medium agar plates at pH 9.) and that of the colony, only 24 isolates, which exhibited the highest ratio (> 3 mm), were selected for further assays pertaining to protease or keratinase production in liquid media. The two bacterial strains that displayed the highest extracellular protease and keratinase activity were termed as strain CBS (from the CBS bacterial strain collection) and strain AB1 (from Algerian soil samples) and retained for all subsequent experimental assays.

Identification and Molecular Phylogeny Of The Microorganisms

The two newly isolated bacterial strains, CBS and AB1, were submitted to identification and typing by molecular and catabolic techniques. The data from the morphological, biochemical and physiological characterization tests, performed on the isolates in accordance with the methods described in the Bergey's Manual of Systematic Bacteriology, showed that the CBS and AB1 strains appeared in a bacilli and filamentous form, respectively, that are aerobic, endospore-forming, Gram-positive, catalase+, oxydase+ and motile rod-shaped. The findings from API 50 CH gallery tests revealed that the CBS isolate metabolized l -arabinose, d-tagatose, ribose, and mannitol in addition to several other simple sugars. The AB1 strain, on the other hand, could use galactose, sucrose, maltose, cellobiose, fucose, raffinose, d-xylose, l-arabinose, and d-ribose, but not lactose, starch, l-rhamnose, erythritol, adonitol, and inositol. The results from API ZYM tests revealed that strain AB1 also exhibited alkaline phosphatase, esterase lipase (C8), leucine arylamidase and valine arylamidase activities, but no lipase (C14), trypsin, α-chymotrypsin, N-acetyl-β-lucosamidase, β -glucuronidase, α-mannosidase, and α-fucosidase ones. Taken together, the data obtained with regard to the physiological and biochemical properties of the two isolates strongly confirmed that the strains CBS and AB1 belonged to the *Bacillus*and *Streptomyces* genera, respectively.

A molecular approach was used to establish further support for the identification of the CBS and AB1 isolates. Two 16S rRNA gene fragments, namely 1,497 bp (Jaouadi et al., 2009) and 1541 bp (Jaouadi et al., 2010a), were amplified from the genomic DNA of the CBS and AB1 isolates, respectively, and then cloned and sequenced on both strands. The 16S rRNA gene sequences obtained were subjected to GenBank BLAST search analyses, which yielded strong homologies of up to 98 and 99% with those of several cultivated strains of *Bacillus* and *Streptomyces*, respectively. The nearest *Bacillus* and*Streptomyces* strains identified by the BLAST analysis were the *Bacillus pumilus*, with the accession numbers of DQ988522, AM292995, AY548955, AB195283, and EF173329, and the *Streptomyces rochei* strains of A-1 (GQ392058) and NRRL B-1559 (EF626598) as well as the *Streptomyces* sp. Strain B5W22-2 (EF114310), respectively. Those sequences were imported into the ARB and MEGA software packages, respectively, and then aligned. After that, the phylogenetic trees were constructed using neighbour-joining methods and Jukes-Cantor distance matrices (Fig. 1). Phylogenetic analyses confirmed that the CBS and AB1 strains were closely related to the five isolated *Bacillus* and three isolated *Streptomyces* strains mentioned earlier. In conclusion, all the results obtained strongly supported the assignment of the CBS and AB1 isolates to the *Bacillus pumilus* strain CBS and *Streptomyces* sp. strain AB1, respectively.

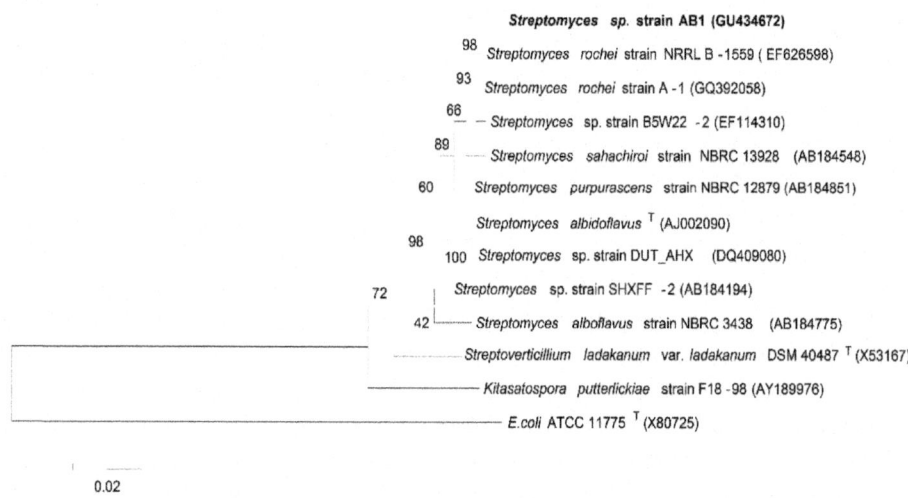

Figure 1: Example of the phylogenetic tree of *Streptomyces sp.* strain AB1. Phylogenetic and molecular evolutionary analyses were conducted using MEGA version 4.1. Reference type-strain organisms are included and sequence accession numbers are given in parentheses. Bootstrap values, expressed as percentage of 100 replications, are shown in branching points and bar indicated 2 substitutions per 100 nt. The out-group used in the analysis, *E. coli* (X80725), was chosen arbitrarily.

PRODUCTION, PURIFICATION AND BIOCHEMICAL CHARACTERIZATION OF SAPB AND KERAB ENZYMES

Sapb and Kerab Production

Different carbon and nitrogen sources and trace elements were assayed to optimize the culture growth conditions for the production of the enzymes. In the medium containing (g/l): gelatin 10, yeast extract 5, $CaCl_2$ 1, K_2HPO_4 1, and KH_2PO_4 1, the addition of 0.1% (v/v) trace elements [composed of (g/l): $ZnCl_2$, 0.4; $FeSO_4$ $7H_2O$, 2; H_3BO_3, 0.065; and $MoNa_2O_4$ $2H_2O$, 0.135] at pH 10 was noted to bring about a significant enhancement of 1.32 folds in SAPB production, which reached 6,500 U/ml under the optimal conditions used (pH 10.6 and 65°C), after 24 h of incubation at 37°C and 250 rpm (Jaouadi et al., 2009). In medium containing trace salts with feather as carbon and nitrogen source (g/l): NaCl, 0.5; KH_2PO_4, 0.5; K_2HPO_4, 0.5; KCl, 0.1; $MgSO_4$ $7H_2O$, 1; and chicken feather meal, 10; at pH 9, KERAB production was observed to undergo a significant improvement, reaching a maximum of 9,500 U/ml under the optimal conditions used (pH 11.5 and 75°C) after 96 h of incubation at 30°C and 200 rpm (Jaouadi et al., 2010a). Under these particular conditions, the production of the SAPB and KERAB enzymes started after a 6- and 10-h lag phase, respectively. These productions were then noted to increase exponentially and concomitantly with the increase of cellular growth and to reach the maxima within 24 h of cultivation for SAPB (Fig. 2) and 96 h for KERAB (data not shown).

Compared to the production yields obtained in flask cultivations, the use of a 7-litre fermentor containing the optimized medium after 24-h cultivation at 37°C, an aeration of 1.5 vvm, and an agitation of 600 rpm was noted to improve SAPB production by about 4-folds. It is worth noting here that the cell densities obtained in both cases (Rotary flask and fermentor) were almost the same (about O.D. = 10.9). Based on this particular finding, it was possible to infer that the improvement of enzyme production was related not only to the cell's growth but also to the stability of fermentation parameters (pH and pO_2).

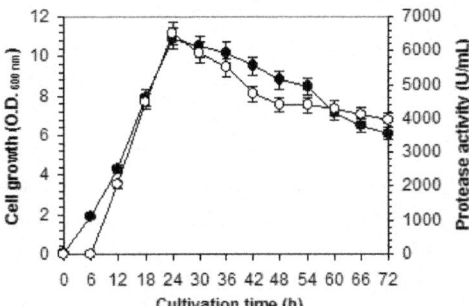

Figure 2: Time course of *B. pumilus* strain CBS cell growth (●) and SAPB production (○). The culture was carried out under the submerged shaking flask conditions at 37°C for 72 h with an agitation rate of 250 rpm in broth medium containing (g/l): gelatin 10, yeast extract 5, $CaCl_2$ 1, K_2HPO_4 1, KH_2PO_4 1, and trace elements 0.1% (v/v) at pH 10. Cell growth was monitored by measuring the O.D. at 600 nm.

Sapb and Kerab Purification And Characterization

The purification protocols used for the purification of each enzyme were conducted at temperatures not exceeding 4°C. Five-hundred ml of 24 h and 96 h cultures of *B. pumilus* strain CBS and *Streptomyces*sp. strain AB1, respectively, were centrifuged to remove microbial cells. Ammonium sulfate was added to each supernatant to a final concentration of 270 g/l. In the case of SAPB, the precipitate formed was collected by centrifugation, dissolved in a minimum amount of 50 mM Tris-HCl (pH 7.5) supplemented with 2 mM $CaCl_2$ and 0.05% Triton X-100 (Buffer A). In the case of KERAB, the precipitate was suspended in 50 mM bicarbonate-NaOH buffer and supplemented with 5 mM $MgSO_4$ at pH 11.5 (Buffer B) containing 10 mM NaCl (Buffer C), and then dialyzed overnight against repeated changes of the buffer A and C, respectively.

Purification to homogeneity was achieved for SAPB by HPLC using Shodex Protein WK 802-5 column. The analysis indicated that enzyme achieved a degree of purity that was about 38-fold greater than that of the crude extract. Under the optimal assay conditions used, the purified enzyme preparation exhibited a yield of about 12% with a specific activity of 25,500 U/mg (Jaouadi et al., 2008). As far as KERAB was concerned, the insoluble material was then removed by centrifugation. The supernatant obtained was incubated for 1 h at 50°C and insoluble material was removed by centrifugation. The supernatant was loaded on a Sephacryl S-200 column equilibrated with buffer B. The elution of protease was performed with the same buffer. The fractions containing keratinase activity were then pooled and applied to a Q-Sepharose column equilibrated in buffer D. The column was rinsed with 500 ml of the same buffer and the adsorbed material was eluted with a linear NaCl gradient. At the final purification step, Keratinase activity was eluted between 0.15 and 0.3 M NaCl. The purity of the enzyme was estimated to be about 86-fold greater than that of the crude extract. The purified enzyme preparation contained about 24% of the total activity of the crude enzyme and had a specific activity of 67,000 U/mg (Jaouadi et al., 2010a). These preparations were homogeneous enzymes with high purity as they exhibited single protein bands on native PAGE and unique elution symmetrical peaks on gel filtration chromatography.

For determination of their molecular weight, enzyme preparations were treated with 1 mM PMSF prior to electrophoresis to inhibit possible autolysis during electrophoresis. Electrophoresis under denaturing conditions (SDS-PAGE) also revealed single bands with molecular masses estimated as 34 kDa for SAPB (Jaouadi et al., 2008) and 30 kDa for KERAB (Jaouadi et al., 2010a). The exact molecular masses obtained for the purified SAPB and KERAB were confirmed by MALDI-TOF mass spectrometry as being 34598.19 and 29850.17 Da, respectively. Zymogram activity staining also revealed two clear zones of proteolytic activity at 34 and 30 kDa for the SAPB and KERAB, respectively. These observations indicated that SAPB extracted from the newly isolated bacterium *B. pumilus* CBS was a monomeric holoenzyme comparable to those previously reported for other proteases from *B. pumilus* strains (Han and Damodaran, 1998; Huang et al., 2003; Kumar, 2002; Miyaji et al., 2006; Yasuda et al., 1999). They also showed that KERAB was a monomeric protein comparable to those previously reported for other proteases from *Streptomyces* strains (Syed et al., 2009; Tatineni et al., 2008).

The molecular mass of SAPB determined by SDS-PAGE (~ 34000 Da) and conducted by MALDI-TOF mass spectrometry (34598.19 Da) were not close to that calculated from the primary sequence of the mature polypeptide (27789 Da), which strongly suggested that the protein underwent noteworthy post-translational changes that were presumably pertaining to glycosylation. Similar differences between experimental and theoretical determinations were previously observed for several *B. pumilus* proteases, including those from *B. pumilus* TYO-67 (Yasuda et al., 1999), *B. pumilus* UN-31-C-42 (Huang et al., 2003), and *B. pumilus* MS-1 (Miyaji et al., 2006).

Physico-Chemical and Kinetic Properties Of Sapb And Kerab

Phenylmethanesulfonyl fluoride (PMSF) and diiodopropyl fluorophosphates (DIFP) were noted to strongly inhibit SAPB and KERAB, which indicated that both enzymes belonged to the serine proteases family. While the optimal pH and temperature values of 10.6 and 65°C were determined for SAPB using casein as a substrate, those obtained for KERAB were 11.5 and 75°C with keratin azure as substrate. The thermoactivity and thermostability of KERAB were also demonstrated to be enhanced in the presence of 5 mM Mg^{2+} against 2 mM Ca^{2+} for SAPB. One of the distinguishing properties of SAPB was its catalytic efficiency (*kcat/Km*) which was 4.77, 2.73, and 2.11 times higher than those of Subtilisin Carlsberg, Subtilisin BPN', and Subtilisin 309, respectively. The catalytic efficiency of KERAB was higher than those of SAPB, nattokinase and subtilisin Carlsberg.

Substrate Specificity of Sapb And Kerab

The activity of the purified SAPB and KERAB enzymes towards various natural and modified protein substrates is summarized in Table 1. Among the proteinaceous substrates tested, casein and keratin were most efficiently hydrolyzed by SAPB and KERAB, respectively. When SAPB and KERAB activities against casein and keratin were taken as 100%, the hydrolysis rates of gelatine and casein were 95 and 92%, respectively. Poor BSA hydrolysis rates were, however, noted in both cases. A similarly low hydrolysis level was also observed with gluten and egg albumin. Using modified proteins as substrates, the highest activities observed for SAPB and KERAB were with azocasein and keratin azure, respectively. Previous reports also showed that alkaline serine proteases from *B. stearotermophilus* FI (Rahman et al., 1994) and *Bacillus pumilus* A1 (Fakhfakh-Zouari et al., 2010) exhibited highest activities towards casein and keratin, respectively. Interestingly, no collagenase activities were detected for SAPB and KERAB on collagen types I and II, which suggests the potential utility of both enzymes for hair removal in the leather industry.

The cleavage specificities of SAPB and KERAB toward various oligopeptidyl and ester substrates were also investigated. The findings revealed that SAPB exhibited both esterase and amidase activities on oligopeptides, with Tyr or Phe at position P_1 (the amino acid residue at the N-terminal side of the scissile peptide bond). This included *N*-benzol-l-arginine ethyl ester (BTEE) or *N*-acetyl-l-tyrosine ethyl ester monohydrate (ATEE) and *N*-succinyl-l-Ala-l-Ala-l-Pro-l-Phe-*p*-nitroanilide (AAPF), which are specific substrates for chymotrypsin-like proteases (DelMar et al., 1979; Walsh, 1970). In fact, however, the activity of SAPB on AAPF did not necessarily mean that it was a chymotrypsin-like enzyme. Firstly, most of the microbial members of the Subtilisin family are reported to have specificity that is somewhat similar to that of chymotrypsin (Rawlings and Barrett, 1977). Moreover, SAPB was not observed to show sensitivity to *N*α-p-tosyl l-phenylalanine chloromethyl ketone (TPCK), which is an inhibitor of chymotrypsin-like enzymes (Schoellman and Shaw, 1963). Last but not least, SAPB showed neither esterase nor amidase activity on synthetic substrates with P_1 = Arg, such as *N*-benzol-l-arginine ethyl ester (BAEE) and -benzoyl-l-tyrosine *p*-nitroanilide (BAPNA), which are substrates for trypsin-like proteases (Rick, 1995). In contrast, the purified KERAB exhibited esterase and amidase activities on BAEE and BAPNA, but not on BTEE and ATEE.

In the same way, the purified KERAB was noted to exhibit a preference for aromatic and hydrophobic amino acid residues, such as Phe, Leu, Ala, and Val, at the carboxyl side of the splitting point in the P1 position. KERAB was, therefore, active against leucine peptide bonds. When Suc-(Ala)$_n$-*p*NA was used as the synthetic oligopeptide substrate, a minimum length of three residues was necessary for hydrolysis. Enzymatic activity was observed to depend mainly on secondary enzyme substrate contacts with amino acid residues (P2, P3, etc.)

more distant from the scissile bond, as illustrated by the differences observed between the kinetic parameters of Suc-(Ala)$_2$-Val-pNA and those of Suc-Tyr-Leu-Val-pNA. The highest hydrolysis levels achieved by KERAB and SAPB were 100% for AAPF and Suc-Tyr-Leu-Val-pNA, respectively.

Table 1: Substrate specificity of SAPB and KERAB

Substrate	Concentration	Relative activity (%)*	
		SAPB	KERAB
Natural protein[a]			
Keratin	10 g/l	65 ± 1.4	100 ± 3.0
Casein	20 g/l	100 ± 2.5	92 ± 2.0
Gelatine	20 g/l	95 ± 2.4	79 ± 1.4
BSA	20 g/l	52 ± 1.3	66 ± 1.4
Albumin (egg)	20 g/l	15 ± 0.7	26 ± 0.9
Gluten (wheat)	20 g/l	20 ± 0.8	11 ± 0.8
Modified protein[b]			
Keratin azure	10 g/l	63 ± 1.4	100 ± 3.0
Azo-casein	20 g/l	100 ± 2.5	94 ± 2.5
Collagen type I[c]	1 mg/ml	0 ± 0.0	0 ± 0.0
Collagen type I[c]	1 mg/ml	0 ± 0.0	0 ± 0.0
Ester[d]			
BAEE	4 mM	0 ± 0.0	100 ± 3.0
BTEE	3 mM	71 ± 1.4	10 ± 1.5
ATEE	3 mM	100 ± 2.5	20 ± 0.9
Synthetic peptide[e]			
Suc-Tyr-Leu-Val-pNA	2 mM	30 ± 1.1	100 ± 3.0
Suc-(Ala)$_2$-Pro-Phe-pNA	3 mM	100 ± 2.5	17 ± 0.9
Suc-(Ala)$_2$-Pro-Leu-pNA	3 mM	45 ± 1.3	13 ± 0.8
Suc-(Ala)$_2$-Val-Ala-pNA	3 mM	39 ± 1.2	50 ± 1.8
Suc-(Ala)$_2$-Val-pNA	3 mM	25 ± 0.9	56 ± 2.0
Suc-(Ala)$_3$-pNA	2 mM	10 ± 0.4	67 ± 2.2
Suc-(Ala)$_2$-Phe-pNA	2 mM	17 ± 0.5	89 ± 2.5
BAPNA	2 mM	0 ± 0.0	66 ± 1.3

[a]The activity of these natural protein substrates were assessed by measuring

absorbance at 660 nm following the previously reported Folin-Ciocalteu method (Jaouadi et al., 2010b).

[b]The activity of these modified protein substrates were determined by measuring absorbance at 440 nm following the method of Riffel and Brandelli, (2002).

[c]The collagenolytic activity was determined by measuring absorbance at 490 nm as described in the protocol of Sigma Co.

[d]The esterase and amidase activities of these substrates were determined by measuring absorbance at 253 nm as described in the method of Walsh (Walsh, 1970).

[e]The activity of these synthetic oligopeptide substrates was determined by measuring absorbance at 410 nm according to the method of DelMar et al., (1979).

[*]Values represent the mean of four replicates and standard errors are reported.

MOLECULAR CLONING OF *SAPB* GENE AND ENGINEERING OF MORE EFFICIENT SAPB MUTANT ENZYMES

The *sap*B gene encoding SAPB was cloned, sequenced, and over-expressed in *Escherichia coli*. The purified recombinant enzyme, called rSAPB, exhibited the same biochemical properties of the native enzyme (Jaouadi and Bejar, 2008). An additional study by the authors further investigated the implications of five amino acid residues (L31, T33, N99, F159, and G182) on the pH and temperature behavior as well as kinetic parameters of the enzyme using site-directed mutagenesis and 3D-modeling approaches (Jaouadi et al., 2010b). Seven more efficient SAPB mutant enzymes, particularly L31I/T33S/N99Y, were generated. The latter had an optimal of pH of 12 and an optimal temperature of 70°C. It was also noted to exhibit a high specific activity that was approximately 2-fold higher than that of the wild-type enzyme and a prominent increase in its *kcat/Km* value that was 42-fold higher than that of the wild-type enzyme.

POTENTIAL AND PROSPECTS FOR SAPB AND KERAB IN DETERGENT FORMULATIONS

Effect of Detergents On The Activity And Stability Of Sapb And Kerab

With the aim of evaluating the performance of the purified proteases in real life-like detergents, SAPB and KERAB were pre-incubated at 40°C and in the presence of several commercially available laboratory non-ionic surfactants, denaturing agents or anionic surfactants, and bleach agents for 24 and 72 h, respectively. The residual activity was determined at pH 10.6 and 65°C (for SAPB) and pH 11.5 and 75°C (for KERAB). The findings revealed that the SAPB enzyme exhibited high stability at 10% of oxidizing agents (Tween 60 or Triton X-100) as well as against strong anionic surfactants, particularly sodium dodecyl sulphate(SDS) and linear alkylbenzene sulfonate (LAS) (Jaouadi et al., 2008). In fact, SAPB retained its activity upon treatment with 0.8% SDS and 0.5% LAS. In addition, 80 and 65% residual activity were obtained after incubation with 1.5% SDS and 1% LAS, respectively. The SAPB and KERAB enzymes were also highly stable against bleaching agents for they retained 110 and 115% of their initial activity after treatment with 15% hydrogen peroxide, respectively. This is an important behaviour of SAPB and KERAB because oxidant-, surfactant-, and bleach-stable wild-type enzymes are rarely reported.

By way of comparison, the alkaline protease from alkalophilic *Bacillus* sp. JB-99 lost 25% activity during treatment with 0.5% SDS for only 1 h of incubation at 40°C (Johnvesly and Naik, 2001) while two other alkaline proteases (FI and FII) from *Vibrio fluvialis* TKU005 were activated by 1% SDS (Wang et al., 2007b). The present native SAPB and KERAB enzymes showed inherent stability in the presence of high concentrations of detergent compounds, especially Tween 60 at 10%, SDS at 1.5%, and hydrogen peroxide at 15%. In addition, their enzymatic activity and stability were observed to improve in the presence of high concentrations of 1% perfume and anti-redeposition agents, particularly 100 mM Na_2 CMC, and of cationic (TTAB, CTAB) and zwitterionic (Zwittergent 3-12, CHAPS) detergent agents (Table 2). This stability is of interest since only few wild-type proteases have so far been reported to be oxidant, surfactant and bleach stable. These include those reported by Gupta et al., (1999) and Haddar et al., (2009). Bleach stability was also attained through protein engineering (Pillai and Archana, 2008; Radha and Gunasekaran, 2008). These findings suggest the potential strong candidacy of SAPB and KERAB for application as cleaning additives in detergent formulations to facilitate the release of proteinacious materials in tough stains caused by blood, chocolate, grime, milk, etc.

Table 2: Effect of various detergents on SAPB and KERAB activity and stability. The non-incubated enzymes were considered as 100%. The activity is expressed as a percentage of the activity level in the absence of additives. Values represent the mean of three replicates and standard errors are reported

Detergent additive	Concentration	Relative activity (%)		Residual activity (%)	
		SAPB	KERAB	SAPB	KERAB
None	–	100 ± 2.5	100 ± 2.5	100 ± 2.5	100 ± 2.5
H$_{20}$2	15%	140 ± 3.7	155 ± 3.8	110 ± 2.6	115 ± 2.6
Sodium perborate	2% (w/v)	85 ± 2.2	110 ± 2.6	55 ± 2.0	85 ± 2.2
SDS	1.5%	110 ± 2.6	125 ± 3.0	80 ± 2.2	109 ± 2.6
LAS	1% (w/v)	79 ± 2.2	120 ± 3.2	65 ± 2.1	103 ± 2.5
Sulfobetaine	30 mM	105 ± 2.5	130 ± 3.3	90 ± 2.3	113 ± 2.6
Tween 40	5% (v/v)	111 ± 2.6	135 ± 3.4	101 ± 2.5	119 ± 2.8
Tween 60	10% (v/v)	120 ± 3.0	126 ± 3.3	105 ± 2.5	111 ± 2.6
Triton X-100	10% (v/v)	101 ± 2.5	132 ± 3.5	94 ± 2.3	112 ± 2.7
TAED	10% (w/v)	115 ± 2.8	128 ± 3.5	103 ± 2.5	117 ± 2.7
Na$_2$·CMC	5% (w/v)	109 ± 2.6	137 ± 3.5	101 ± 2.5	112 ± 2.6
Zeolithe	1% (w/v)	99 ± 2.5	100 ± 2.5	94 ± 2.3	95 ± 2.3
STPP	1% (w/v)	88 ± 2.3	90 ± 2.3	80 ± 2.2	82 ± 2.2
Perfume	1% (v/v)	115 ± 2.6	116 ± 2.6	103 ± 2.5	104 ± 2.5
Na$_2$·C$_0$3	100 mM	50 ± 2.0	113 ± 2.4	42 ± 1.8	100 ± 2.5
Zwittergent 3-12	10 mM	107 ± 2.5	116 ± 2.6	100 ± 2.5	109 ± 2.5
CHAPS	15 mM	121 ± 3.0	133 ± 3.1	106 ± 2.5	115 ± 2.6
CTAB	25 mM	104 ± 2.5	107 ± 2.3	95 ± 2.4	100 ± 2.5
TTAB	25 mM	99 ± 2.4	105 ± 2.6	90 ± 2.3	98 ± 2.3

Sulfobetaine: *N*-dodecyl-N-N'-dimethyl-3-ammonio-1-propane sulfonate; Tween: poly (oxyethylene) sorbitan monolaurate; Triton: octyphenolpoly (ethylene glycolether); TAED: tetraacetylethylenediamine; STPP: sodium tripolyphosphate; CHAPS: 3-[(3-cholamidopropyl) dimethylammonio]-1-propane sulfonate; CTAB: hexadecyltrimethylammonium bromide; TTAB: tetradecyl trimethylammonium bromides.

Compatibility of Sapb And Kerab Enzymes With Various Commercial Laundry Detergents

To check the compatibility and stability of the alkaline proteases towards detergents, the enzymes were pre-incubated in the presence of various commercial laundry detergents of different compositions for 1 h at 40°C. The laundry detergents were diluted in tap water to a final concentration of 7 mg/ml to simulate washing conditions. The endogenous proteases were inactivated by incubating the diluted detergents for 1 h at 65°C, prior to the addition of the SAPB and KERAB enzymes, or the SB 309 commercial enzyme, which was used for comparison (Table 3). The findings showed that SAPB and KERAB were relatively more stable and compatible with some commercial liquid detergents than the commercial enzyme. In fact, while SB 309 retained 100, 85, 70 and 90% of its initial activity in the presence of Axion, Dinol, Nadhif, and Lav+, SAPB retained about 100, 95, 94, and 85% and KERAB about 87, 90, 75, and 95% of their initial activities, respectively. The SAPB and SB 309 enzymes were, however, less stable in the presence of Axion, where they were totally active. Furthermore, SAPB and KERAB showed excellent stability and compatibility in the presence of some commercial solid detergents, namely OMO, New Det, and Skip, with SAPB retaining about 96, 82, and 69% of its initial activity, and KERAB about 88, 93, and 95%, respectively. SAPB and KERAB were, however, less stable in the presence of Ariel, retaining about 55 and 51% of their initial activities, respectively. Nevertheless, the compatibility and stability exhibited by SAPB and KERAB were much more significant than that of SB 309, which retained only 70, 68, and 84% of its initial activity in the presence of OMO, New Det, and Skip, respectively. Incubated in the same conditions in the presence of New Det, the NH1 protease was reported to retain 60% of its initial activity (Hadj-Ali et al., 2007) and, in the presence of Ariel, the VM10 (Venugopal and Saramma, 2006) and SSR1 (Singh et al., 2001) proteases were reported to retain only 42 and 37% of their initial activities, respectively. Overall, the results obtained clearly indicated the superior performance of SAPB and KERAB enzymes in detergents compared to currently commercialized or previously described proteases. A minor discordance was, however, reported as present with regards this performance, which was presumably correlated to the nature and concentration of the laundry detergent compounds used.

Table 3: Stability of the purified SAPB and KERAB proteases in the presence of various commercial laundry detergents. The non-incubated enzyme was considered as 100%. The activity is expressed as a percentage of the activity level in the absence of organic solvent

Laundry detergent (7 mg/ml)	Relative activity (%)			Residual activity (%)		
	SAPB	KERAB	SB 309	SAPB	KERAB	SB 309
None	100 ± 2.5	100 ± 2.5	100 ± 2.5	100 ± 2.5	100 ± 2.5	100 ± 2.5
Liquid detergent						
Dinol	95 ± 2.4	90 ± 2.2	85 ± 2.2	81 ± 1.4	80 ± 2.2	77 ± 2.0
Lav+	85 ± 2.2	95 ± 2.4	90 ± 2.2	75 ± 2.0	81 ± 2.2	75 ± 2.0
Nadhif	94 ± 2.4	75 ± 2.0	70 ± 2.0	77 ± 2.0	62 ± 1.8	60 ± 1.7
Axion	100 ± 2.5	87 ± 2.1	100 ± 2.5	91 ± 2.3	66 ± 1.9	85 ± 2.2
Solid detergent						
New Det	99 ± 2.5	94 ± 2.4	68 ± 2.0	82 ± 2.2	93 ± 2.4	58 ± 1.7
Skip	75 ± 2.0	100 ± 2.5	84 ± 2.2	69 ± 2.0	95 ± 2.4	72 ± 2.2
Ariel	65 ± 1.8	60 ± 1.7	61 ± 1.7	55 ± 1.6	51 ± 1.5	50 ± 1.5
OMO	100 ± 2.5	95 ± 2.4	70 ± 2.2	96 ± 2.4	88 ± 2.1	61 ± 1.7

Values represent the mean of three replicates and standard errors are reported.

WASH PERFORMANCE ANALYSIS OF SAPB

In order to evaluate the performance of SAPB in terms of ability to remove harsh stains, namely those caused by chocolate or human blood, several pieces of stained cotton cloth were incubated at different conditions (Fig. 3). The findings from these assays revealed that the blood and chocolate stain removal levels achieved with the use of SAPB alone were more effective than the ones obtained with detergent (Det) alone. In fact, SAPB facilitated the release of proteinacious materials in a much easier way than the commercialized SB 309 protease (Jaouadi et al., 2009). Furthermore, the combination of SAPB and the Det detergent resulted in complete stain removal (Fig. 3). In fact, a similar study has previously reported on the usefulness of alkaline proteases from *Spilosoma obliqua* (Anwar and Saleemuddin, 1997) and *B. brevis* (Banerjee et al., 1999) in the assistance of blood stain removal from cotton cloth both in the presence and absence of detergents, but, in terms of reported results, the SAPB enzyme was more effective.

Figure 3: Example of washing performance analysis test of SAPB. Stained cloth pieces with blood (I) or chocolate (II). (A) Control: untreated stained cloth pieces; or stained cloth pieces washed with: (B) distilled water, (C) Det detergent (7 mg/ml), (D) SAPB (500 U/ml), (E) SB 309 (commercial enzyme, 500 U/ml), and (F) SAPB (500 U/ml) + Det detergent (7 mg/ml).

Storage Stability of The Spray-Dried And Lyophilized SAPB

Table 4: Stability of the spray-dried and lyophilized SAPB with and without xylitol at 1% during storage at room temperature and during prolonged storage within Det solid detergent. The activity of each treated SAPB before incubation was taken as 100% and the residual activity was determined at regular intervals.

Values are the means of three independent experiments.

	Condition	t = 2 months	t = 12 months
		Residual activity (%)	
Spray-died	SAPB alone	76	55
	SAPB + Xylitol	88	70
	SAPB + Det	64	50
	SAPB + Det + Xylitol	78	70
Lyophilized	SAPB alone	74	55
	SAPB + Xylitol	80	68
	SAPB + Det	61	50
	SAPB + Det + Xylitol	75	65

The findings indicated that spray-dried SAPB, from fermentor culture, lost about 3% of its original activity; lyophilized SAPB lost about 10% (Jaouadi et al., 2009). Several of the additives used during the spray drying and lyophilizing processes were noted to improve SAPB stability (Table 4). However, the best results were actually obtained with 1% of xylitol, maltodextrin, and PEG 8000, which preserved about 100, 99 and 97% of its proteolytic activity, respectively.

The stability of the spray-dried and lyophilized SAPB during subsequent storage in the presence of 1% xylitol showed that, after incubation at room temperature for 12 months, the enzymes lost only about 20 and 25% of their original activity, respectively, against 35% for the control without additives. The non-treated enzyme was rapidly inactivated, losing about 50% of its initial activity after 2 months of incubation. Moreover, compared to the treated enzyme and in the absence of additives, 1% xylitol clearly enhanced SAPB stability during storage within the Det solid detergent (Jaouadi et al., 2009). In fact, after being incubated for 12 months at room temperature in the presence and absence of xylitol, the spray-dried enzyme retained 68 and 55% of its initial activity, respectively. The level of stability enhancement achieved for the lyophilized SAPB by xylitol was, on the other hand, less pronounced, since the enzyme retained only 65% of its initial activity.

POTENTIAL AND PROSPECTS FOR SAPB AND KERAB IN THE LEATHER PROCESSING INDUSTRY

Keratin-Degradation Profile of *Bacillus Pumilus* CBS and *Streptomycessp.* Ab1

Keratinacious substrates, such as keratin and keratin azure, were previously reported to be significantly hydrolyzed by SAPB (Jaouadi et al., 2008) and KERAB (Jaouadi et al., 2010a). It was also demonstrated that the *B. pumilus* strain CBS was able to grow in an optimized medium containing 10 g/l of feather-meal, chicken feather (Fig. 4A), goat hair, bovine hair, and sheep wool (as a sole carbon and nitrogen source) instead of gelatin and yeast extract, reaching an absorbance at 600 nm of 6 to 10 after 48 h-culture (Jaouadi et al., 2009). Of the 5 keratin substrates tested, feather-meal was the most strongly degraded (98.5%), followed by chicken feather (92%), goat hair (80%), and bovine hair (68%), with sheep wool showing a relatively low degradation rate (12%).

The feather-meal degradation rate achieved by *B. pumilus* CBS was higher than those of *B. pumilis* F3-4 (97%) (Son et al., 2008) and *Streptomyces albidoflavus* (67%) (Bressollier et al., 1999). The maximum release of protein obtained with the *B. pumilus* strain CBS occurred in the feather-meal medium, which was followed by the chicken feather medium. Moreover, while feather-meal and chicken feather gave the best SAPB production yields of 4,800 and 4,512 U/ml, respectively, sheep wool supported very low keratinolytic activity (1250 U/ml) (Jaouadi et al., 2009). Hence, its full-grown and intense Feather-Degrading (FD) activity could be achieved, in 24 h, at the range of 30 - 37°C, and with initial pH adjusted from 8 to 9. This profile contrasts with previously reported results stipulating that *B. pumilus*FH9 solubilize feather in 72 h at 55°C with pH 9 (El-Refai et al., 2005) while *B. pumilis* F3-4 show intense FD activity in 168 h at 30°C with pH 7.5 (Son et al., 2008).

An increase simultaneous to keratin degradation was noted in protein levels and sulfhdryl groups (Jaouadi et al., 2009). Higher levels of keratin degradation resulted in high sulfhydryl group formation. The results obtained, therefore, suggested that *B. pumilus* CBS had a disulfide bond-reducing ability. Moreover, the processing of data from amino acid analysis following keratin degradation revealed a marked increase in the release of free amino acids after 12 h of incubation. The profile suggested that phenylalanine, tryptophan, leucine, isoleucine, valine, and alanine were the major amino acids liberated, whereas the untreated keratin (control) did not release any free amino acids. In fact, this amino acid profile matched well with the one described for the keratinolytic serine-enzyme produced by *B. licheniformis* PWD-1 (Williams et al., 1990). When SAPB was shaking-incubated with a white feather, a partial degradation was observed after 24 h with a simultaneous increase in protein concentration and sulfhydryl group formation, whereas no degradation was noticed with the control (Fig. 4B). These results confirmed that SAPB alone could accomplish the whole dehairing process.

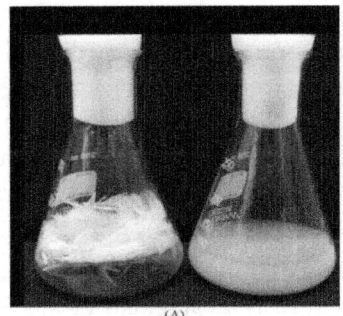 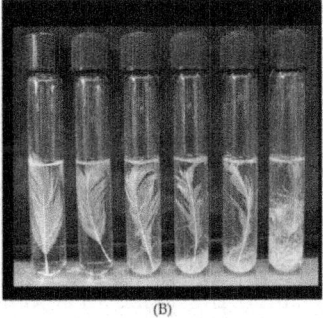

(A)   (B)

Figure 4: Keratin(feather)-degradation by *B. pumilus* strain CBS and SAPB. (A) Feathers were incubated for 24 h at 37°C under shake culture condition with 2.8×10^8 cells/ml as an initial inoculum density of the strain CBS (right flask) and with autoclaved inoculum as control (left flask). (B) SABP was incubated for 24 h at 37°C with chicken feather.

The *Streptomyces* sp. strain AB1 was able to grow after 2 days of culture in a mineral salt medium containing 30 g/l of intact chicken feathers as sole carbon, nitrogen, and sulfur sources instead of 10 g/l of feather meal. Intense feather-degrading activity was achieved at 30 °C and initial pH 9 (Jaouadi et al., 2010a). Interestingly, a nearly complete feather degradation was achieved, including the delamination of the rachis. A simultaneous increase of protein concentration and sulfhydryl group formation followed by a higher disulfide bond-reducing activity of KERAB were also observed. In contrast, no degradation was noted with the control. These results, therefore, suggested that the *Streptomyces* sp. strain AB1 had a disulfide bond-reducing ability. Furthermore, when KERAB was incubated with native chicken feathers, total degradation was observed after 24 h with a simultaneous increase of protein concentration and sulfhydryl group formation; whereas no degradation was noted with the control (Jaouadi et al., 2009).

The use of enzymatic and/or microbiological methods for the hydrolysis of feathers is an attractive alternative to the currently used methods of feather meal preparation which involve high temperature and pressure treatments that result in the loss of essential amino acids (Hess and FitzGerald, 2007). The ability of the *B. pumilus* strain CBS and *Streptomyces* sp. strain AB1 to grow and produce appreciable levels of protease and keratinase using feather as a substrate could open new opportunities for the achievement of efficient biodegradation and valorization processes of keratin-containing wastes and, thereby, help reduce the environmental impact of such biowaste.

Dehairing Utility Of Sapb

The incubation of the SAPB protease with skin from goat (Fig. 5), bovine (Jaouadi et al., 2009), and sheep (Jaouadi et al., 2009) for dehairing showed that after 24 h-incubation at 37°C, hair was removed very easily for all skins, compared to the corresponding controls, with no observable damage on the collagen. Therefore, the dehaired skins obtained exhibited clean hair pore and clear grain structure (data not shown). Again, these results confirmed that SAPB alone could accomplish the whole dehairing process.

The dehairing operation in leather processing is generally carried out under a relatively high pH value of about 8 - 10 (Dayanandan et al., 2003). This criterion was also satisfied by SAPB. In fact, approximately similar results were reached with the *A. tamarri* alkaline protease on goat skin at pH 9 - 11 and temperature of 30-37°C (Dayanandan et al., 2003). Likewise, similar results were obtained with the *Vibrio* sp. strain Kr2 but at pH values ranging from 6 to 8 and temperature of 30°C (Grazziotin et al., 2007). Other alkaline proteases from *B. pumilus* with high keratinolytic activity were also reported to accomplish alone the dehairing process on bovine hair (Kumar et al., 2008), cowhides (Wang et al., 2007a) and goatskins (Huang et al., 2003). However, with the higher dehairing ability and FD activities reported for SAPB, the latter could be considered a potential strong candidate for application in biotechnological bioprocesses involving the dehairing of hides or skins or the conversion of feather-rich wastes into economically useful feather-meal.

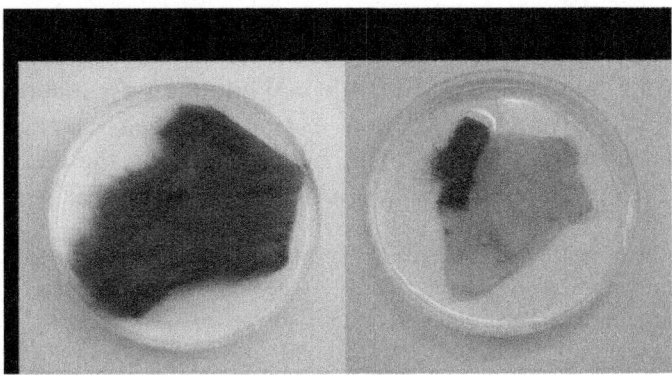

Figure 5: Dehairing function of SAPB. SABP was incubated for 24 h at 37°C with, goat hair. (Left = control, Right = test).

EFFECT OF ORGANIC SOLVENTS ON THE ACTIVITY AND STABILITY OF SAPB AND KERAB

In addition to the key areas of application discussed for proteases above, the latter constitute a highly resourceful class of enzymes for various industrial sectors. They are, for instance, necessary in the biocatalysis of various peptide coupling reactions, which are of an extremely pharmaceutical and nutritional interest, namely those involved in the synthesis of several drug precursors such as the enkephalin (Kimura et al., 1990) and aspartame precursors (Nakanishi et al., 1990). However, the ultimate application of proteases in the synthesis of peptides has often been curtailed by the poor levels of specificity and instability in the presence of organic solvents so far reported in the literature. Accordingly, various water-miscible organic solvents and alcohols at final concentrations of 50% were assayed for their effect on SAPB and KERAB activity at pH 10 and 60°C. Buthanol, acetonitrile, and ethyl acetate had significant inhibitory effects on the activity of both enzymes (Table 5). By contrast, dimethylformamide (DMF), DMSO, and hexane were noted to enhance the activity and stability of both enzymes while isopropanol and ethanol enhanced those of SAPB and KERAB, respectively (Table 5). Hence, good stability rates of 115, 97, 90 and 85% were exhibited by SAPB in the presence of DMF, hexane, isopropanol and DMSO, respectively. Equally good stability rates of 150, 125, 115 and 105% were displayed by KERAB in the presence DMSO, DMF, ethanol, and haxane, respectively. Acetonitrile, however, exerted a considerably negative effect on enzyme stability. Compared to SAPB, NH1 (Hadj-Ali et al., 2007) seemed less efficient for it exhibited only 181.5 and 94.5% of its initial activity and stability in the presence of 25% DMSO, respectively. The exception was observed with the organic solvent-tolerant protease BG1 (Ghorbel-Frikha et al., 2005; Ghorbel et al., 2003), which showed a half-life of 50 days of its activity in the presence of 25% DMSO. While the only report available to date on organic solvent protease from *B. pumilus* 115b (Rahman et al., 2007) showed that it exhibited 134% of its initial activity in the presence of 25% hexane as opposed to the 190% for SAPB and 145% for KERAB.

Table 5: Effect organic solvents on SAPB and KERAB activity and stability. The non-incubated enzyme was considered as 100%. The activity is expressed as a percentage of the activity level in the absence of organic solvent

Organic solvent (50%)	Relative activity (%)		Residual activity (%)	
	SAPB	KERAB	SAPB	KERAB
None	100 ± 2.5	100 ± 2.5	100 ± 2.5	100 ± 2.5
Methanol	100 ± 2.1	80 ± 2.1	85 ± 2.2	75 ± 2.0
Ethanol	75 ± 2.0	132 ± 3.2	55 ± 2.6	115 ± 2.6
Buthanol	50 ± 1.4	79 ± 2.0	38 ± 1.4	63 ± 1.5
Isopropanol	115 ± 2.6	25 ± 0.5	90 ± 2.3	15 ± 0.8
Acetonitrile	25 ± 1.0	20 ± 1.0	10 ± 0.8	0 ± 0.1
Ethyl acetate	85 ± 2.2	66 ± 1.5	72 ± 2.0	58 ± 1.5
DMF	200 ± 5.0	155 ± 3.7	115 ± 3.0	125 ± 3.0
DMSO	150 ± 3.7	195 ± 4.9	85 ± 2.2	150 ± 3.7
Hexane	170 ± 4.0	160 ± 3.8	97 ± 2.5	105 ± 2.5

Values represent the mean of three replicates and standard errors are reported.

A combination of high esterase and low amidase activities is necessary for several synthetic applications of proteases, including peptides coupling (Plettner et al., 1999). In addition to demonstrating its organic tolerance, the findings presented above show that both SAPB and KERAB exhibited powerful esterase activities on BTEE and on BAEE. Furthermore, no amidase activity was detected for SAPB and KERAB on BAEE with P_1= Arg and ATEE with P1 = Tyr, respectively. These findings, in addition of the observed activity and stability in certain organic solvents strongly suggested that SAPB and KERAB are potential strong candidates for use in peptide synthesis reactions in low water systems.

CONCLUSION

This chapter described the valuable advantages inherent in proteases and the promising opportunities they offer for the enhancement of a variety of industrial and consumer product applications. This was illustrated by an overview on the purification and characterization of two extracellular extremozyme serine alkaline proteinases, namely SAPB and KERAB, which were isolated from *B. pumilus* strain CBS and *Streptomyces* sp. strain AB1, respectively. These pure enzymes were significantly tolerant and stable in the presence of the various laundry detergents tested, which strongly supported their suitablity for liquid and solid laundry detergents. Furthermore, and in comparison with the standard enzyme, namely SB 309, both SAPB and KERAB turned to be more effective under alkaline and high temperature conditions. Furthermore, the *B. pumilus* strain CBS and *Streptomyces* sp. strain AB1 proved suitable for the degradation of avian feathers and feather-meal, showing strong potential for application in future biotechnological processes. More interestingly, SAPB demonstrated powerful dehairing abilities against various skins with minimal damage on collagen. Last but not least, these enzymes showed high esterase and low amidase activities as well a good tolerance for several organic solvents. Overall, the findings presented in this chapter strongly suggest that both enzymes, SAPB and KERAB, offer new and promising opportunities for prospective application in biotechnological bioprocesses, particularly those involving the synthesis of detergent formulations, dehairing during leather processing, and peptide biocatalysis in non-aqueous environments.

ACKNOWLEDGEMENTS

This work was funded by the Tunisian Ministry of Higher Education and Scientific Research (contract program CBS-LEMP, grant no. RL02CBS01) and the Algerian Ministry of Higher Education and Scientific Research (CNEPRU project grant no. JO100420070004). The authors wish to express their sincere gratitude to Pr. Anouar Smaoui, from the English department at the Sfax Faculty of Science for carefully structuring, proofreading, and polishing the language and format of the present book chapter.

REFERENCES

1. Anonyme 2007 World enzymes to 2011 (2229). Focus on Catalysts 2007 22 .

2. A. Anwar, M. Saleemuddin, 1997 Alkaline-pH-acting digestive enzymes of Polyphagous brevis and its characterization as a laundry detergent additive. Process Biochem. 35 213216 .

3.   W. M. Awad, Jr , A. R. Soto, S. Siegel, W. E. Skiba, G. G. Bernstrom, M. S. Ochoa, 1972 The proteolytic enzymes of the K-1 strain of Streptomyces griseus obtained from a commercial preparation (Pronase). I. Purification of four serine endopeptidases. J Biol Chem. 247 41444154 .

4.   A. Badis, F. Z. Ferradji, A. Boucherit, D. Fodil, H. Boutoumi, 2010 Characterization and biodegradation of soil humic acids and preliminary identification of decolorizing actinomycetes at Mitidja plain soils (Algeria). Afr J Microbiol Res. 3 9971007 .

5.   U. Banerjee, R. Sani, W. Azmi, R. K. Sani, 1999 Thermostable alkaline protease from Bacillus brevis and its characterisation as a laundry detergent additive. Process Biochem. 35 213219 .

6.   Q. Beg, R. Gupta, 2003 Purification and characterization of an oxidation stable, thiol-dependent serine alkaline protease from Bacillus mojavensis. Enzyme Microb Technol. 32 294304 .

7.   A. Bertsch, N. Coello, 2005 A biotechnological process for treatment and recycling poultry feathers as a feed ingredient. Bioresour Technol. 96 17031708 .

8.   A. Brandelli, D. J. Daroit, A. Riffel, 2010 Biochemical features of microbial keratinases and their production and applications. Appl Microbiol Biotechnol. 85 17351750 .

9.   P. Bressollier, F. Letourneau, M. Urdaci, B. Verneuil, 1999 Purification and characterization of a keratinolytic serine proteinase from Streptomyces albidoflavus. Appl Environ Microbiol. 65 25702576 .

10. R. R. Chitte, V. K. Nalawade, S. Dey, 1999 Keratinolytic activity from the broth of a feather-degrading thermophilic Streptomyces thermoviolaceus strain SD8. Lett Appl Microbiol. 28 131136 .

11. P. A. Coulombe, M. B. Omary, 2002 'Hard' and 'soft' principles defining the structure, function and regulation of keratin intermediate filaments. Curr Opin Cell Biol. 14 110122 .

12. A. Dayanandan, J. Kanagaraj, L. Sounderraj, R. Govindaraju, G. S. Rajkumar, 2003 Application of an alkaline protease in leather processing: An ecofriendly approach. Journal Clean Prod. 11 533536 .

13. E. G. Del Mar, C. Largman, J. W. Brodrick, M. C. Geokas, 1979 A sensitive new substrate for chymotrypsin. Anal Biochem. 99 316320 .

14. H. A. El -Refai, Naby. M. A. Abdel, A. Gaballa, M. H. El -Araby, A. F. Abdel Fattah, 2005 Improvement of the newly isolated Bacillus pumilus FH9 keratinolytic activity. Process Biochem. 40 23252332 .

15. J. P. Essien, A. A. Umoh, E. J. Akpan, S. I. Eduok, A. Umoiyoho, 2009 Growth, keratinolytic proteinase activity and thermotolerance of dermatophytes associated with alopecia in Uyo, Nigeria. Acta Microbiol Immunol Hung. 56 6169 .

16. N. Fakhfakh-Zouari, N. Hmidet, A. Haddar, S. Kanoun, M. Nasri, 2010 A novel serine metallokeratinase from a newly isolated Bacillus pumilus A1 grown on chicken feather meal: biochemical and molecular characterization. Appl Biochem Biotechnol. 162 329344 .

17. B. Ghorbel-Frikha, A. Sellami-Kamoun, N. Fakhfakh, A. Haddar, L. Manni, M. Nasri, 2005 Production and purification of a calcium-dependent protease from Bacillus cereus BG1. J Ind Microbiol Biotechnol. 32 186194 .

18. B. Ghorbel, A. Sellami-Kamoun, M. Nasri, 2003 Stability studies of protease from Bacillus cereus BG1. Enzyme Microb Technol. 32 513518 .

19. A. Grazziotin, F. A. Pimentel, S. Sangali, E. V. de Jong, A. Brandelli, 2007 Production of feather protein hydrolysate by keratinolytic bacterium Vibrio sp. kr2. Bioresour Technol. 98 31723175 .

20. R. Gupta, Q. K. Beg, P. Lorenz, 2002 Bacterial alkaline proteases: Molecular approaches and industrial applications. Appl Microbiol Biotechnol. 59 1532 .

21. R. Gupta, K. Gupta, R. Saxena, S. Khan, 1999 Bleach-stable alkaline protease from & sp. Biotechnol Lett. 21 135138 .

22. R. Gupta, R. K. Saxena, P. Chaturvedi, J. S. Virdi, 1995 Chitinase production by Streptomyces viridificans: Its potential in fungal cell wall lysis. J Appl Bacteriol. 78 378383 .

23. A. Haddar, R. Agrebi, A. Bougatef, N. Hmidet, A. Sellami-Kamoun, M. Nasri, 2009 Two detergent stable alkaline serine-proteases from Bacillus mojavensis A21: Purification, characterization and potential application as a laundry detergent additive. Bioresour Technolo.100 33663373 .

24. N. E. Hadj-Ali, R. Agrebi, B. Ghorbel-Frikha, A. Sellami-Kamoun, S. Kanoun, M. Nasri, 2007 Biochemical and molecular characterization of a detergent stable alkaline serine-protease from a newly isolated Bacillus licheniformis NH1. Enzyme Microb Technol. 40 515523 .

25. X. Han, Q. , S. Damodaran, 1998 Purification and Characterization of Protease Q: A detergent- and urea-stable serine endopeptidase from Bacillus pumilus. J Agric Food Chem. 46 35963603 .

26. J. F. Hess, Gerald. P. G. Fitz, 2007 Treatment of keratin intermediate filaments with sulfur mustard analogs. Biochem Biophys Res Commun.

359 616621 .

27. Q. Huang, Y. Peng, X. Li, H. Wang, Y. Zhang, 2003 Purification and characterization of an extracellular alkaline serine protease with dehairing function from Bacillus pumilus. Curr Microbiol. 46 169173 .

28. Z. Ignatova, A. Gousterova, G. Spassov, P. Nedkov, 1999 Isolation and partial characterisation of extracellular keratinase from a wool degrading thermophilic actinomycete strain Thermoactinomyces candidus. Can J Microbiol. 45 217222 .

29. B. Jaouadi, B. Abdelmalek, D. Fodil, F. Z. Ferradji, H. Rekik, N. Zaraî, S. Bejar, 2010a Purification and characterization of a thermostable keratinolytic serine alkaline proteinase from Streptomyces sp. strain AB1 with high stability in organic solvents. Bioresour Technol. 101 83618369 .

30. B. Jaouadi, N. Aghajari, R. Haser, S. Bejar, 2010b Enhancement of the thermostability and the catalytic efficiency of Bacillus pumilus CBS protease by site-directed mutagenesis. Biochimie. 92 360369 .

31. B. Jaouadi, S. Bejar, 2008 Characterization of an original serine alkaline proteinase from Bacillus pumilus CBS. J Biotechnol.136 (Suppl.) S305 EOF .

32. B. Jaouadi, S. Ellouz-Chaabouni, Ali. M. Ben, Messaoud. E. Ben, B. Naili, A. Dhouib, S. Bejar, 2009 Excellent laundry detergent compatibility and high dehairing ability of the Bacillus pumilus CBS alkaline proteinase (SAPB). Biotechnol Bioprocess Eng. 14 503512 .

33. B. Jaouadi, S. Ellouz-Chaabouni, M. Rhimi, S. Bejar, 2008 Biochemical and molecular characterization of a detergent-stable serine alkaline protease from Bacillus pumilus CBS with high catalytic efficiency. Biochimie. 90 1291305 .

34. P. Johnson, L. B. Smillie, 1974 The amino acid sequence and predicted structure of Streptomyces griseus protease A. FEBS Lett. 47 16 .

35. B. Johnvesly, G. Naik, 2001 Studies on production of thermostable alkaline protease from thermophilic and alkaliphilic Bacillus sp. JB-99 in a chemically defined medium. Process Biochem. 37 139144 .

36. Y. Kimura, K. Nakanishi, R. Matsuno, 1990 Enzymatic synthesis of the precursor of Leu-enkephalin in water-immiscible organic solvent systems. Enzyme Microb Technol. 12 272280 .

37. O. Kirk, T. V. Borchert, C. C. Fuglsang, 2002 Industrial enzyme applications. Curr Opin Biotechnol. 13 345351 .

38. K. Kitadokoro, H. Tsuzuki, E. Nakamura, T. Sato, H. Teraoka, 1994

Purification, characterization, primary structure, crystallization and preliminary crystallographic study of a serine proteinase from Streptomyces fradiae ATCC 14544. Eur J Biochem. 220 5561 .

39. A. G. Kumar, S. Swarnalatha, S. Gayathri, N. Nagesh, G. Sekaran, 2008 Characterization of an alkaline active-thiol forming extracellular serine keratinase by the newly isolated Bacillus pumilus. J Appl Microbiol. 104 411420 .

40. C. G. Kumar, 2002 Purification and characterization of a thermostable alkaline protease from alkalophilic Bacillus pumilus. Lett Appl Microbiol. 34 1317 .

41. K. H. Maurer, 2004 Detergent proteases. Curr Opin Biotechnol. 15 330334 .

42. T. Miyaji, Y. Otta, T. Nakagawa, T. Watanabe, Y. Niimura, N. Tomizuka, 2006 Purification and molecular characterization of subtilisin-like alkaline protease BPP-A from Bacillus pumilus strain MS-1. Lett Appl Microbiol. 42 242247 .

43. K. Nakanishi, A. Takeuchi, R. Matsuno, 1990 Long-term continuous synthesis of aspartame precursor in a column reactor with an immobilized thermolysin. Appl Microbiol Biotechnol. 32 633636 .

44. A. A. Onifade, N. A. Al-Sane, A. A. Al-Musallam, S. Al-Zarban, 1998 A review: Potentials for biotechnological applications of keratin-degrading microorganisms and their enzymes for nutritional improvement of feathers and other keratins as livestock feed resources. Bioresour Technol. 66 111 .

45. P. Pillai, G. Archana, 2008 Hide depilation and feather disintegration studies with keratinolytic serine protease from a novel Bacillus subtilis isolate. Appl Microbiol Biotechnol. 78 643650 .

46. E. Plettner, G. De Santis, R. Stabile, M. , J. B. Jones, 1999 Modulation of esterase and amidase activity of subtilisin bacillus lentus by chemical modification of cysteine mutants. J Am Chem Soc. 1214977 .

47. S. Radha, P. Gunasekaran, 2008 Sustained expression of keratinase gene under PxylA and PamyL promoters in the recombinant Bacillus megaterium MS941. Bioresour Technol. 99 55285537 .

48. R. N. Rahman, S. Mahamad, A. B. Salleh, M. Basri, 2007 A new organic solvent tolerant protease from Bacillus pumilus 115b. J Ind Microbiol Biotechnol. 34 509517 .

49. R. N. Z. A. Rahman, C. N. Razak, K. Ampon, M. Basri, W. M. Zin, W. Yunus, A. B. Salleh, 1994 Purification and characterization of a heat-

stable alkaline protease from Bacillus stearothermophilus F1. Appl Microb Biotechnol. 40 822827 .

50. M. B. Rao, A. M. Tanksale, M. S. Ghatge, V. V. Deshpande, 1998 Molecular and biotechnological aspects of microbial proteases. Microbiol Mol Biol Rev. 62 597635 .

51. D. N. Rawlings, A. J. Barrett, 1994 Families of serine peptidases. Methods Enzymol. 244 1961 .

52. W. Rick, In. H. U. Trypsin, Editor. Bergmeyer, 2 English edition.), Methods of Enzymatic Analysis. 2 Verlag Chemie, Weinheim (1974), 10131024 Academic Press, New York, NY USA.

53. A. Riffel, A. Brandelli, 2002 Isolation and characterization of a feather-degrading bacterium from the poultry processing industry. J Ind Microbiol Biotechnol. 29 255258 .

54. G. Schoellman, E. Shaw, 1963 Direct evidence for the presence of histidine in the active center of chympotrypsin. Biochemistry. 2 252255 .

55. J. Singh, N. Batra, R. C. Sobti, 2001 Serine alkaline protease from a newly isolated Bacillus sp. SSR1. Process Biochem. 36 781785 .

56. H. J. Son, H. C. Park, H. S. Kim, C. Y. Lee, 2008 Nutritional regulation of keratinolytic activity in Bacillus pumilis. Biotechnol Lett. 30 461465 .

57. D. G. Syed, J. C. Lee, W. J. Li, C. J. Kim, D. Agasar, 2009 Production, characterization and application of keratinase from Streptomyces gulbargensis. Bioresour Technol. 100 18681871 .

58. I. Szabo, A. Benedek, Szabo. I. B. Mihaly, G. Y. , 2000 Feather degradation with a thermotolerant Streptomyces graminofaciens strain. World J Microbiol Biotechnol. 16 153255 .

59. R. Tatineni, K. K. Doddapaneni, R. C. Potumarthi, R. N. Vellanki, M. T. Kandathil, N. Kolli, L. N. Mangamoori, 2008 Purification and characterization of an alkaline keratinase from Streptomyces sp. Bioresour Technol. 99 15961602 .

60. M. Venugopal, A. V. Saramma, 2006 Characterization of alkaline protease from Vibrio fluvialis strain VM10 isolated from a mangrove sediment sample and its application as a laundry detergent additive. Process Biochem. 41 12391243 .

61. K. A. Walsh, 1970 Trypsinogens and trypsins of varoius species. Methods Enzymol. 19 4163 .

62. H. Y. Wang, D. M. Liu, Y. Liu, C. F. Cheng, Q. Y. , Q. Huang, Y. Z. Zhang, 2007a Screening and mutagenesis of a novel Bacillus pumilus

strain producing alkaline protease for dehairing. Lett Appl Microbiol. 44 16 .

63.   S. L. Wang, Y. H. Chio, Y. H. Yen, C. L. Wang, 2007b Two novel surfactant-stable alkaline proteases from Vibrio fluvialis TKU005 and their applications. Enzyme Microb Technol. 40 12131220 .

64.   S. L. Wang, W. T. Hsu, T. W. Liang, Y. H. Yen, C. L. Wang, 2008 Purification and characterization of three novel keratinolytic metalloproteases produced by Chryseobacterium indologenes TKU014 in a shrimp shell powder medium. Bioresour Technol. 99 56795686 .

65.   C. M. Williams, C. S. Richter, J. M. Mackenzie, J. C. Shih, 1990 Isolation, identification, and characterization of a feather-degrading bacterium. Appl Environ Microbiol. 56 15091515 .

66.   F. Xie, Y. Chao, X. Yang, J. Yang, Z. Xue, Y. Luo, S. Qian, 2010 Purification and characterization of four keratinases produced by Streptomyces sp. strain 16 in native human foot skin medium. Bioresour Technol.101 344350 .

67.   M. Yasuda, M. Aoyama, M. Sakaguchi, K. Nakachi, N. Kobamoto, 1999 Purification and characterization of a soybean-milk-coagulating enzyme from Bacillus pumilus TYO-67. Appl Microbiol Biotechnol. 51 474479

Chapter 12

TRYPTOPHAN BIOCHEMISTRY: STRUCTURAL, NUTRITIONAL, METABOLIC, AND MEDICAL ASPECTS IN HUMANS

Lionella Palego,[1] Laura Betti,[2,3] Alessandra Rossi,[1] and Gino Giannaccini[2,3]

[1]Department of Clinical and Experimental Medicine, University of Pisa, 56126 Pisa, Italy

[2]Department of Pharmacy, University of Pisa, 56126 Pisa, Italy

[3]Interdepartmental Center of "Nutraceutical Research and Food for Health", University of Pisa, 56124 Pisa, Italy

ABSTRACT

L-Tryptophan is the unique protein amino acid (AA) bearing an indole ring: its biotransformation in living organisms contributes either to keeping this chemical group in cells and tissues or to breaking it, by generating in both cases a variety of bioactive molecules. Investigations on the biology of Trp highlight the pleiotropic effects of its small derivatives on homeostasis processes. In addition to protein turn-over, in humans the pathways of Trp indole derivatives cover the synthesis of the neurotransmitter/hormone serotonin (5-HT), the pineal gland melatonin (MLT), and the trace amine tryptamine. The breakdown of the Trp indole ring defines instead the "kynurenine shunt" which produces cell-response adapters as L-kynurenine, kynurenic and quinolinic acids, or the coenzyme nicotinamide adenine dinucleotide (NAD+). This review aims therefore at tracing a "map" of the main molecular effectors in human tryptophan (Trp) research, starting from the chemistry of this AA, dealing then with its biosphere distribution and nutritional value for humans, also focusing on some proteins responsible for its tissue-dependent uptake and biotransformation. We will thus underscore the role of Trp biochemistry in the pathogenesis of human complex diseases/syndromes primarily involving the gut, neuroimmunoendocrine/stress responses, and the CNS, supporting the use of -Omics approaches in this field.

INTRODUCTION

L-Tryptophan (L-Trp) is a large neutral amino acid (LNAA) present in living organisms, precisely one of the 20 L-amino acids (AAs) incorporated in proteins during the process of mRNA translation. All Trp residues in protein and peptide sequences are conventionally indicated with the alphabetic letter W. The AA L-Trp, discovered by the English chemist F. Hopkins in 1901, is also one of the 9 essential AAs for humans which cannot be endogenously synthesized and need to be supplied with aliments, as revealed through diet manipulation studies [1]. Besides being an intermediate of protein/peptide synthesis and turn-over, Trp is the object of scientific investigations in human biological research since decades because of its transformation, after absorption, into a series of small bioactive, pleiotropic compounds, each capable of influencing a number of cell metabolic pathways and physiological responses. Hence, alterations of L-Trp-deriving compounds can be found associated with a variety of metabolic diseases and syndromes affecting those systems and organs responsible for maintaining the chemical, cellular, and behavioural homeostasis: the gut-liver apparatus and the neuroendocrine and immune systems along with the CNS. In particular, an imbalanced metabolism of this AA can interfere with the ability of these systems to interact with as well as discriminate, during development, stressors and stimuli, exogenous and endogenous antigens, and nutrients and xenobiotics.

Amongst Trp-derived compounds produced in the human body, there is the ancient neurotransmitter serotonin (5-hydroxy-tryptamine, 5-HT), a biogenic amine which is known to regulate, in the human CNS, the main adaptive reactions and responses to environmental changes, such as mood-anxiety, cognition, nociception, impulsivity, aggressiveness, libido, feeding behaviour, and body temperature [2, 3]. Next to its role as a neurotransmitter, 5-HT also modulates the activity of peripheral districts, in particular the gut function, the immune and inflammatory responses, the differentiation process of blood stem cells, and the hemodynamic function [3]. Indeed, an altered 5-HT transmission has been found associated with mood-affective disorders [4], autism and cognitive deficit [5, 6], anorexia or bulimia nervosa and obesity [6], and other syndromes presenting peripheral symptoms, such as fibromyalgia, chronic fatigue syndrome, and irritable bowel syndrome (IBS) [7]. Moreover, 5-HT is in turn the precursor of the circadian regulators N-acetyl-5-HT (NAS) and melatonin (MLT), primarily produced in the pineal gland but also in periphery where the two indoleamines act as scavenger compounds [8]. In vertebrates and humans, another main metabolic pathway of Trp is the indole ring breakdown, through the so-called "kynurenine shunt" which produces a number of molecules involved in inflammation, immune response, excitatory neurotransmission, and many other functions.

Only a very small amount of endogenous/dietary L-Trp is converted into 5-HT, suggesting that the bioavailability of this AA and/or changes in the regulation of its metabolism in tissues might be critical for maintaining a healthy balance between all its different paths and destinies [9, 10]. Even though the various components of L-Trp metabolism have been studied since a long time, their regulatory mechanisms in humans have been explored in a lesser extent, especially concerning developmental and/or gender-dependent aspects. Without claiming to provide herein an exhaustive vision of the complexity of Trp research in living organisms and humans, we will start this review by highlighting the impact of the chemistry of this molecule for life, its distribution in the alimentary chain, and nutritional value for human diet and then presenting some among the main tissue-dependent mechanisms of Trp uptake/metabolism. We will then underpin those molecular players in Trp biochemistry which are considered or are possible vulnerability markers in the physiopathology of human complex diseases, trying to point out their regulation. At the same time, we will briefly introduce some Trp research targets actually under investigation for therapeutic strategies in human pathology as well as the utility of -Omics approaches.

TRYPTOPHAN: A PIVOTAL CHEMICAL STRUCTURE FOR LIVING ORGANISMS

The molecular evolution of life in Earth has selected the chemical structure of –R groups of the 20 L-AAs as the most suitable for building proteins. L-Trp is the only AA in proteins deriving from indole, a bicyclic ring formed by a benzene and a pyrrole group (Figure 1(a)), linked to the α-carbon by a –CH_2- group. The presence of the indole ring in the chemical structure of Trp gives high hydrophobic features to this molecule among all protein AAs. Several AAs could be theoretically synthesized starting from indole (Figure 1(b)), but, amongst these, only L-Trp has been "retained" as a constituent of proteins in living organisms, presumably being the most simple structure of all possible indole AAs. In fact, Trp is the AA at the highest number of C atoms (C_{11}) and the presence of other C atoms or substituent groups would be unnecessary. The advantage to keep indole in life chemistry derives either from the possibility to exploit its C_{11} skeleton in metabolism or to utilize it as –R residue in proteins and peptides to promote and stabilize their structure. Also, Trp is metabolized to produce biologically active indole compounds which have great impact on life functions. In fact, beside being present in the chemical structure of the neurotransmitter 5-HT and, in turn, in the circadian molecules NAS and MLT in animals and humans (Figure 1(c)), the indole ring of Trp can be transformed into bioactive compounds also by plants: for instance, the plant hormone indole-3-acetic acid (IAA) or auxin, the defense compounds indolyl glucosinolates [11], and the indole alkaloid and natural hallucinogen dimethyltryptamine.

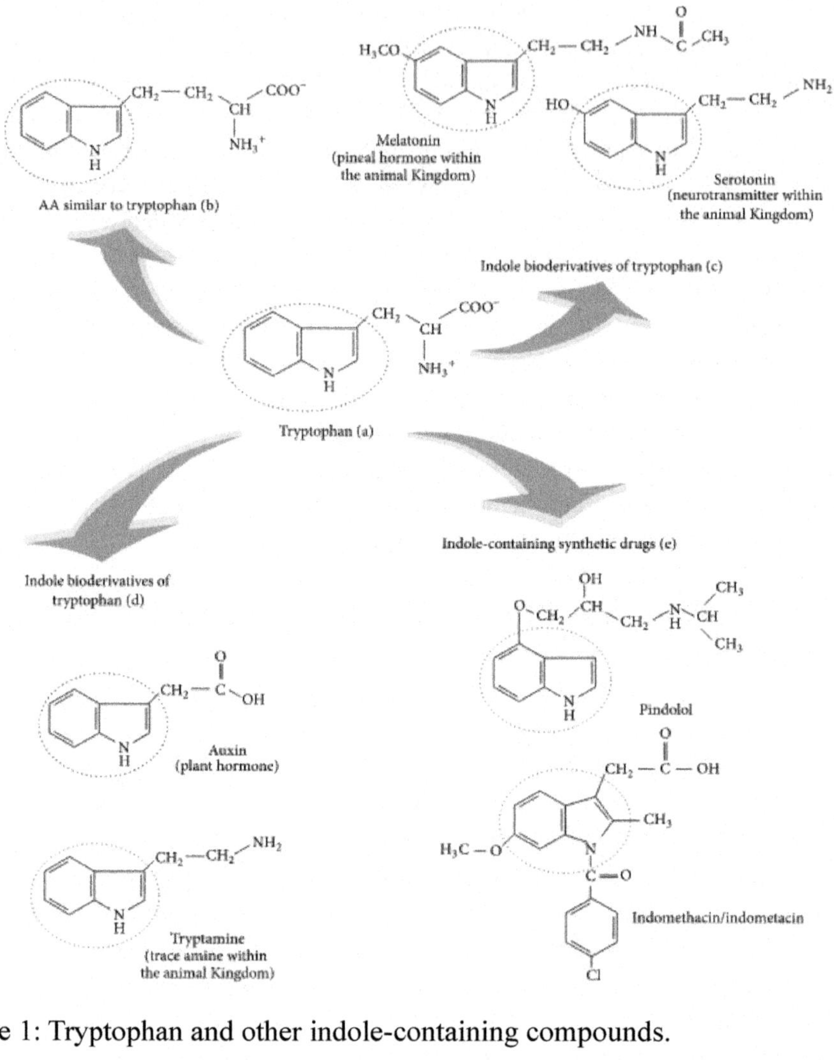

Figure 1: Tryptophan and other indole-containing compounds.

In particular, the plant hormone auxin has been found linked to a specific Trp metabolism pathway involved in plant photoperception and development [11, 12]. Interestingly, indoleamines as 5-HT and MLT have been detected also in plants where their function is under investigation [13]. Similarly, tryptamine (Figure 1(e)) and derivatives, as dimethyltryptamine, have been found in mammalian CNS where they act as neuromodulators, the so-called "trace amines" [14]. Thus, the physiological significance of indole derivatives and their pathways in the evolution of living organisms pivotally involves development, the response to light and environmental variations, and defence against viruses, bacteria, parasites, and/or toxic substances.

Moreover, the study of the metabolic fates of Trp and the other aromatic AAs (AAAs) phenylalanine (Phe) and tyrosine (Tyr) in living organisms has revealed that these pathways represent a "signature" of the cellular evolutionary stages, starting from prokaryote unicellular organisms, passing through eukaryote ancestor cells, arriving then to distinct vegetal or animal eukaryote cells with specific subcellular compartments and trophic features [15]. If Trp is essential for animals, bacteria or other eukaryotes as fungi and plants are instead able to synthesize it from chorismic (also the precursor of Tyr and Phe) and anthranilic (Trp path only) acids.

In bacteria, fungi, and plants, the biosyntheses of Trp, Tyr, and Phe are linked together by the shikimate pathway: phosphoenolpyruvate (PEP) and erythrose 4-phosphate (E-4P), deriving from glycolysis and the pentose phosphate shunt, enter into a series of reactions involving the activity of seven enzymes, whose final product is chorismate, the common precursor for the synthesis of the other two main metabolites, prephenate and anthranilate, the first generating Tyr and Phe, the last producing Trp. Prephenate derives from chorismate through the activity of the enzyme chorismate mutase; in turn, prephenate enters into a 3-branch path producing Tyr and Phe. The biosynthesis of Trp in bacteria shares common genes and chemical reactions with plants or fungi: chorismate is recognized by the enzyme anthranilate synthase which transfers to it an amino group from the AA glutamine generating anthranilate and pyruvate; anthranilate is then transformed into Trp via 5 subsequent enzymatic steps. Bacteria and plants or fungi follow, however, different regulatory mechanisms of this metabolic path. In bacteria, chorismate produces Trp under the control of one of the most studied models of gene expression regulation in prokaryote organisms, the Trp operon, activated or repressed depending upon the intracellular concentrations of this AA: operon genes are not constitutively expressed but are induced only by the absence of Trp [16, 17]. The regulation of Trp formation then diverges from the bacterial one because of the different gene organization in eukaryotes. In the absence of the operon, the biosynthetic path of this AA is essentially regulated through a negative feedback of the final product on anthranilate synthase, whose α-subunit is specifically recognized by Trp [15]. Additionally, Trp formation in plants has been found tightly regulated by several transcriptional factors acting on gene expression of the enzymes of the shikimate pathway and AAA metabolism, as evidenced in the Brassicaceae Arabidopsis. Many of these transcriptional factors have been identified, each differentially stimulated by diverse stressors, as infections by pathogens, trauma, or light. In particular, different stressors can selectively induce the expression of genes involved in Tyr/Phe or Trp biosynthesis, suggesting a distinct physiological significance of the two branches of AAA metabolism in plants [18, 19]. In the common ancestor of animal and vegetable cells, the shikimate path was localized in the cytosol whereas in higher plants this metabolic shunt occurs inside plastids [15]. Thus, genes of Trp biosynthesis would have been lost in plastid-lacking animal eukaryote cells. These considerations lead to reasonably think that AAAs metabolism and production of their bioactive derivatives occupy a central position in the early stages of the evolution of living organisms and trophism lineages. It is not therefore surprising that AAAs represent foremost compounds for human nutrition and health. The importance of maintaining intact the indole ring in Trp derivatives is mirrored by natural therapeutic agents: plant indole alkaloids, as vinblastine and related compounds, exert in fact antitumor properties. Also, indole derivatives have been used in pharmacological research as the starting point for the synthesis of therapeutically relevant compounds. These comprise the nonsteroidal anti-inflammatory compound indomethacin and the antihypertension drug pindolol, a β-adrenoceptor blocker acting also on 5-HT transmission (Figure 1(e)).

TRYPTOPHAN RESIDUES IN PROTEINS AND PEPTIDES

The presence of Trp residues in polypeptides, as previously introduced, deserves a specific mention: in fact, the Trp indole ring in –R residues gives unique properties to proteins and peptides promoting protein-protein, protein-peptide, or protein-biomolecule structural hydrophobic interactions. The Trp indole ring is able to stabilize structures, domains, and interactions through Van der Waals forces while the indole-N shows propensity as a hydrogen bond donor evidencing a role of this AA also in protein binding and recognition. The presence of Trp –R groups in precise domains, for instance, in transmembrane domains of membrane-bound proteins, is fundamental for the protein stability/assemblage to the phospholipid bilayer [20]. Hydrophobic interactions between proteins and peptides or between these and other biologically active molecules have great importance in cell physiology. Some reviews in the current literature show interesting investigations focusing on these structural aspects of Trp residues: these works are relevant in the study of both cellular and synthetic (peptidomimetics) peptides, with the purpose of evaluating the specific function of secondary and tertiary conformational structures [21]. It is noteworthy that protein hot-spots relevant as therapeutic targets are frequently localized in Trp-rich β-hairpin regions [21]. Residues of Trp in the AA sequence of small bioactive peptides, as endogenous anti-inflammatory/antiobesity melanocortin peptides or defense antimicrobial peptides of innate immunity [22, 23], need to be further explored in the field of peptides' structure-activity relationships.

TRYPTOPHAN REQUIREMENT AND CONTENT IN FOOD

A main consideration deriving from previous paragraphs is that Trp is precious for life: its biosynthetic pathway is in fact energetically expensive and requires the expression of several enzymes and substrates either for Trp operon in bacteria or for the shikimate and chorismate paths in plants. This probably explains why L-Trp is an AA scarcely represented in the alimentary chain [21, 24–26] and its presence in animal cells and tissues must be tightly regulated. The frequency of Trp residues in proteins is, on average, 1-2% with respect to 5% of other AAs and 9% of leucine, the most abundant AA [27]. The recommended dietary Trp daily doses for human adults ranges from 250 to 425 mg/day, corresponding to 3.5–6 mg kg^{-1} (meanly 4 mg kg^{-1}) body weight per day [27, 28]. As for other essential AA, new-borns and children require from the diet much higher Trp levels than adults, about 12 mg kg^{-1} bw day [28]. Together with cysteine (Cys), Trp is the essential AA required in lesser amount in human diet [27, 29]. This apparent paradox, due to the variety of important enzymatic reactions in the body and production of crucial metabolites deriving from Trp, would suggest that just a "right" amount of this AA is necessary for humans, without a need to be accumulated: its chemistry is necessary for health but not its accrual in tissues.

The AA Trp is introduced along with all other AAs in the body with protein-rich foods, mainly of vegetal or animal origin. Aliments at higher Trp content include animal origin: milk, cheese, and dairy products, eggs (white), meat, and seafood (fish and crustaceous) and vegetal origin: potatoes, chickpeas, soybeans, cocoa beans, and nuts (walnuts, hazelnuts, and cashew). Lower Trp amounts can be found in some varieties of cereals and maize. Thus, a normal, varied, and balanced diet, as in developed countries, can largely ensure the daily Trp requirement. A main nutritional impact of Trp for human diet is represented by chronic exposure to a diet low in niacin (vitamin B$_3$) and Trp, which produces pellagra, a metabolic dysfunction defined by severe alterations of the skin, gut, and brain activity [30]. Pellagra was frequent in past centuries, in people eating almost exclusively low-niacin/Trp maize varieties. In fact, niacin is classified as a vitamin, but this compound can be produced through the metabolic transformations of L-Trp into its precursor quinolinic acid; this explains why L-Trp exerts a protective action against the onset of pellagra symptoms in low-niacin diets. Thus, in economically disadvantaged countries, Trp content in foods, together with other essential AAs, can be of great importance. The analysis of the composition of nutrients, vitamins, essential elements, and AAs represents the basis for good health and children development in these countries. Besides, the amount of Trp in diet represents a challenge for human health and nutritional status worldwide, especially as concerns the regulation of its concentration in plasma as well as its uptake to tissues and brain. The role of the gut microbiome is also an interesting aspect that is emerging as a link between nutrition, gut absorption, Trp fates, and health.

TRYPTOPHAN ABSORPTION, TRANSPORT, AND UPTAKE: REGULATION OF PLASMA LEVELS BY DIET, HORMONES, AND CARRIERS

On the whole, plasma levels of Trp undergo regulatory mechanisms comparable to those operating for other protein L-AAs: AA uptake occurs in all tissues and cells according to the need for protein synthesis or degradation, with gut, liver, and muscle tissue primarily involved in its modulation. Once introduced with food, all AAs, including Trp, are absorbed by the gut, pass into the bloodstream, are transported to all main tissue districts, overall muscles, and liver, and are finally taken by cells to be part of the AA pool used for the synthesis and turn-over of proteins. Proteolysis and protein catabolism inside cells regenerate, in part, the intracellular reserve of AAs (and Trp) for subsequent protein synthesis and, in part, provoke their release in the bloodstream. Insulin, glucagon, and cortisol are the regulatory hormones of endogenous protein turn-over: insulin blocks the proteolytic activity and promotes recovery of AAs from the bloodstream for protein synthesis in the tissues (overall muscle), while glucagon reduces plasmatic AAs, as alanine (Ala), glycine (Gly), and proline (Pro), for use in the synthesis of glucose in the liver. Cortisol increases the AAs plasma levels (efflux from muscle), shifting the balance towards proteolysis [31]. At the same time, each AA can undergo its own regulation originating its own cell in- and out-flow, in relation to AA composition of both endogenous proteins and those derived from the diet; these last at more variable content [32]. Also, multiple factors as age, gender, or physical activity concur to affect plasma concentrations of AAs [32].

Differently from nonessential AAs, for which, in addition to diet, the rate of de novo synthesis is able to control the homeostatic balance of endogenous contents, essential AAs and Trp plasma concentrations are more directly related to their amount in diet. Specifically as concerns dietary Trp, it must be also pointed out that the degree of its relative contribution to protein synthesis/degradation remains unclear [33]: the "paradox" of its rarity in the biosphere and its concomitant worth for life influences its absorption, transport in the bloodstream, tissue uptake and, as a consequence, its destinies.

A foremost and intriguing aspect of human Trp biology is in fact defined by the observation that diet and the type of meal can change its plasma levels as well as its uptake by different cell types. After food digestion, for gut absorption this AAA shares its passage across enterocytes with other neutral AAs through two distinct carrier molecules: the first, expressed at the level of apical membranes of the gut epithelium, is a Na^+-dependent transmembrane protein codified by the SLC6A19 gene, while the second, named TAT1, is codified by the SLC16A10 gene and is localized on basolateral epithelial membranes, controlling in particular the absorption of AAAs [33]. Tryptophan has also the lowest affinity for the apical carrier than other competitive NAAs, except lysine (Lys), confirming that tissues require defined amounts of Trp and suggesting that gut absorption is a regulated step for the subsequent transport and biotransformation of this AAA.

A widely studied model of Trp uptake mechanism is that regulating its transport across the blood-brain barrier (BBB). For that, insulin and other large neutral AAs, valine (Val), leucine (Leu), isoleucine (Ileu), Tyr, and Phe, have been found to play a chief role: in fact, LNAAs compete with each other for the same transporter system across the BBB, under the control of insulin. This explains why a protein-rich meal increases Trp plasma levels but not its uptake to the brain. Trp uptake to CNS is thus rather favoured by carbohydrate-rich meals. After a carbohydrate meal, 5-HT biosynthesis in raphe nuclei is increased. This mechanism has been extensively studied in mammals: carbohydrate ingestion increases insulin secretion and the clearance of AAs from plasma, in particular of branched-chain AAs (BCAAs: Val, Ileu, and Leu), transported from the bloodstream to muscles, thus increasing Trp availability for CNS uptake and, as described later, to 5-HT synthesis. The 5-HT release at the hypothalamic level activates specific 5-HT receptor subtypes devolved to inhibit appetite brain nuclei [34]. Thus, meal composition, palatable food, and poor protein foods all contribute to Trp uptake across the BBB in favor of 5-HT synthesis. Protein-rich foods in fact contain Trp, but at lower levels than other LNAAs, which, on the whole, rather provoke inhibition of Trp brain uptake. Briefly: a protein-rich meal increases the availability of AAs and Trp, but LNAA competition for transport to the brain reduces Trp entering into the brain in comparison with the amount crossing the BBB after a low protein diet. On the other side, some proteins containing a higher Trp/LNAAs ratio than others, as the milk-derived α-lactalbumin, can on the contrary elevate Trp uptake into the brain. Thus, conclusively, proteins can enhance 5-HT synthesis but in relation to their low or high content in Trp [35, 36]. Another important uptake regulatory aspect is represented by the fact that Trp is highly lipophilic and scarcely soluble in aqueous solutions at the physiological pH, so that its transport in blood requires plasma albumin binding: Trp is the only AA transported by albumin. Therefore, a finely regulated equilibrium between free and bound Trp levels exists in plasma, an argument of actual scientific interest and debate.

Next to nutritional considerations, for a deeper understanding of Trp uptake, the transport proteins across tissues and the BBB are currently under investigation. The molecular complex acting as the Trp and other LNAAs transporter across the BBB is formed by a protein which belongs to the superfamily of AA carriers of type "L," LAT1, and an accessory protein, the cell surface antigen CD98 (heavy chain 4F2), highly expressed in the barrier capillary endothelium [37, 38]. Regulatory mechanisms of this LNAA carrier complex are therefore important for Trp passage into SNC and availability for brain metabolism. The ratio between free and albumin-bound Trp has been also found to modulate Trp passage into the brain: under various conditions, including intense sport activity, activation of nervous sympathetic system, lipolysis, and increased plasma levels of free fatty acids (NEFA), the intracerebral Trp uptake is facilitated by the displacement of the Trp-albumin bound provoking, as a result, the increased availability of the free AA [39]; the Trp-albumin bound is also displaced through interactions with the capillary vessel endothelium (glycocalyx) [40]. Lastly, once it crossed the BBB, Trp can enter inside SNC cells by transport carriers for AAs, which have not been fully characterized. Transport proteins candidates for this role are proteins of the superfamily of G transporters or "ATP-binding cassette transporters," highly expressed by 5-HTergic neurons. If transporters of types "L" and "G" display an affinity for AAs ranging from 10 and 100 µM, other carriers show a greater affinity for Trp, <1 µM: these proteins are type "T" carriers of pinealocytes or macrophages. This reveals that Trp uptake follows a tissue-dependent regulation based upon a molecular heterogeneity of Trp protein carriers in various tissues [41].

TRYPTOPHAN AND METABOTROPIC G-PROTEIN COUPLED RECEPTORS FOR AROMATIC AAS

Some AAs, as γ-aminobutyric acid (GABA) or glutamate, are neurotransmitters through the activation of specific subtypes of G-protein coupled receptors (GPCR), particularly relevant for brain function. Research in metabotropic receptors has evidenced a surprising diversity of these proteins and their ligand specificity, also involving elements, nutrients, and metabolites [42]. An interesting aspect needing to be deepened, linked to the topic of this review, is that AAAs and, in particular, L-Trp and L-Phe recognize and activate a class of Ca^{2+} "sensor" and taste metabotropic receptors whose physiological role is under investigation [43]. Beside their primary action on Ca^{2+} regulation, the high expression of these receptors in the gut implies a main action on feeding, food choice, nutrient absorption, and gastrointestinal function [44–47]. It can be supposed that the extensive study of their localization, gene expression, and function within the body would provide useful information and clinical application.

METABOLIC FATES OF TRP

After its uptake into the various districts, tissues, and cells, Trp is introduced into protein metabolism and synthesis or can enter into various metabolic paths depending upon the tissue expression of specific enzyme activities. Figure 2 summarizes Trp transport in the bloodstream, its uptake to different tissues, and its main metabolic fates. In substance, beside protein turn-over, Trp metabolism can be divided into two main branches: one, limited to approximately the 3–10% of these Trp biotransformations, which keeps the indole ring intact while producing chemical messengers as the indoleamines 5-HT, NAS, and MLT and the trace amine tryptamine and derivatives and the other, the prevalent one (about 90% or more), which breaks the indole ring generating the kynurenine path, kynurenines, nicotinic acid, and the nicotinamide adenine dinucleotide (NAD^+) synthesis. We thus will follow herein this schema for describing the Trp metabolic paths which generate low-molecular weight derivatives. The limiting enzymatic reaction for 5-HT biosynthesis is Trp-hydroxylase, TPH, which is active in specialized tissues: 5-HTergic neurons of mesencephalic raphe nuclei; pinealocytes in the pineal gland; blood cells (lymphomonocytes, macrophages, and mast-cells); enterochromaffin cells; neuroendocrine epithelial cells in the lung [41] and other emerging tissues. The synthesis of 5-HT occurs in two enzymatic steps: the first consists in the C-5 hydroxylation of Trp at the level of the benzene ring of the indole (cofactors: O_2 and tetrahydrobiopterin, TBH_4) by TPH leading to 5-hydroxy-Trp; the second one is the decarboxylation of 5-hydroxy-Trp to 5-HT, a reaction catalyzed by the enzyme L-amino acid aromatic decarboxylase (cofactor: pyridoxal-5′-phosphate, P5P). This last enzyme is ubiquitous. Newly synthesized 5-HT can enter into storage vesicles to be released as a neurotransmitter in CNS or a modulator in periphery; after its release, excess 5-HT is internalized again through 5-HT reuptake (5-HT transporter, SERT), degraded to 5-hydroxy-acetaldehyde by monooxygenase activities (MAO-A) on mitochondrial outer membrane and then oxidized into 5-hydroxyindoleacetic acid (5-HIAA) by aldehyde dehydrogenase (cofactor: NAD^+). This last compound is excreted in urine. Two main TPH isoforms exist, TPH_1 and TPH_2, codified by distinct genes [48]. The TPH_1 activity is prevalent in periphery and pinealocytes, while TPH_2 is expressed in raphe nuclei. Both isoforms are partially saturated in tissues, so that the rate of 5-HT production depends on Trp levels in the SNC and periphery. Intriguingly, in pinealocytes, specialized in the production and secretion of the circadian hormone MLT, the precursor 5-HT can be obtained either by its uptake through a pineal 5-HT transporter (SERT) [49] or by its synthesis from Trp, transported by type "T" protein carriers [50]. Then, 5-HT is acetylated by the enzyme aryl-alkyl-amine-N-acetyl transferase (AANAT) producing NAS which is in turn converted by the enzyme hydroxyl-indole-methyl-transferase (HIOMT) and cosubstrate S-adenosyl-methionine (SAM) into MLT. Melatonin synthesis can also occur in peripheral tissues where this molecule acts as a paracrine/scavenger effector. Another metabolic fate which maintains the indole ring is the formation, by Trp direct decarboxylation, of the trace amine tryptamine, a compound with a physiological meaning which has not been fully understood.

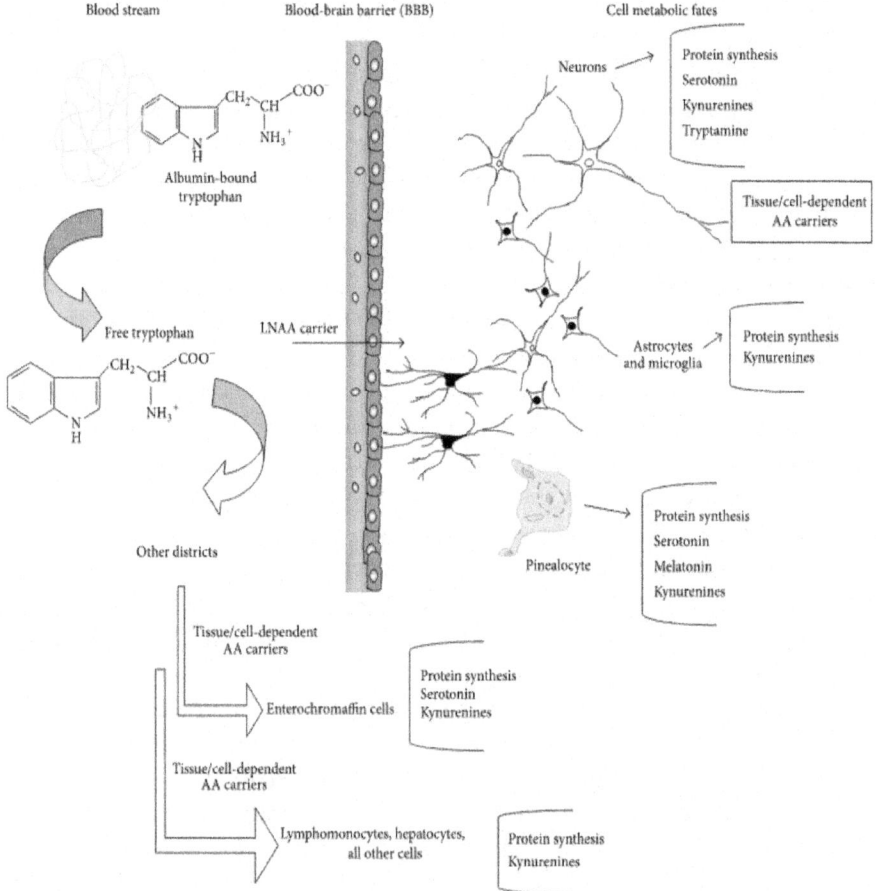

Figure 2: Tryptophan uptake to tissues and main metabolic destinies in either physiological or pathological states.

Trace amines, present in mammalian tissues at very low, nanomolar, concentrations, can be divided into those deriving from Trp (5-HT-related) and those deriving from phenylalanine and tyrosine (catecholamine-related) and are thought to regulate monoamine transmission [51]. A class of metabotropic, G-protein coupled receptors (GPCR) specific for trace amines recognition (trace amine associated receptors, TAARs) has been discovered [52]. TAARs have been widespread localized in mammalian brain, prevalently in the amygdale region; they are highly expressed in populations of nonmonoaminergic neurons colocalized with monoaminergic neurons [52], implying that trace amines can exert a GPCR-mediated regulation of monoamine neurotransmission. Interestingly, in humans, TAARs genes have been located in chromosome 6, within a DNA region linked to schizophrenia and bipolar disorder [53]; these receptors (TAAR1), which activate adenylate cyclase and cAMP via a protein, have been also found to exert a chief role in drug addictions.

The most active metabolic path of Trp is the indole-breaking pathway, the so-called L-kynurenine shunt. Genes codifying for enzymes of the kynurenine shunt have been prevalently found in eukaryote animal cells. Aerobic bacteria also express enzymes of this path [54], implying the ancient origin of this biotransformation. The majority of studies concerning Trp breakdown have been conducted in animals: the kynurenine pathway is active in almost all tissues, being almost rate-limited by the first reaction, the opening, by oxidation, of the indole ring of this AA: the reaction can be catalyzed by two types of heme-containing enzymes differently located in tissues: (a) the Trp-2,3-dioxygenase or Trp pyrrolase (TDO) mainly expressed in the liver but also in brain, prevalently in populations of astrocytes and (b) the indoleamine 2,3-dioxygenase (IDO) expressed in most peripheral tissues, in immune system cells and in the SNC, prevalently in microglia [41, 55, 56]. If TDO is substrate-specific, IDO recognizes and disrupts indole also from D-Trp, 5-HT, MLT, 5-HIAA, and tryptamine. Metabolites deriving from all the other reactions of the kynurenine shunt are considered either cytoprotective (kynurenic acid) or cytotoxic/proepilepsy (3-hydroxy-kynurenine, 3-hydroxy-anthranilic, quinolinic, and nicotinic acids). The kynurenic acid is an antagonist of the excitatory neurotransmitter N-methyl-D-aspartic acid (NMDA) modulating the synthesis of antioxidant species and MLT; the nicotinic path, formed by NMDA agonists, concurs to NAD^+ synthesis and can also generate free radicals. In effect, an adaptive orchestration of each of these derivatives is physiologically relevant: gene expression of IDO, TDO, or other enzymes of this shunt can be in fact differentially and subtly modified/regulated in various physiological conditions, following a dynamic model. The deregulation or malfunctioning of this shunt, for instance, due to a significant enzyme hypo- or hyperactivation, can underlie a pathological state. Also, manipulations of this path can underscore significant cell responses. Secretion of cytokines and proinflammatory factors as γ-interferon (γ-INF) and tumor necrosis factor-α (TNF-α) induces IDO gene expression, whereas antidepressant drugs, tricyclic or selective 5-HT reuptake inhibitors (SSRIs), are able to inhibit TDO, while increasing Trp plasma levels and 5-HT synthesis [41, 57, 58]. The balance between the kynurenine shunt, formation of quinolinic acid or other species potentially producing free radicals, and production of antioxidant metabolites is under the control of the purine system. Purines, as NMDA antagonists, counteract Trp catabolism towards the production of prooxidant species under physiological and pathological conditions [59].

REGULATION OF TRYPTOPHAN METABOLISM AND HUMAN DISEASES

Both exogenous and endogenous factors finely regulate Trp biotransformation. Diet represents one amongst those main factors which influence Trp availability, distribution, and metabolism in different anatomical districts: Trp diet manipulations have revealed brain, digestive apparatus, and gut as "critical" tissues. Dietary Trp manipulation and acute Trp depletion (TD) have contributed to identify patients' vulnerability to depression or other mood symptoms linked to dysfunctional monoaminergic systems: TD strongly modifies Trp plasma levels, decreasing 5-HT synthesis and mood tonus in subjects with a familiarity or history of mood-affective disorders, similarly to what is observed, even if with opposite effects, for sleep deprivation/restriction [60]. This suggests that depressed patients would be less capable of compensating dietary Trp variations with respect to healthy subjects. Tryptophan depletion has also shown to affect gut motility in IBS [61]. An altered homeostasis of the Trp/LNAAs ratio in plasma has been considered the origin of the so-called "carbohydrate craving" syndrome, linked to some types of human obesity and/or feeding behavior disorders, often in comorbidity with depression and bipolar disorder [62–64]. These findings relate Trp availability and metabolism with severe illnesses as metabolic syndrome and diabetes. Other human diseases linked to Trp are those generated by defective carriers at the level of gut absorption, as Hartnup disease, showing pellagra-like symptoms, or malabsorption conditions [65]: fructose malabsorption is characterized by low Trp plasma levels and mild depressive symptoms, evidencing a link between gut dysfunction, Trp availability, and mood. In addition to (i) the influence of diet or alimentary intolerances and/or (ii) Trp content in protein and control exerted by insulin/glucagon, Trp absorption/metabolism in the different cells and tissues can be specifically regulated by many other factors involving the stress neuroendocrine axis, the immune response, and inflammation. A main aspect needing to be better understood is the variation and adaptation of Trp metabolism during the lifespan: the relative contribution of each destiny (5-HT synthesis, tryptamine, kynurenine shunt, and protein synthesis) is supposed to vary during development and aging, in relation to gender, individual vulnerability, and/or lifestyle. The competition between 5-HT synthesis and the kynurenine shunt in cells expressing both enzymatic pools is in fact under the control of various peptides, neurotransmitters, hormones, and cytokines, also in response to stressors of different nature and to variations of the quality of life, an issue still presenting unsolved questions [66]. The enzymatic pool of TPH, TDO, and IDO can be specifically modulated by all these factors: environmental factors and gene polymorphisms/vulnerability can concurrently modify Trp fates. The enzyme TDO is under the control of the arousal response, by cortisol and prolactin [41, 67], implying that, during prolonged stress, 5-HT requirements could be not compensated. The hyperactivation of the glucocorticoid-insensitive IDO can be provoked by cytokines, chemokines, and inflammation mediators. Immune system and inflammation are the main targets of unbalanced IDO activity [68–70]: the role of this enzyme of Trp metabolism consists in the ability of preventing/restraining excessive tissue damage due to cytotoxic immune actions as well as creating a Trp-poor microenvironment as a host defense mechanism. In addition, the activity of IDO is involved in the regulation of Trp levels during development, for instance, to establish immune tolerance and the discrimination of self/non-self-antigens during pregnancy ensuring the homeostasis of individual tissue identity and defense functions [71, 72]. The IDO fine tuning and the γ-interferon mediated switching of its isoforms, IDO1 and IDO2, modulate the balance between immune response suppression or activation [73]. It is worth noting that the regulation of these two IDO activities is disturbed in cancer cells, participating in immune tolerance against tumor antigens [73].

Intriguingly, IDO has also a particular relevance in mood disorders and in neurodegenerative SNC diseases, brain aging, and Alzheimer or Parkinson disease [74, 75]. Oxidative stress and TDO and IDO activities have been found impaired in children with autism [76, 77]. Other factors which influence the availability and metabolism of Trp are aging and gender-related mediators, sustaining the prevalence of many pathological conditions in the elderly or in one of the two sexes [78–82]. Feeding disorders, obesity, and depression are more frequent in women and TD has revealed gender-dependent mood vulnerabilities. The homeostatic interrelation between 5-HT-glucocorticoids can vary in different phases of relational/social life, as adolescence in the two sexes, pregnancy, premenstrual or pre- and postmenopausal periods in women, or aging in man and women separately [83]. The onset of feeding behavior disturbances as bulimia or anorexia nervosa is prevalent in women occurring during adolescence, as sustained in homozygote twin studies [84]. Exposition to chronic stress can modify appetite and macronutrient choice. In feeding disorders, the evaluation of Trp plasma levels versus the other LNAAs can provide useful information on either the nutritional state or patient metabolism, as well as the AA availability of brain uptake [85]. Other new potential molecular targets for these disorders are the metabotropic aromatic AA GPCRs [45, 86] or circulating tryptamine levels and TAARs [87]. The measure of the levels of Trp versus metabolites of the kynurenine shunt could permit "visualizing" the adrenal axis activation, proinflammatory cytokine secretion together with levels of the excitatory AA glutamate. On the other side, the genetic research has confirmed the importance of Trp metabolism as a support for a better clinical response and tolerance to treatments, for instance, in clinical psychiatry and SNC pathology: if polymorphisms of the SERT gene and 5-HT or other monoamine receptor subtypes are directly involved in these disorders and their treatment, the presence of polymorphisms of the brain Trp uptake G transporter as well as those related to TPH_2 has been observed in mood disorders [88, 89], often in comorbidity with bulimia or anorexia. Genetic polymorphisms of the SLC6A14 carrier transport for basic and neutral AAs as Trp have been related with human obesity [90]. Moreover, Trp metabolism components have been linked to gender-dependent disturbances associated with oxidative stress and cognitive processes [91, 92]. In summary, disturbances of Trp metabolism occupy a key position in human pathology and complex, multifactorial diseases.

THERAPEUTIC STRATEGIES AND DRUG DEVELOPMENT

Therapeutic strategies based on Trp chemical properties are in progress. As already reported, synthetic indole derivatives have been used for treating various human diseases (Figure 1(e)). New compounds are currently studied for human health. For instance, Trp-containing peptides or Trp-containing agents are explored as therapeutic agents against protein aggregation in neurodegeneration processes [21, 93, 94]. Alternative strategies in treating CNS diseases consist instead in targeting the kynurenine shunt and its modulation, acting therefore in the core of the balance of Trp fates: since kynurenine derivatives are related to both NMDA agonism or antagonism and nicotinic acid paths, their metabolism can be evaluated for treating cognitive deficits, dementia, and other severe neuropsychiatric conditions [95]. As reported before, Trp metabolism plays a relevant role in cancer: as regards IDO-related immune tolerance for cancer antigens, methyl-Trp derivatives and IDO inhibitors are promising compounds for therapy against tumor growth and metastasis formation [96]. New hydroxyl-indole derivatives are also appraised for their lactate dehydrogenase (LDH) inhibitory effect and anticancer properties [97].

BEYOND TRYPTOPHAN RESEARCH: THE USEFULNESS OF THE -OMICS TECHNIQUES

L-Tryptophan biochemistry lies in the heart of converging nutritional, neuroendocrine, and immune paths, through a variety of molecular effectors, each presumably contributing to relevant, complex, and severe diseases and syndromes, as reported in previous paragraphs. Advances in technologies of applied biochemistry and molecular biology have much improved the study of Trp metabolism and its implications in clinical research and medical genetics. Actually, new perspectives are emerging: in particular, it appears increasingly evident that pathologies at unclear aetiology/pathogenesis need multidisciplinary and multifactorial proceedings. This would allow defining groups of patients within the same disease showing common and distinct symptoms or responses to treatment correlated with specific biochemical patterns. For instance, the identification of biochemical clusters within neuropsychiatric disorders or other complex diseases would further support the notion that these illnesses are not "single," "fixed" pathological entities but rather spectrum disorders [98]. The targeting of Trp biochemistry in the context of other metabolic pathways is included in such a methodological advance. This approach would enable explaining symptoms' overlaps such as chronic fatigue and depression, pain disturbances, and IBS [7], sustaining the confluence of vulnerability factors for a complex disease and/or for diverse treatment responses. This research field would also provide useful information for cancer research and therapy. A valuable and consistent help seems to come from high-dimensional biology [99], involving the -Omics tools, as, primarily, the genomics, transcriptomics, proteomics, and metabolomics ones [100]. Metabolomics techniques permit in particular appraising multiple pathways and a number of metabolic intermediates, known and unknown, following an opposite approach of the classical one. This would allow the tracing of metabolic signatures of patients, to evaluate epigenetic and genetic factors which define a pathological condition as well as identify new correlates of pharmacological responses or new susceptibility traits of disease. Another main advantage of these approaches is that they can implement the use of personalized pharmacological therapies, also with respect to patients' age, gender, and lifestyle, by possibly considering diet habits and supplementation with specific nutrients and elements.

CONCLUSIONS

The essential AA tryptophan displays a peculiar chemistry among all other protein AAs and its derivatives are conserved in all living organisms, being linked to stress/environmental adaptive response. In humans, the molecular effectors of its indole-conserving or indole-disrupting fates are up- and downregulated by multiple factors which can play a role in many human complex diseases and syndromes. Molecular biology techniques and genetics are investigating components of Trp pathways while the application of high-dimensional biology and -Omics techniques is supposed to provide more insights about the regulation of Trp content in cells, its availability for human nutrition, and its role in the pathogenesis of disease. Other perspectives in Trp research concern the efficacy, monitoring, and personalization of pharmacological treatments as well as the development of new therapeutic compounds.

Conflict of Interests

The authors declare that there is no conflict of interests regarding the publication of this paper.

REFERENCES

1. W. C. Rose, "II. The sequence of events leading to the establishment of the amino acid needs of man," American Journal of Public Health and the Nation's Health, vol. 58, no. 11, pp. 2020–2027, 1968.

2. A. Frazer and J. G. Hensler, "Serotonin involvement in physiological function and behaviour," in Basic Neurochemistry: Molecular, Cellular and Medical Aspects, G. J. Siegel, B. W. Agranoff, R. W. Albers, et al., Eds., Lippincott-Raven, Philadelphia, Pa, USA, 6th edition, 1999.

3. M. Berger, J. A. Gray, and B. L. Roth, "The expanded biology of serotonin," Annual Review of Medicine, vol. 60, pp. 355–366, 2009.

4. J. A. Gingrich and R. Hen, "Dissecting the role of the serotonin system in neuropsychiatric disorders using knockout mice," Psychopharmacology, vol. 155, no. 1, pp. 1–10, 2001.

5. P. M. Whitaker-Azmitia, "Serotonin and brain development: role in human developmental diseases," Brain Research Bulletin, vol. 56, no. 5, pp. 479–485, 2001.

6. G. Giannaccini, L. Betti, L. Palego et al., "The expression of platelet serotonin transporter (SERT) in human obesity," BMC Neuroscience, vol. 14, article 128, 2013.

7. J. I. Hudson and H. G. Pope Jr., "The management of treatment-resistant depression in disorders on the interface of psychiatry and medicine," Psychiatric Clinics, vol. 19, no. 2, pp. 351–369, 1996.

8. G. Oxenkrug and R. Ratner, "N-Acetylserotonin and aging-associated cognitive impairment and depression," Aging and Disease, vol. 3, no. 4, pp. 330–338, 2012.

9. H. M. van Praag and C. Lemus, "Monoamine precursors in the treatment of psychiatric disorders," inNutrition and the Brain, R. J. Wurtman and J. J. Wurtman, Eds., pp. 89–139, Raven Press, New York, NY, USA, 1986.

10. D. A. Bender, "Biochemistry of tryptophan in health and disease," Molecular Aspects of Medicine, vol. 6, no. 2, pp. 101–197, 1983.

11. J. Normanly, "Approaching cellular and molecular resolution of auxin biosynthesis and metabolism,"Cold Spring Harbor Perspectives in Biology, vol. 2, no. 1, Article ID a001594, 2010.

12. Y. Tao, J.-L. Ferrer, K. Ljung et al., "Rapid synthesis of auxin via a new tryptophan-dependent pathway is required for shade avoidance in plants," Cell, vol. 133, no. 1, pp. 164–176, 2008.

13. M. B. Arnao and J. Hernández-Ruiz, "The physiological function of melatonin in plants," Plant Signaling and Behavior, vol. 1, no. 3, pp. 89–95, 2006.

14. L. Lindemann and M. C. Hoener, "A renaissance in trace amines inspired by a novel GPCR family,"Trends in Pharmacological Sciences, vol. 26, no. 5, pp. 274–281, 2005.

15. V. Tzin and G. Galili, "New Insights into the shikimate and aromatic amino acids biosynthesis pathways in plants," Molecular Plant, vol. 3, no. 6, pp. 956–972, 2010.

16. M. C. Mackey, M. Santillán, and N. Yildirim, "Modeling operon dynamics: the tryptophan and lactose operons as paradigms," Comptes Rendus—Biologies, vol. 327, no. 3, pp. 211–224, 2004.

17. G. Xie, N. O. Keyhani, C. A. Bonner, and R. A. Jensen, "Ancient origin of the tryptophan operon and the dynamics of evolutionary change," Microbiology and Molecular Biology Reviews, vol. 67, no. 3, pp. 303–342, 2003.

18. Y. Chen, X. Zhang, W. Wu, Z. Chen, H. Gu, and L.-J. Qu, "Overexpression of the wounding-responsive gene AtMYB15 activates the shikimate pathway in Arabidopsis," Journal of Integrative Plant Biology, vol. 48, no. 9, pp. 1084–1095, 2006.

19. S. Ferrari, R. Galletti, C. Denoux, G. de Lorenzo, F. M. Ausubel, and J. Dewdney, "Resistance to Botrytis cinerea induced in Arabidopsis by elicitors is independent of salicylic acid, ethylene, or jasmonate signaling but requires PHYTOALEXIN DEFICIENT3," Plant Physiology, vol. 144, no. 1, pp. 367–379, 2007.

20. A. J. de Jesus and T. W. Allen, "The role of tryptophan side chains in membrane protein anchoring and hydrophobic mismatch," Biochimica et Biophysica Acta (BBA)—Biomembranes, vol. 1828, no. 2, pp. 864–876, 2013.

21. C. M. Santiveri and M. A. Jiménez, "Tryptophan residues: scarce in proteins but strong stabilizers of β-hairpin peptides," Biopolymers, vol. 94, no. 6, pp. 779–790, 2010.

22. P. Grieco, M. Cai, A. V. Mayorov, D. Trivedi, and V. J. Hruby, "Structure-activity studies of new melanocortin peptides containing an aromatic amino acid at the N-terminal position," Peptides, vol. 27, no. 2, pp. 472–481, 2006.

23. B. Mojsoska and H. Jenssen, "Peptides and peptidomimetics for antimicrobial drug design,"Pharmaceuticals, vol. 8, no. 3, pp. 366–415, 2015.

24. G. Hrazdina and R. A. Jensen, "Spatial organization of enzymes in plant metabolic pathways," Annual Review of Plant Physiology and Plant Molecular Biology, vol. 43, no. 1, pp. 241–267, 1991.

25. E. R. Radwanski and R. L. Last, "Tryptophan biosynthesis and metabolism: biochemical and molecular genetics," Plant Cell, vol. 7, no. 7, pp. 921–934, 1995.

26. D. M. Richard, M. A. Dawes, C. W. Mathias, A. Acheson, N. Hill-Kapturczak, and D. M. Dougherty, "L-tryptophan: basic metabolic functions, behavioral research and therapeutic indications,"International Journal of Tryptophan Research, vol. 2, no. 1, pp. 45–60, 2009. ·

27. World Health Organization, "Protein and amino acid requirements in human nutrition. Report of a joint WHO/FAO/ expert consultation," WHO Technical Report Series 935, World Health Organization (WHO), Geneva, Switzerland, 2007.

28. D. J. Millward, "The nutritional value of plant-based diets in relation to human amino acid and protein requirements," Proceedings of the Nutrition Society, vol. 58, no. 2, pp. 249–260, 1999.

29. V. R. Young and P. L. Pellett, "Plant proteins in relation to human protein and amino acid nutrition,"American Journal of Clinical Nutrition, vol. 59, no. 5, supplement, pp. 1203S–1212S, 1994.

30. J. Hegyi, R. A. Schwartz, and V. Hegyi, "Pellagra: dermatitis, dementia, and diarrhea," International Journal of Dermatology, vol. 43, no. 1, pp. 1–5, 2004.

31. P. S. Simmons, J. M. Miles, J. E. Gerich, and M. W. Haymond, "Increased proteolysis. An effect of increases in plasma cortisol within the physiologic range," The Journal of Clinical Investigation, vol. 73, no. 2, pp. 412–420, 1984.

32. L. A. Cynober, "Plasma amino acid levels with a note on membrane transport: characteristics, regulation, and metabolic significance," Nutrition, vol. 18, no. 9, pp. 761–766, 2002.

33. D. Keszthelyi, F. J. Troost, and A. A. M. Masclee, "Understanding the role of tryptophan and serotonin metabolism in gastrointestinal function," Neurogastroenterology and Motility, vol. 21, no. 12, pp. 1239–1249, 2009.

34. R. J. Wurtman and J. D. Fernstrom, "Control of brain monoamine synthesis by diet and plasma amino acids," The American Journal of Clinical Nutrition, vol. 28, no. 6, pp. 638–647, 1975.

35. R. J. Wurtman, "Non-nutritional uses of nutrients," European Journal of Pharmacology, vol. 668, supplement 1, pp. S10–S15, 2011.

36. J. D. Fernstrom, "Effects and side effects associated with the non-nutritional use of tryptophan by humans," Journal of Nutrition, vol. 142, no. 12, pp. 2236S–2244S, 2012.

37. Y. Kanai, H. Segawa, K.-I. Miyamoto, H. Uchino, E. Takeda, and H. Endou, "Expression cloning and characterization of a transporter for large neutral amino acids activated by the heavy chain of 4F2 antigen (CD98)," The Journal of Biological Chemistry, vol. 273, no. 37, pp. 23629–23632, 1998.

38. J. Chillarón, R. Roca, A. Valencia, A. Zorzano, and M. Palacín, "Heteromeric amino acid transporters: biochemistry, genetics, and physiology," The American Journal of Physiology—Renal Physiology, vol. 281, no. 6, pp. F995–F1018, 2001.

39. G. Curzon, J. Friedel, and P. J. Knott, "The effect of fatty acids on the binding of tryptophan to plasma protein," Nature, vol. 242, no. 5394, pp. 198–200, 1973.

40. W. M. Pardridge and G. Fierer, "Transport of tryptophan into brain from the circulating, albumin-bound pool in rats and in rabbits," Journal of Neurochemistry, vol. 54, no. 3, pp. 971–976, 1990.

41. J. P. Ruddick, A. K. Evans, D. J. Nutt, S. L. Lightman, G. A. Rook, and

C. A. Lowry, "Tryptophan metabolism in the central nervous system: medical implications," Expert Reviews in Molecular Medicine, vol. 8, no. 20, pp. 1–27, 2006.

42. D. M. Rosenbaum, S. G. F. Rasmussen, and B. K. Kobilka, "The structure and function of G-protein-coupled receptors," Nature, vol. 459, no. 7245, pp. 356–363, 2009.

43. P. Wellendorph and H. Bräuner-Osborne, "Molecular cloning, expression, and sequence analysis of GPRC6A, a novel family C G-protein-coupled receptor," Gene, vol. 335, no. 1-2, pp. 37–46, 2004.

44. A. D. Conigrave, H. C. Mun, and H. C. Lok, "Aromatic L-amino acids activate the calcium-sensing receptor 1–3," The Journal of Nutrition, vol. 137, pp. 1524S–1527S, 2007.

45. A. D. Conigrave, H.-C. Mun, and S. C. Brennan, "Physiological significance of L-amino acid sensing by extracellular Ca^{2+}-sensing receptors," Biochemical Society Transactions, vol. 35, no. 5, pp. 1195–1198, 2007.

46. S. Nakajima, T. Hira, and H. Hara, "Calcium-sensing receptor mediates dietary peptide-induced CCK secretion in enteroendocrine STC-1 cells," Molecular Nutrition and Food Research, vol. 56, no. 5, pp. 753–760, 2012.

47. A. M. San Gabriel, "Taste receptors in the gastrointestinal system," Flavour, vol. 4, article 14, 2015. ·

48. S. A. Sakowski, T. J. Geddes, D. M. Thomas, E. Levi, J. S. Hatfield, and D. M. Kuhn, "Differential tissue distribution of tryptophan hydroxylase isoforms 1 and 2 as revealed with monospecific antibodies," Brain Research, vol. 1085, no. 1, pp. 11–18, 2006.

49. M. Boldrini, M. Castagna, I. Nardi, A. Giromella, A. Martini, I. Pampaloni I, et al., "Serotonin transporter binding density changes seasonally in the human pineal gland," in Proceedings of the 42nd Annual Meeting of the American College of Neuropsychopharmacology (ACNP ‹03), San Juan, Puerto Rico, December 2003.

50. C. I. Gutiérrez, M. Urbina, F. Obregion, J. Glykys, and L. Lima, "Characterization of tryptophan high affinity transport system in pinealocytes of the rat. Day-night modulation," Amino Acids, vol. 25, no. 1, pp. 95–105, 2003.

51. M. D. Berry, "Mammalian central nervous system trace amines. Pharmacologic amphetamines, physiologic neuromodulators," Journal of Neurochemistry, vol. 90, no. 2, pp. 257–271, 2004.

52. R. Zucchi, G. Chiellini, T. S. Scanlan, and D. K. Grandy, "Trace amine-associated receptors and their ligands," British Journal of Pharmacology, vol. 149, no. 8, pp. 967–978, 2006.

53. L. Lindemann, M. Ebeling, N. A. Kratochwil, J. R. Bunzow, D. K. Grandy, and M. C. Hoener, "Trace amine-associated receptors form structurally and functionally distinct subfamilies of novel G protein-coupled receptors," Genomics, vol. 85, no. 3, pp. 372–385, 2005.

54. O. Kurnasov, L. Jablonski, B. Polanuyer, P. Dorrestein, T. Begley, and A. Osterman, "Aerobic tryptophan degradation pathway in bacteria: novel kynurenine formamidase," FEMS Microbiology Letters, vol. 227, no. 2, pp. 219–227, 2003.

55. Y. Murakami, M. Hoshi, Y. Imamura, Y. Arioka, Y. Yamamoto, and K. Saito, "Remarkable role of indoleamine 2,3-dioxygenase and tryptophan metabolites in infectious diseases: potential role in macrophage-mediated inflammatory diseases," Mediators of Inflammation, vol. 2013, Article ID 391984, 9 pages, 2013.

56. B. M. Campbell, E. Charych, A. W. Lee, and T. Möller, "Kynurenines in CNS disease: regulation by inflammatory cytokines," Frontiers in Neuroscience, vol. 8, pp. 1–22, 2014.

57. A. A.-B. Badawy and M. Evans, "Inhibition of rat liver tryptophan pyrrolase activity and elevation of brain tryptophan concentration by administration of antidepressants," Biochemical Pharmacology, vol. 30, no. 11, pp. 1211–1216, 1981.

58. C. M. Robinson, P. T. Hale, and J. M. Carlin, "The role of IFN-γ and TNF-α-responsive regulatory elements in the synergistic induction of indoleamine dioxygenase," Journal of Interferon and Cytokine Research, vol. 25, no. 1, pp. 20–30, 2005.

59. T. W. Stone, C. M. Forrest, G. M. Mackay, N. Stoy, and L. G. Darlington, "Tryptophan, adenosine, neurodegeneration and neuroprotection," Metabolic Brain Disease, vol. 22, no. 3-4, pp. 337–352, 2007.

60. A. Neumeister, N. Praschak-Rieder, B. Heßelmann et al., "Effects of tryptophan depletion in drug-free depressed patients who responded to total sleep deprivation," Archives of General Psychiatry, vol. 55, no. 2, pp. 167–172, 1998.

61. T. O. C. Kilkens, A. Honig, M. A. van Nieuwenhoven, W. J. Riedel, and R.-J. M. Brummer, "Acute tryptophan depletion affects brain-gut responses in irritable bowel syndrome patients and controls,"Gut, vol. 53, no. 12, pp. 1794–1800, 2004.

62. B. Caballero, N. Finer, and R. J. Wurtman, "Plasma amino acids and insulin levels in obesity: response to carbohydrate intake and tryptophan supplements," Metabolism, vol. 37, no. 7, pp. 672–676, 1988.

63. C. R. Markus, C. Firk, C. Gerhardt, J. Kloek, and G. F. Smolders, "Effect of different tryptophan sources on amino acids availability to the brain and mood in healthy volunteers," Psychopharmacology, vol. 201, no. 1, pp. 107–114, 2008.

64. M. C. W. Kroes, G. A. van Wingen, J. Wittwer, M. H. Mohajeri, J. Kloek, and G. Fernández, "Food can lift mood by affecting mood-regulating neurocircuits via a serotonergic mechanism," NeuroImage, vol. 84, pp. 825–832, 2014.

65. M. Ledochowski, B. Widner, C. Murr, B. Sperner-Unterweger, and D. Fuchs, "Fructose malabsorption is associated with decreased plasma tryptophan," Scandinavian Journal of Gastroenterology, vol. 36, no. 4, pp. 367–371, 2001.

66. G. F. Oxenkrug, "Genetic and hormonal regulation of tryptophan-kynurenine metabolism: implications for vascular cognitive impairment, major depressive disorder, and aging," Annals of the New York Academy of Sciences, vol. 1122, pp. 35–49, 2007.

67. R. T. Rubin, "Adrenal cortical activity changes in manic-depressive illness. Influence on intermediary metabolism of tryptophan," Archives of General Psychiatry, vol. 17, no. 6, pp. 671–679, 1967.

68. U. Grohmann, F. Fallarino, and P. Puccetti, "Tolerance, DCs and tryptophan: much ado about IDO,"Trends in Immunology, vol. 24, no. 5, pp. 242–248, 2003.

69. A. L. Mellor and D. H. Munn, "IDO expression by dendritic cells: tolerance and tryptophan catabolism," Nature Reviews Immunology, vol. 4, no. 10, pp. 762–774, 2004.

70. A. Heitger, "Regulation of expression and function of IDO in human dendritic cells," Current Medicinal Chemistry, vol. 18, no. 15, pp. 2222–2233, 2011.

71. P. Ligam, U. Manuelpillai, E. M. Wallace, and D. Walker, "Localisation of indoleamine 2,3-dioxygenase and kynurenine hydroxylase in the human placenta and decidua: implications for role of the kynurenine pathway in pregnancy," Placenta, vol. 26, no. 6, pp. 498–504, 2005.

72. A. L. Mellor, J. Sivakumar, P. Chandler et al., "Prevention of T cell-driven complement activation and inflammation by tryptophan catabolism during pregnancy," Nature Immunology, vol. 2, no. 1, pp. 64–68, 2001.

73. H. Soliman, M. Mediavilla-Varela, and S. Antonia, "Indoleamine 2,3-dioxygenase is it an immune suppressor?" Cancer Journal, vol. 16, no. 4, pp. 354–359, 2010.

74. A.-M. Myint and Y.-K. Kim, "Network beyond IDO in psychiatric disorders: revisiting neurodegeneration hypothesis," Progress in Neuro-Psychopharmacology and Biological Psychiatry, vol. 48, pp. 304–313, 2014.

75. R. Schwarcz, J. P. Bruno, P. J. Muchowski, and H.-Q. Wu, "Kynurenines in the mammalian brain: when physiology meets pathology," Nature Reviews Neuroscience, vol. 13, no. 7, pp. 465–477, 2012.

76. S. J. James, S. Melnyk, S. Jernigan et al., "Metabolic endophenotype and related genotypes are associated with oxidative stress in children with autism," American Journal of Medical Genetics Part B, vol. 141, no. 8, pp. 947–956, 2006.

77. L. Boccuto, C.-F. Chen, A. R. Pittman et al., "Decreased tryptophan metabolism in patients with autism spectrum disorders," Molecular Autism, vol. 4, no. 1, article 16, 2013.

78. S. Nishizawa, C. Benkelfat, S. N. Young et al., "Differences between males and females in rates of serotonin synthesis in human brain," Proceedings of the National Academy of Sciences of the United States of America, vol. 94, no. 10, pp. 5308–5313, 1997.

79. D. Marazziti, A. Rossi, L. Palego et al., "Effect of aging and sex on the [³H]-paroxetine binding to human platelets," Journal of Affective Disorders, vol. 50, no. 1, pp. 11–15, 1998.

80. L. Palego, A. Giromella, M. R. Mazzoni et al., "Gender and age-related variation in adenylyl cyclase activity in the human prefrontal cortex, hippocampus and dorsal raphe nuclei," Neuroscience Letters, vol. 279, no. 1, pp. 53–56, 2000.

81. D. Marazziti, S. Baroni, I. Masala et al., "Impulsivity, gender, and the platelet serotonin transporter in healthy subjects," Neuropsychiatric Disease and Treatment, vol. 6, no. 1, pp. 9–15, 2010.

82. C. Hammen, P. A. Brennan, D. Keenan-Miller, N. A. Hazel, and J. M. Najman, "Chronic and acute stress, gender, and serotonin transporter gene-environment interactions predicting depression symptoms in youth," Journal of Child Psychology and Psychiatry and Allied Disciplines, vol. 51, no. 2, pp. 180–187, 2010.

83. N. Goel and T. L. Bale, "Sex differences in the serotonergic influence on the hypothalamic-pituitary-adrenal stress axis," Endocrinology, vol. 151, no. 4, pp. 1784–1794, 2010.

84. W. H. Kaye, "Neurobiology of anorexia and bulimia nervosa," Physiology and Behavior, vol. 94, no. 1, pp. 121–135, 2008.

85. W. H. Kaye, K. A. Gendall, M. H. Fernstrom, J. D. Fernstrom, C. W. McConaha, and T. E. Weltzin, "Effects of acute tryptophan depletion on mood in bulimia nervosa," Biological Psychiatry, vol. 47, no. 2, pp. 151–157, 2000.

86. F. Reimann, G. Tolhurst, and F. M. Gribble, "G-Protein-coupled receptors in intestinal chemosensation," Cell Metabolism, vol. 15, no. 4, pp. 421–431, 2012.

87. D. Narang, S. Tomlinson, A. Holt, D. D. Mousseau, and G. B. Baker, "Trace Amines and their relevance to psychiatry and neurology: a brief overview," Klinik Psikofarmakoloji Bulteni, vol. 21, no. 1, pp. 73–79, 2011.

88. M. Nakamura, S. Ueno, A. Sano, and H. Tanabe, "Polymorphisms of the human homologue of theDrosophila white gene are associated with mood and panic disorders," Molecular Psychiatry, vol. 4, no. 2, pp. 155–162, 1999.

89. A. Rotondo, C. Mazzanti, L. Dell›Osso et al., "Catechol O-methyltransferase, serotonin transporter, and tryptophan hydroxylase gene polymorphisms in bipolar disorder patients with and without comorbid panic disorder," American Journal of Psychiatry, vol. 159, no. 1, pp. 23–29, 2002.

90. E. Suviolahti, L. J. Oksanen, M. Öhman et al., "The SLC6A14 gene shows evidence of association with obesity," The Journal of Clinical Investigation, vol. 112, no. 11, pp. 1762–1772, 2003.

91. G. F. Oxenkrug, "Tryptophan metabolism as a new target for the treatment of schizophrenia," US Psychiatry Review, pp. 38–39, 2007, (Touch Briefings).

92. D. Marazziti, S. Baroni, M. Picchetti, A. Piccinni, S. Silvestri, and L. Dell›Osso, "New developments on the serotonin hypothesis of depression: shunt of tryptophan," Rivista di Psichiatria, vol. 48, no. 1, pp. 23–34, 2013.

93. A. Frydman-Marom, M. Rechter, I. Shefler, Y. Bram, D. E. Shalev, and E. Gazit, "Cognitive-performance recovery of Alzheimer›s disease model mice by modulation of early soluble amyloidal assemblies," Angewandte Chemie—International Edition, vol. 48, no. 11, pp. 1981–1986, 2009.

94. S. A. Funke and D. Willbold, "Peptides for therapy and diagnosis of Alzheimer›s disease," Current Pharmaceutical Design, vol. 18, no. 6, pp. 755–767, 2012.

95. T. W. Stone, N. Stoy, and L. G. Darlington, "An expanding range of targets for kynurenine metabolites of tryptophan," Trends in Pharmacological Sciences, vol. 34, no. 2, pp. 136–143, 2013.

96. D.-Y. Hou, A. J. Muller, M. D. Sharma et al., "Inhibition of indoleamine 2,3-dioxygenase in dendritic cells by stereoisomers of 1-methyl-tryptophan correlates with antitumor responses," Cancer Research, vol. 67, no. 2, pp. 792–801, 2007.

97. C. Granchi, S. Roy, C. Giacomelli et al., "Discovery of N-hydroxyindole-based inhibitors of human lactate dehydrogenase isoform A (LDH-A) as starvation agents against cancer cells," Journal of Medicinal Chemistry, vol. 54, no. 6, pp. 1599–1612, 2011.

98. L. Palego, L. Betti, and G. Giannaccini, "Sulfur metabolism and sulfur-containing amino acids derivatives—part II: autism spectrum disorders, schizophrenia and fibromyalgia," Biochemical Pharmacology, vol. 4, article 159, 2015.

99. G. A. Evans, "Designer science and the 'omic' revolution," Nature Biotechnology, vol. 18, article 127, 2000.

100. L. Atzori, R. Antonucci, L. Barberini, J. L. Griffin, and V. Fanos, "Metabolomics: a new tool for the neonatologist," Journal of Maternal-Fetal and Neonatal Medicine, vol. 22, no. 3, pp. 50–53, 2009.

CITATION

CHAPTER 1

Shaniko Shini Asad Sultan and Wayne L. Bryden; Selenium Biochemistry and Bioavailability: Implications for Animal Agriculture; doi:10.3390/agriculture5041277

CHAPTER 2

Lewbart GA, Hirschfeld M, Denkinger J, Vasco K, Guevara N, et al. (2014) Blood Gases, Biochemistry, and Hematology of Galapagos Green Turtles (Chelonia Mydas). PLoS ONE 9(5): e96487. doi:10.1371/journal.pone.0096487

CHAPTER 3

Mustafa Türk, Zakir M. O. Rzayev, Gülcihan Kurucu, Bioengineering functional copolymers. XII. Interaction of boron-containing and PEO branched derivatives of poly(MA-alt-MVE) with HeLa cells, doi:10.4236/health.2010.21009.

CHAPTER 4

M. Türk, G. Kahraman, S. Khalilova, Z. Rzayev and S. Oguztüzün, "Bioengineering Functional Copolymers. XVII. Interaction of Organoboron Amide-Ester Branched Derivatives of Poly(Acrylic Acid) with Cancer Cells," Journal of Cancer Therapy, Vol. 2 No. 2, 2011, pp. 266-275. doi: 10.4236/jct.2011.22034.

CHAPTER 5

Laura Mejía-Teniente, Angel María Chapa-Oliver, Moises Alejandro Vazquez-Cruz, Irineo Torres-Pacheco and Ramón Gerardo Guevara-González (2011). Aflatoxins Biochemistry and Molecular Biology - Biotechnological Approaches for Control in Crops, Aflatoxins - Detection, Measurement and Control, Dr Irineo Torres-Pacheco (Ed.), ISBN: 978-953-307-711-6, InTech, DOI: 10.5772/25663.

CHAPTER 6

Umberto Lucia; Bioengineering thermodynamics of biological cells; doi: 10.1186/s12976-015-0024-z

CHAPTER 7

Pedro Fernandes, "Enzymes in Food Processing: A Condensed Overview on Strategies for Better Biocatalysts," Enzyme Research, vol. 2010, Article ID 862537, 19 pages, 2010. doi:10.4061/2010/862537

CHAPTER 8

Heping Cao; Bioengineering Recombinant Diacylglycerol Acyltransferases; DOI: 10.5772/18994

CHAPTER 9

Pontini, M.M. Sfriso, M.I. Buompensiere, V. Vindigni and F. Bassetto; Bioengineering of Vascular Conduits; DOI: 10.5772/59148

CHAPTER 10

Odhiambo C, Oyaro B, Odipo R, Otieno F, Alemnji G, Williamson J, et al. (2015) Evaluation of Locally Established Reference Intervals for Hematology and Biochemistry Parameters in Western Kenya. PLoS ONE 10(4): e0123140. doi:10.1371/journal.pone.0123140

CHAPTER 11

Bassem Jaouadi, Badis Abdelmalek, Nedia Zaraî Jaouadi and Samir Bejar (2011). The Bioengineering and Industrial Applications of Bacterial Alkaline Proteases: the Case of SAPB and KERAB, Progress in Molecular and Environmental Bioengineering - From Analysis and Modeling to Technology Applications, Prof. Angelo Carpi (Ed.), ISBN: 978-953-307-268-5, InTech, DOI: 10.5772/2385

CHAPTER 12

Lionella Palego, Laura Betti, Alessandra Rossi, and Gino Giannaccini, "Tryptophan Biochemistry: Structural, Nutritional, Metabolic, and Medical Aspects in Humans," Journal of Amino Acids, vol. 2016, Article ID 8952520, 13 pages, 2016. doi:10.1155/2016/8952520

INDEX